# An Introduction to
# *Wavelets*

# Wavelet Analysis and Its Applications

The subject of wavelet analysis has recently drawn a great deal of attention from mathematical scientists in various disciplines. It is creating a common link between mathematicians, physicists, and electrical engineers. This book series will consist of both monographs and edited volumes on the theory and applications of this rapidly developing subject. Its objective is to meet the needs of academic, industrial, and governmental researchers, as well as to provide instructional material for teaching at both the undergraduate and graduate levels.

This first volume is intended to be introductory in nature. It is suitable as a textbook for a beginning course on wavelet analysis and is directed towards both mathematicians and engineers who wish to learn about the subject. Specialists may use this volume as supplementary reading to the vast literature that has already emerged in this field.

This is a volume in
**WAVELET ANALYSIS AND ITS APPLICATIONS**

CHARLES K. CHUI, SERIES EDITOR
*Texas A&M University, College Station, Texas*

A list of titles in this series appears at the end of this volume.

# An Introduction to
# *Wavelets*

CHARLES K. CHUI
*Department of Mathematics*
*Texas A&M University*
*College Station, Texas*

ACADEMIC PRESS, INC.
*Harcourt Brace Jovanovich, Publishers*

Boston   San Diego   New York
London   Sydney   Tokyo   Toronto

Cover designed by Camille Pecoul.

ACADEMIC PRESS, INC.
1250 Sixth Avenue, San Diego, CA 92101

*United Kingdom Edition published by*
ACADEMIC PRESS LIMITED
24–28 Oval Road, London NW1 7DX

Library of Congress Catalog Card Number: 91-58831

Printed in the United States of America

93 94 95    9 8 7 6 5 4

*To Margaret*

# Contents

**Preface** ................................................................. ix

**1. An Overview** .......................................................... 1

    1.1. From Fourier analysis to wavelet analysis ........................... 1
    1.2. The integral wavelet transform and time-frequency analysis ........ 6
    1.3. Inversion formulas and duals ...................................... 9
    1.4. Classification of wavelets ....................................... 13
    1.5. Multiresolution analysis, splines, and wavelets ................. 16
    1.6. Wavelet decompositions and reconstructions ..................... 18

**2. Fourier Analysis** .................................................... 23

    2.1. Fourier and inverse Fourier transforms .......................... 23
    2.2. Continuous-time convolution and the delta function ............. 27
    2.3. Fourier transform of square-integrable functions ................ 32
    2.4. Fourier series ................................................... 36
    2.5. Basic convergence theory and Poisson's summation formula ....... 43

**3. Wavelet Transforms and Time-Frequency Analysis** .............. 49

    3.1. The Gabor transform ............................................. 50
    3.2. Short-time Fourier transforms and the Uncertainty Principle ...... 54
    3.3. The integral wavelet transform .................................. 60
    3.4. Dyadic wavelets and inversions .................................. 64
    3.5. Frames .......................................................... 68
    3.6. Wavelet series .................................................. 74

**4. Cardinal Spline Analysis** ........................................... 81

    4.1. Cardinal spline spaces .......................................... 81
    4.2. $B$-splines and their basic properties .......................... 85
    4.3. The two-scale relation and an interpolatory
          graphical display algorithm ..................................... 90
    4.4. $B$-net representations and computation of cardinal splines ....... 95
    4.5. Construction of spline approximation formulas .................. 100
    4.6. Construction of spline interpolation formulas .................. 109

**5. Scaling Functions and Wavelets** ................................ 119

    5.1. Multiresolution analysis ......................................... 120

    5.2. Scaling functions with finite two-scale relations ................. 128

    5.3. Direct-sum decompositions of $L^2(\mathbb{R})$ ........................... 140

    5.4. Wavelets and their duals ....................................... 146

    5.5. Linear-phase filtering ......................................... 159

    5.6. Compactly supported wavelets ................................. 168

**6. Cardinal Spline-Wavelets** ........................................ 177

    6.1. Interpolatory spline-wavelets .................................. 177

    6.2. Compactly supported spline-wavelets ........................... 182

    6.3. Computation of cardinal spline-wavelets ....................... 187

    6.4. Euler-Frobenius polynomials .................................. 195

    6.5. Error analysis in spline-wavelet decomposition ................. 199

    6.6. Total positivity, complete oscillation, zero-crossings ............. 207

**7. Orthogonal Wavelets and Wavelet Packets** ...................... 215

    7.1. Examples of orthogonal wavelets ............................... 215

    7.2. Identification of orthogonal two-scale symbols .................. 220

    7.3. Construction of compactly supported orthogonal wavelets ....... 229

    7.4. Orthogonal wavelet packets .................................... 236

    7.5. Orthogonal decomposition of wavelet series .................... 240

**Notes** ............................................................. 245

**References** ....................................................... 251

**Subject Index** .................................................... 257

**Appendix** ........................................................ 265

# Preface

Fourier analysis is an established subject in the core of pure and applied mathematical analysis. Not only are the techniques in this subject of fundamental importance in all areas of science and technology, but both the integral Fourier transform and the Fourier series also have significant physical interpretations. In addition, the computational aspects of the Fourier series are especially attractive, mainly because of the orthogonality property of the series and of its simple expression in terms of only two functions: $\sin x$ and $\cos x$.

Recently, the subject of "wavelet analysis" has drawn much attention from both mathematicians and engineers alike. Analogous to Fourier analysis, there are also two important mathematical entities in wavelet analysis: the "integral wavelet transform" and the "wavelet series". The integral wavelet transform is defined to be the convolution with respect to the dilation of the reflection of some function $\tilde{\psi}$, called a "basic wavelet", while the wavelet series is expressed in terms of a single function $\psi$, called an "$\mathcal{R}$-wavelet" (or simply, a wavelet) by means of two very simple operations: binary dilations and integral translations. However, unlike Fourier analysis, the integral wavelet transform with a basic wavelet $\tilde{\psi}$ and the wavelet series in terms of a wavelet $\psi$ are intimately related. In fact, if $\tilde{\psi}$ is chosen to be the "dual" of $\psi$, then the coefficients of the wavelet series of any square-integrable function $f$ are precisely the values of the integral wavelet transform, evaluated at the dyadic positions in the corresponding binary dilated scale levels. Since the integral wavelet transform of $f$ simultaneously localizes $f$ and its Fourier transform $\hat{f}$ with the zoom-in and zoom-out capability, and since there are real-time algorithms for obtaining the coefficient sequences of the wavelet series, and for recovering $f$ from these sequences, the list of applications of wavelet analysis seems to be endless. On the other hand, polynomial spline functions are among the simplest functions for both computational and implementational purposes. Hence, they are most attractive for analyzing and constructing wavelets.

This is an introductory treatise on wavelet analysis with an emphasis on spline-wavelets and time-frequency analysis. A brief overview of this subject, including classification of wavelets, the integral wavelet transform for time-frequency analysis, multiresolution analysis highlighting the important properties of splines, and wavelet algorithms for decomposition and reconstruction of functions, will be presented in the first chapter. The objective of this chapter is not to go into any depth but only to convey a general impression of what wavelet analysis is about and what this book aims to cover.

This monograph is intended to be self-contained. The only prerequisite is a basic knowledge of function theory and real analysis. For this reason, preliminary material on Fourier analysis and signal theory is covered in Chapters 2 and 3, and an introductory study of cardinal splines is included in Chapter 4. It must be pointed out, however, that Chapters 3 and 4 also contribute as an integral part of wavelet analysis. In particular, in Chapter 3, the notion of "frames", and more generally "dyadic wavelets", is introduced in the discussion of reconstruction of functions from partial information of their integral wavelet transforms in time-frequency analysis.

The common theme of the last three chapters is "wavelet series". Hence, a general approach to the analysis and construction of scaling functions and wavelets is discussed in Chapter 5. Spline-wavelets, which are the simplest examples, are studied in Chapter 6. The final chapter is devoted to an investigation of orthogonal wavelets and wavelet packets.

The writing of this monograph was greatly influenced by the pioneering work of A. Cohen, R. Coifman, I. Daubechies, S. Mallat, and Y. Meyer, as well as my joint research with X. L. Shi and J. Z. Wang. In learning this fascinating subject, I have benefited from conversations and correspondence with many colleagues, to whom I am very grateful. In particular, I would like to mention P. Auscher, G. Battle, A. K. Chan, A. Cohen, I. Daubechies, D. George, T. N. T. Goodman, S. Jaffard, C. Li, S. Mallat, Y. Meyer, C. A. Micchelli, E. Quak, X. L. Shi, J. Stöckler, J. Z. Wang, J. D. Ward, and R. Wells. Among my friends who have read portions of the manuscript and made many valuable suggestions, I am especially indebted to C. Li, E. Quak, X. L. Shi, and N. Sivakumar. As usual, I have again enjoyed superb assistance from Robin Campbell, who TEXed the entire manuscript, and from Stephanie Sellers and my wife, Margaret, who produced the manuscript in camera-ready form. Finally, to the editorial office of Academic Press, and particularly to Charles Glaser, who has complete confidence in me, I wish to express my appreciation of their efficient assistance and friendly cooperation.

College Station, Texas            Charles K. Chui
October, 1991


### Preface to the second printing

The second printing gave me an opportunity to make some corrections and append two tables of weights for implementing spline-wavelet reconstruction and decomposition. The inclusion of these numerical values was suggested by David Donoho to whom I am very grateful. I would also like to thank my student Jun Zha for his assistance in producing these two tables.

April, 1992            C. K. C.

# 1 An Overview

"Wavelets" has been a very popular topic of conversations in many scientific and engineering gatherings these days. Some view wavelets as a new basis for representing functions, some consider it as a technique for time-frequency analysis, and others think of it as a new mathematical subject. Of course, all of them are right, since "wavelets" is a versatile tool with very rich mathematical content and great potential for applications. However, as this subject is still in the midst of rapid development, it is definitely too early to give a unified presentation. The objective of this book is very modest: it is intended to be used as a textbook for an introductory one-semester course on "wavelet analysis" for upper-division undergraduate or beginning graduate mathematics and engineering students, and is also written for both mathematicians and engineers who wish to learn about the subject. For the specialists, this volume is suitable as complementary reading to the more advanced monographs, such as the two volumes of *Ondelettes et Opérateurs* by Yves Meyer, the edited volume of *Wavelets–A Tutorial in Theory and Applications* in this series, and the forthcoming CBMS volume by Ingrid Daubechies.

Since wavelet analysis is a relatively new subject and the approach and organization in this book are somewhat different from that in the others, the goal of this chapter is to convey a general idea of what wavelet analysis is about and to describe what this book aims to cover.

## 1.1. From Fourier analysis to wavelet analysis

Let $L^2(0, 2\pi)$ denote the collection of all measurable functions $f$ defined on the interval $(0, 2\pi)$ with

$$\int_0^{2\pi} |f(x)|^2 dx < \infty.$$

For the reader who is not familiar with the basic Lebesgue theory, the sacrifice is very minimal by assuming that $f$ is a piecewise continuous function. It will always be assumed that functions in $L^2(0, 2\pi)$ are extended periodically to the real line

$$\mathbb{R} := (-\infty, \infty),$$

namely: $f(x) = f(x - 2\pi)$ for all $x$. Hence, the collection $L^2(0, 2\pi)$ is often called the space of $2\pi$-periodic square-integrable functions. That $L^2(0, 2\pi)$ is

1

a vector space can be verified very easily. Any $f$ in $L^2(0, 2\pi)$ has a Fourier series representation:

$$f(x) = \sum_{n=-\infty}^{\infty} c_n e^{inx}, \tag{1.1.1}$$

where the constants $c_n$, called the Fourier coefficients of $f$, are defined by

$$c_n = \frac{1}{2\pi} \int_0^{2\pi} f(x) e^{-inx} dx. \tag{1.1.2}$$

The convergence of the series in (1.1.1) is in $L^2(0, 2\pi)$, meaning that

$$\lim_{M,N \to \infty} \int_0^{2\pi} \left| f(x) - \sum_{n=-M}^{N} c_n e^{inx} \right|^2 dx = 0.$$

There are two distinct features in the Fourier series representation (1.1.1). First, we mention that $f$ is decomposed into a sum of infinitely many mutually orthogonal components $g_n(x) := c_n e^{inx}$, where orthogonality means that

$$\langle g_m, g_n \rangle^* = 0, \quad \text{for all} \quad m \neq n, \tag{1.1.3}$$

with the *"inner product"* in (1.1.3) being defined by

$$\langle g_m, g_n \rangle^* := \frac{1}{2\pi} \int_0^{2\pi} g_m(x) \overline{g_n(x)} dx. \tag{1.1.4}$$

That (1.1.3) holds is a consequence of the important, yet simple fact that

$$w_n(x) := e^{inx}, \qquad n = \cdots, -1, 0, 1, \ldots, \tag{1.1.5}$$

is an orthonormal (o.n.) basis of $L^2(0, 2\pi)$. The second distinct feature of the Fourier series representation (1.1.1) is that the o.n. basis $\{w_n\}$ is generated by *"dilation"* of a single function

$$w(x) := e^{ix} ; \tag{1.1.6}$$

that is, $w_n(x) = w(nx)$ for all integers $n$. This will be called *integral dilation*.

Let us summarize this remarkable fact by saying that *every $2\pi$-periodic square-integrable function is generated by a "superposition" of integral dilations of the basic function* $w(x) = e^{ix}$.

We also remark that from the o.n. property of $\{w_n\}$, the Fourier series representation (1.1.1) also satisfies the so-called *Parseval Identity*:

$$\frac{1}{2\pi} \int_0^{2\pi} |f(x)|^2 dx = \sum_{n=-\infty}^{\infty} |c_n|^2. \tag{1.1.7}$$

Let $\ell^2$ denote the space of all square-summable bi-infinite sequences; that is, $\{c_n\} \in \ell^2$ if and only if

$$\sum_{n=-\infty}^{\infty} |c_n|^2 < \infty.$$

Hence, if the square-root of the quantity on the left of (1.1.7) is used as the "norm" for the measurement of functions in $L^2(0, 2\pi)$, and similarly, the square-root of the quantity on the right of (1.1.7) is used as the norm for $\ell^2$, then the function space $L^2(0, 2\pi)$ and the sequence space $\ell^2$ are "*isometric*" to each other. Returning to the above mentioned observation on the Fourier series representation (1.1.1), we can also say that *every $2\pi$-periodic square-integrable function is an $\ell^2$-linear combination of integral dilations of the basic function* $w(x) = e^{ix}$.

We emphasize again that the basic function

$$w(x) = e^{ix} = \cos x + i \sin x,$$

which is a "*sinusoidal wave*", is the *only* function required to generate all $2\pi$-periodic square-summable functions. For any integer $n$ with large absolute value, the wave $w_n(x) = w(nx)$ has high "*frequency*", and for $n$ with small absolute value, the wave $w_n$ has low frequency. So, every function in $L^2(0, 2\pi)$ is composed of waves with various frequencies.

We next consider the space $L^2(\mathbb{R})$ of measurable functions $f$, defined on the real line $\mathbb{R}$, that satisfy

$$\int_{-\infty}^{\infty} |f(x)|^2 dx < \infty.$$

Clearly, the two function spaces $L^2(0, 2\pi)$ and $L^2(\mathbb{R})$ are quite different. In particular, since (the local average values of) every function in $L^2(\mathbb{R})$ must "decay" to zero at $\pm\infty$, the sinusoidal (wave) functions $w_n$ do not belong to $L^2(\mathbb{R})$. In fact, if we look for "waves" that generate $L^2(\mathbb{R})$, these waves should decay to zero at $\pm\infty$; and for all practical purposes, the decay should be very fast. That is, we look for small waves, or "*wavelets*", to generate $L^2(\mathbb{R})$. As in the situation of $L^2(0, 2\pi)$, where one single function $w(x) = e^{ix}$ generates the entire space, we also prefer to have a single function, say $\psi$, to generate all of $L^2(\mathbb{R})$. But if the wavelet $\psi$ has very fast decay, how can it cover the whole real line? The obvious way is to shift $\psi$ along $\mathbb{R}$.

Let $\mathbb{Z}$ denote the set of integers:

$$\mathbb{Z} = \{\ldots, -1, 0, 1, \ldots\}.$$

The simplest way for $\psi$ to cover all of $\mathbb{R}$ is to consider all the *integral shifts* of $\psi$, namely:

$$\psi(x - k), \qquad k \in \mathbb{Z}.$$

Next, as in the sinusoidal situation, we must also consider waves with different frequencies. For various reasons which will soon be clear to the reader, we do not wish to consider "single-frequency" waves, but rather, waves with frequencies partitioned into consecutive "octaves" (or frequency bands). For computational efficiency, we will use integral powers of 2 for frequency partitioning; that is, we now consider the small waves

$$\psi(2^j x - k), \qquad j, k \in \mathbb{Z}. \tag{1.1.8}$$

Observe that $\psi(2^j x - k)$ is obtained from a single "wavelet" function $\psi(x)$ by a *binary dilation* (i.e. dilation by $2^j$) and a *dyadic translation* (of $k/2^j$).

So, we are interested in "wavelet" functions $\psi$ whose binary dilations and dyadic translations are enough to represent all the functions in $L^2(\mathbb{R})$. For simplicity, let us first consider an orthogonal basis generated by $\psi$. Later in this chapter (see Section 1.4), we will introduce the more general "wavelet series".

Throughout this book, we will use the following notations for the *inner product* and *norm* for the space $L^2(\mathbb{R})$:

$$\langle f, g \rangle := \int_{-\infty}^{\infty} f(x) \overline{g(x)} dx; \tag{1.1.9}$$

$$\|f\|_2 := \langle f, f \rangle^{1/2}, \tag{1.1.10}$$

where $f, g \in L^2(\mathbb{R})$. Note that for any $j, k \in \mathbb{Z}$, we have

$$\|f(2^j \cdot - k)\|_2 = \left\{ \int_{-\infty}^{\infty} |f(2^j x - k)|^2 dx \right\}^{1/2}$$
$$= 2^{-j/2} \|f\|_2.$$

Hence, if a function $\psi \in L^2(\mathbb{R})$ has unit length, then all of the functions $\psi_{j,k}$, defined by

$$\psi_{j,k}(x) := 2^{j/2} \psi(2^j x - k), \qquad j, k \in \mathbb{Z}, \tag{1.1.11}$$

also have unit length; that is,

$$\|\psi_{j,k}\|_2 = \|\psi\|_2 = 1, \qquad j, k \in \mathbb{Z}. \tag{1.1.12}$$

In this book, the Kronecker symbol

$$\delta_{j,k} := \begin{cases} 1 & \text{for} \quad j = k; \\ 0 & \text{for} \quad j \neq k, \end{cases} \tag{1.1.13}$$

defined on $\mathbb{Z} \times \mathbb{Z}$, will be often used.

**Definition 1.1.** *A function $\psi \in L^2(\mathbb{R})$ is called an orthogonal wavelet (or o.n. wavelet), if the family $\{\psi_{j,k}\}$, as defined in (1.1.11), is an orthonormal basis of $L^2(\mathbb{R})$; that is,*

$$\langle \psi_{j,k}, \psi_{\ell,m} \rangle = \delta_{j,\ell} \cdot \delta_{k,m}, \qquad j, k, \ell, m \in \mathbb{Z}, \tag{1.1.14}$$

and every $f \in L^2(\mathbb{R})$ can be written as

$$f(x) = \sum_{j,k=-\infty}^{\infty} c_{j,k}\psi_{j,k}(x), \qquad (1.1.15)$$

where the convergence of the series in (1.1.15) is in $L^2(\mathbb{R})$, namely:

$$\lim_{M_1,N_1,M_2,N_2\to\infty} \left\| f - \sum_{j=-M_2}^{N_2} \sum_{k=-M_1}^{N_1} c_{j,k}\psi_{j,k} \right\|_2 = 0.$$

The simplest example of an orthogonal wavelet is the Haar function $\psi_H$ defined by

$$\psi_H(x) := \begin{cases} 1 & \text{for} & 0 \le x < \frac{1}{2}; \\ -1 & \text{for} & \frac{1}{2} \le x < 1; \\ 0 & \text{otherwise.} \end{cases} \qquad (1.1.16)$$

We will give a brief discussion of this function in Sections 1.5 and 1.6. Other o.n. wavelets will be studied in some details in Chapter 7.

The series representation of $f$ in (1.1.15) is called a *wavelet series*. Analogous to the notion of Fourier coefficients in (1.1.2), the wavelet coefficients $c_{j,k}$ are given by

$$c_{j,k} = \langle f, \psi_{j,k} \rangle. \qquad (1.1.17)$$

That is, if we define an integral transform $W_\psi$ on $L^2(\mathbb{R})$ by

$$(W_\psi f)(b,a) := |a|^{-\frac{1}{2}} \int_{-\infty}^{\infty} f(x)\overline{\psi(\frac{x-b}{a})}\,dx, \qquad f \in L^2(\mathbb{R}), \qquad (1.1.18)$$

then the wavelet coefficients in (1.1.15) and (1.1.17) become

$$c_{j,k} = (W_\psi f)\left(\frac{k}{2^j}, \frac{1}{2^j}\right). \qquad (1.1.19)$$

The linear transformation $W_\psi$ is called the "*integral wavelet transform*" relative to the "basic wavelet" $\psi$. Hence, the $(j,k)^{\text{th}}$ *wavelet coefficient of $f$ is given by the integral wavelet transformation of $f$ evaluated at the dyadic position* $b = k/2^j$ *with binary dilation $a = 2^{-j}$*, where the same o.n. wavelet $\psi$ is used to generate the wavelet series (1.1.15) and to define the integral wavelet transform (1.1.18).

The importance of the integral wavelet transform will be discussed in the next section. Here, we only mention that this integral transform greatly enhances the value of the (integral) Fourier transform $\mathcal{F}$, defined by

$$(\mathcal{F}f)(y) := \int_{-\infty}^{\infty} e^{-iyx} f(x)\,dx, \qquad f \in L^2(\mathbb{R}). \qquad (1.1.20)$$

The mathematical treatment of this transform will be delayed to the next chapter. As is well known, the Fourier transform is the other important component of Fourier analysis. Hence, it is interesting to note that while the two components of Fourier analysis, namely: the Fourier series and the Fourier transform, are basically unrelated; the two corresponding components of wavelet analysis, namely: the wavelet series (1.1.15) and the integral wavelet transform (1.1.18), have an intimate relationship as described by (1.1.19).

## 1.2. The integral wavelet transform and time-frequency analysis

The Fourier transform $\mathcal{F}$ defined in (1.1.20) not only is a very powerful mathematical tool, but also has very significant physical interpretations in applications. For instance, if a function $f \in L^2(\mathbb{R})$ is considered as an *analog signal* with *finite energy*, defined by its norm $\|f\|_2$, then the Fourier transform

$$\hat{f}(\omega) := (\mathcal{F}f)(\omega) \qquad (1.2.1)$$

of $f$ represents the *spectrum* of this signal. In signal analysis, analog signals are defined in the *time-domain*, and the spectral information of these signals is given in the *frequency-domain*. To facilitate our presentation, we will allow negative frequencies for the time being. Hence, both the time- and frequency-domains are the real line $\mathbb{R}$. Analogous to the Parseval Identity for Fourier series, the Parseval Identity that describes the relationship between functions in $L^2(\mathbb{R})$ and their Fourier transforms is given by

$$\langle f, g \rangle = \frac{1}{2\pi} \langle \hat{f}, \hat{g} \rangle, \qquad f, g \in L^2(\mathbb{R}). \qquad (1.2.2)$$

Here, the notation of inner product introduced in (1.1.9) is used, and as will be seen in the next chapter, the Fourier transformation $\mathcal{F}$ takes $L^2(\mathbb{R})$ onto itself. As a consequence of (1.2.2), we observe that the energy of an analog signal is directly proportional to its spectral content; more precisely,

$$\|f\|_2 = \frac{1}{\sqrt{2\pi}} \|\hat{f}\|_2, \qquad f \in L^2(\mathbb{R}). \qquad (1.2.3)$$

However, the formula

$$\hat{f}(\omega) = \int_{-\infty}^{\infty} e^{-it\omega} f(t)\,dt \qquad (1.2.4)$$

of the Fourier transform alone is quite inadequate for most applications. In the first place, to extract the spectral information $\hat{f}(\omega)$ from the analog signal $f(t)$ from this formula, it takes an infinite amount of time, using both past and future information of the signal just to evaluate the spectrum at a single frequency $\omega$. Besides, the formula (1.2.4) does not even reflect frequencies that evolve with time. What is really needed is for one to be able to determine the time intervals that yield the spectral information on any desirable range of frequencies (or frequency band). In addition, since the frequency of a signal is directly proportional to the length of its cycle, it follows that for high-frequency spectral information, the time-interval should be relatively small to give better accuracy, and for low-frequency spectral information, the time-interval should be relatively wide to give complete information. In other words, it is important to have a flexible time-frequency window that automatically narrows at high "*center-frequency*" and widens at low center-frequency. Fortunately, the integral wavelet transform $W_\psi$ relative to some "basic wavelet" $\psi$, as introduced in (1.1.18), has this so-called zoom-in and zoom-out capability.

To be more specific, both $\psi$ and its Fourier transform $\widehat{\psi}$ must have sufficiently fast decay so that they can be used as "*window functions*". For an $L^2(\mathbb{R})$ function $w$ to qualify as a window function, it must be possible to identify its "center" and "width", which are defined as follows.

**Definition 1.2.** *A nontrivial function $w \in L^2(\mathbb{R})$ is called a window function if $xw(x)$ is also in $L^2(\mathbb{R})$. The center $t^*$ and radius $\Delta_w$ of a window function $w$ are defined to be*

$$t^* := \frac{1}{\|w\|_2^2} \int_{-\infty}^{\infty} x|w(x)|^2 dx \tag{1.2.5}$$

*and*

$$\Delta_w := \frac{1}{\|w\|_2} \left\{ \int_{-\infty}^{\infty} (x - t^*)^2 |w(x)|^2 dx \right\}^{1/2}, \tag{1.2.6}$$

*respectively; and the width of the window function $w$ is defined by $2\Delta_w$.*

We have not formally defined a "basic wavelet" $\psi$ yet and will not do so until the next section. An example of a basic wavelet is any orthogonal wavelet as already discussed in the previous section. In any case, we will see that any basic wavelet window function must necessarily satisfy:

$$\int_{-\infty}^{\infty} \psi(x)dx = 0, \tag{1.2.7}$$

so that its graph is a *small wave*.

Suppose that $\psi$ is any basic wavelet such that both $\psi$ and its Fourier transform $\widehat{\psi}$ are window functions with centers and radii given by $t^*, \omega^*, \Delta_\psi, \Delta_{\widehat{\psi}}$, respectively. Then in the first place, it is clear that the integral wavelet transform

$$(W_\psi f)(b, a) = |a|^{-\frac{1}{2}} \int_{-\infty}^{\infty} f(t)\overline{\psi\left(\frac{t - b}{a}\right)} dt \tag{1.2.8}$$

of an analog signal $f$, as introduced in (1.1.18), localizes the signal with a "time window"

$$[b + at^* - a\Delta_\psi, \, b + at^* + a\Delta_\psi],$$

where the center of the window is at $b + at^*$ and the width is given by $2a\Delta_\psi$. This is called "time-localization" in signal analysis. On the other hand, if we set

$$\eta(\omega) := \widehat{\psi}(\omega + \omega^*), \tag{1.2.9}$$

then $\eta$ is also a window function with center at 0 and radius given by $\Delta_{\widehat{\psi}}$; and by the Parseval Identity (1.2.2), the integral wavelet transform in (1.2.8) becomes

$$(W_\psi f)(b,a) = \frac{a|a|^{-\frac{1}{2}}}{2\pi} \int_{-\infty}^{\infty} \hat{f}(\omega)e^{ib\omega}\overline{\eta\left(a\left(\omega - \frac{\omega^*}{a}\right)\right)}\,d\omega. \qquad (1.2.10)$$

Hence, with the exception of multiplication by $a|a|^{-\frac{1}{2}}/2\pi$ and a linear phase-shift of $e^{ib\omega}$, determined by the amount of translation of the time-window, the same quantity $(W_\psi f)(b,a)$ also gives localized information of the spectrum $\hat{f}(\omega)$ of the signal $f(t)$, with a "frequency window"

$$\left[\frac{\omega^*}{a} - \frac{1}{a}\Delta_{\widehat{\psi}}, \; \frac{\omega^*}{a} + \frac{1}{a}\Delta_{\widehat{\psi}}\right],$$

whose center is at $\omega^*/a$ and whose width is given by $2\Delta_{\widehat{\psi}}/a$. This is called "frequency-localization". By equating the quantities (1.2.8) and (1.2.10), we now have a "time-frequency window":

$$[b + at^* - a\Delta_\psi, \; b + at^* + a\Delta_\psi] \times \left[\frac{\omega^*}{a} - \frac{1}{a}\Delta_{\widehat{\psi}}, \; \frac{\omega^*}{a} + \frac{1}{a}\Delta_{\widehat{\psi}}\right] \qquad (1.2.11)$$

for time-frequency analysis using the integral wavelet transform relative to a basic wavelet $\psi$ with the window conditions described above.

Several comments are in order. First, since we must eventually consider positive frequencies, the basic wavelet $\psi$ should be so chosen that the center $\omega^*$ of $\hat{\psi}$ is a positive number. In practice, this positive number, along with the positive scaling parameter $a$, is selected in such a way that $\omega^*/a$ is the "center-frequency" of the "frequency band" $\left[\frac{\omega^*}{a} - \frac{1}{a}\Delta_{\widehat{\psi}}, \frac{\omega^*}{a} + \frac{1}{a}\Delta_{\widehat{\psi}}\right]$ of interest. Then the ratio of the center-frequency to the width of the frequency band is given by

$$\frac{\omega^*/a}{2\Delta_{\widehat{\psi}}/a} = \frac{\omega^*}{2\Delta_{\widehat{\psi}}}, \qquad (1.2.12)$$

which is independent of the location of the center-frequency. This is called "constant-$Q$" frequency analysis. The importance of the time-frequency window (1.2.11) is that it narrows for large center-frequency $\omega^*/a$ and widens for small center-frequency $\omega^*/a$ (cf. Figure 1.2.1), although the area of the window is a constant, given by $4\Delta_\psi\Delta_{\widehat{\psi}}$. This is exactly what is most desirable in time-frequency analysis. Details will be studied in Chapter 3.

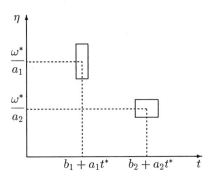

**Figure 1.2.1.** Time-frequency windows, $a_1 < a_2$.

### 1.3. Inversion formulas and duals

The integral wavelet transform $(W_\psi f)(b, a)$ gives the location (in terms of $b + at^*$), the "rate" (in terms of $a$), and the amount (measured by the value $(W_\psi f)(b, a)$) of change of $f$, with the zoom-in and zoom-out capability. This information is extremely valuable in many applications such as time-frequency analysis. For instance, in data compression, the values of $(W_\psi f)(b, a)$ below a certain tolerance level are removed; and in lowpass filtering, $(W_\psi f)(b, a)$ is replaced by zero for small values of $a$. In any case, the (new and modified) function $f$ has to be reconstructed from the values of $(W_\psi f)(b, a)$. Any formula that expresses every $f \in L^2(\mathbb{R})$ in terms of $(W_\psi f)(b, a)$ will be called an "*inverse formula*", and the (kernel) function $\tilde{\psi}$ to be used in this formula will be called a "*dual*" of the basic wavelet $\psi$. Hence, in practice $\psi$ can be used as a *basic wavelet*, only if an inversion formula exists.

In the following, we will study four different situations, to be considered in the order of restrictiveness of the domain of information of $W_\psi f$.

($1°$) *Recovery from* $(W_\psi f)(b, a)$, $a, b \in \mathbb{R}$.

In order to reconstruct $f$ from $W_\psi f$, we need to know the constant

$$C_\psi := \int_{-\infty}^{\infty} \frac{|\widehat{\psi}(\omega)|^2}{|\omega|} d\omega < \infty. \tag{1.3.1}$$

The finiteness of this constant restricts the class of $L^2(\mathbb{R})$ functions $\psi$ that can be used as "basic wavelets" in the definition of the integral wavelet transform. In particular, if $\psi$ must also be a window function, then $\psi$ is necessarily in $L^1(\mathbb{R})$, meaning:

$$\int_{-\infty}^{\infty} |\psi(x)| dx < \infty,$$

so that $\widehat{\psi}$ is a continuous function in $\mathbb{R}$ (see Theorem 2.2 in Chapter 2), and hence, it follows from (1.3.1) that $\widehat{\psi}$ must vanish at the origin; that is,

$$\int_{-\infty}^{\infty} \psi(x)dx = 0. \tag{1.3.2}$$

So, the graph of a basic wavelet $\psi$ is a small wave. With the constant $C_\psi$, we have the following reconstruction formula:

$$f(x) = \frac{1}{C_\psi} \int_{\mathbb{R}^2} \int \{(W_\psi f)(b,a)\} \left\{ |a|^{-\frac{1}{2}} \psi\left(\frac{x-b}{a}\right) \right\} \frac{dadb}{a^2}, \quad f \in L^2(\mathbb{R}), \tag{1.3.3}$$

where $\mathbb{R}^2 = \mathbb{R} \times \mathbb{R}$. Observe that the same kernel,

$$|a|^{-\frac{1}{2}} \psi\left(\frac{x-b}{a}\right),$$

with the exception of complex conjugation, is used to define both the integral wavelet transform in (1.1.18) and its inverse in (1.3.3). Hence, $\overline{\psi}$ may be called a *"dual"* of the basic wavelet $\psi$. Of course, one cannot expect uniqueness of this dual.

$(2°)$ *Recovery from* $(W_\psi f)(b,a)$, $b \in \mathbb{R}$ *and* $a > 0$.

In time-frequency analysis as discussed in the previous section, we use a positive constant multiple of $a^{-1}$ to represent frequency. Hence, since only positive frequency is of interest, we need a reconstruction formula where the integration is over $\mathbb{R} \times (0, \infty)$ instead of $\mathbb{R}^2$. Therefore, we must now consider even a smaller class of basic wavelets $\psi$, namely: wavelets $\psi$ satisfying

$$\int_0^\infty \frac{|\widehat{\psi}(\omega)|^2}{\omega} d\omega = \int_0^\infty \frac{|\widehat{\psi}(-\omega)|^2}{\omega} d\omega = \frac{1}{2} C_\psi < \infty, \tag{1.3.4}$$

where $C_\psi$ is defined in (1.3.1). For instance, any real-valued $\psi$ satisfying (1.3.1) can be used as a basic wavelet in this situation. For any $\psi$ satisfying (1.3.4), we have the following reconstruction formula:

$$f(x) = \frac{2}{C_\psi} \int_0^\infty \left[ \int_{-\infty}^\infty \{(W_\psi f)(b,a)\} \left\{ \frac{1}{\sqrt{a}} \psi\left(\frac{x-b}{a}\right) \right\} db \right] \frac{da}{a^2}, \quad f \in L^2(\mathbb{R}). \tag{1.3.5}$$

With the exception of a factor of 2, this formula is the same as the reconstruction formula (1.3.3). Of course, the basic wavelet $\psi$ in (1.3.5) is more restrictive. As in $(1°)$, we again call the complex conjugate $\overline{\psi}$ of $\psi$ a *"dual"* of the basic wavelet $\psi$ for the situation $(2°)$. Once again there is no reason to expect a unique dual.

$(3°)$ *Recovery from* $(W_\psi f)(b,a)$, $b \in \mathbb{R}, a = \frac{1}{2^j}$; $j \in \mathbb{Z}$.

By restricting our attention to $a = 2^{-j}$, where $j$ runs over all the integers, we can consider time-frequency localization with frequency windows

$$B_j := [2^j \omega^* - 2^j \Delta_{\widehat{\psi}}, 2^j \omega^* + 2^j \Delta_{\widehat{\psi}}], \quad j \in \mathbb{Z}. \tag{1.3.6}$$

In particular, if the center $\omega^*$ of the window function $\widehat{\psi}$ is chosen to be

$$\omega^* = 3\Delta_{\widehat{\psi}},$$

then the frequency bands $B_j$, $j \in \mathbb{Z}$, in (1.3.6) form a disjoint partition of the whole frequency-axis $[0, \infty)$, with the exception of the end-points of the intervals $B_j$. The integral wavelet transform (1.2.8) is used to determine the time-intervals $[b + 2^{-j}t^* - 2^{-j}\Delta_{\psi}, \, b + 2^{-j}t^* + 2^{-j}\Delta_{\psi}]$ on which the spectral content of the signal $f$, with frequencies in the range $B_j$, is of some significance, namely: the value of $|(W_{\psi}f)(b, 2^{-j})|$ is above a certain threshold.

Since only partial information of $W_{\psi}f$ is available, the basic wavelet $\psi$ must again satisfy a stronger condition than (1.3.1) for a reconstruction formula to be available. The condition we impose on $\psi$ is the following so-called "stability condition":

$$A \le \sum_{j=-\infty}^{\infty} |\widehat{\psi}(2^{-j}\omega)|^2 \le B, \tag{1.3.7}$$

where $A$ and $B$, with $0 < A \le B < \infty$, are constants independent of $\omega$. It follows quite easily from (1.3.7) that $\psi$ also satisfies

$$A \ln 2 \le \int_0^{\infty} \frac{|\widehat{\psi}(\omega)|^2}{\omega} d\omega, \, \int_0^{\infty} \frac{|\widehat{\psi}(-\omega)|^2}{\omega} d\omega \le B \ln 2, \tag{1.3.8}$$

which implies that $C_{\psi}$ lies between $2A \ln 2$ and $2B \ln 2$. Details will be discussed in Section 3.4 in Chapter 3. If $\widehat{\psi}$ satisfies (1.3.7), then the basic wavelet $\psi$ has a *"dual"* $\psi^*$ whose Fourier transform is given by

$$\widehat{\psi}^*(\omega) := \frac{\overline{\widehat{\psi}(\omega)}}{\displaystyle\sum_{j=-\infty}^{\infty} |\widehat{\psi}(2^{-j}\omega)|^2}. \tag{1.3.9}$$

The reconstruction formula by using this dual may be stated as follows:

$$f(x) = \sum_{j=-\infty}^{\infty} \int_{-\infty}^{\infty} \{2^{j/2}(W_{\psi}f)(b, 2^{-j})\}\{2^j \psi^*(2^j(x-b))\}db, \quad f \in L^2(\mathbb{R}). \tag{1.3.10}$$

Since basic wavelets $\psi$ for this situation have both theoretical and practical value, they are given the following special name.

**Definition 1.3.** A function $\psi \in L^2(\mathbb{R})$ is called a "dyadic wavelet" if it satisfies the stability condition (1.3.7) for almost all $\omega \in \mathbb{R}$ for some constants $A$ and $B$ with $0 < A \le B < \infty$.

It will be seen in Chapter 3 that even dyadic wavelets do not have unique duals in general. The most interesting examples of dyadic wavelets are probably the so-called "frames", to be introduced in Section 3.5.

(4°) *Recovery from* $(W_\psi f)(b, a), b = \frac{k}{2^j}, a = \frac{1}{2^j}; j, k \in \mathbb{Z}$.

In order to construct efficient algorithms for determining the integral wavelet transform $(W_\psi f)(b, a)$ and for reconstructing $f$ from $(W_\psi f)(b, a)$, only discrete samples are considered. While it is important to partition the frequency axis into frequency bands by using powers of two for the scale parameter $a$, say, as in (3°), it is much more efficient to consider only the samples at the dyadic values $b = k/2^j$ on the time-axis, when $a = 2^{-j}$, $j \in \mathbb{Z}$, instead of all $b \in \mathbb{R}$. In many applications, there is very minimal, if any, sacrifice by using this uniform discrete sampling, and as we shall see later, the mathematical theory of this approach is very attractive.

We first observe that

$$(W_\psi f)\left(\frac{k}{2^j}, \frac{1}{2^j}\right) = \int_{-\infty}^{\infty} f(x)\{\overline{2^{j/2}\psi(2^j x - k)}\}dx \qquad (1.3.11)$$
$$= \langle f, \psi_{j,k}\rangle,$$

where, as in (1.1.11),

$$\psi_{j,k}(x) := 2^{j/2}\psi(2^j x - k), \qquad j, k \in \mathbb{Z}. \qquad (1.3.12)$$

However, in general we do not require $\{\psi_{j,k}\}$ to be an o.n. basis of $L^2(\mathbb{R})$ as in Section 1.1. Indeed a "stable" basis, as defined in the following, is sufficient.

**Definition 1.4.** *A function $\psi \in L^2(\mathbb{R})$ is called an $\mathcal{R}$-function if $\{\psi_{j,k}\}$, as defined in (1.3.12), is a Riesz basis of $L^2(\mathbb{R})$, in the sense that the linear span of $\psi_{j,k}, j, k \in \mathbb{Z}$, is dense in $L^2(\mathbb{R})$ and that positive constants $A$ and $B$ exist, with $0 < A \leq B < \infty$, such that*

$$A\|\{c_{j,k}\}\|_{\ell^2}^2 \leq \left\|\sum_{j=-\infty}^{\infty}\sum_{k=-\infty}^{\infty} c_{j,k}\psi_{j,k}\right\|_2^2 \leq B\|\{c_{j,k}\}\|_{\ell^2}^2, \qquad (1.3.13)$$

*for all doubly bi-infinite square-summable sequences $\{c_{j,k}\}$; that is,*

$$\|\{c_{j,k}\}\|_{\ell^2}^2 := \sum_{j=-\infty}^{\infty}\sum_{k=-\infty}^{\infty} |c_{j,k}|^2 < \infty.$$

Suppose that $\psi$ is an $\mathcal{R}$-function. Then there is a unique Riesz basis $\{\psi^{j,k}\}$ of $L^2(\mathbb{R})$ which is dual to $\{\psi_{j,k}\}$ in the sense that

$$\langle \psi_{j,k}, \psi^{\ell,m}\rangle = \delta_{j,\ell} \cdot \delta_{k,m}, \qquad j, k, \ell, m \in \mathbb{Z}. \qquad (1.3.14)$$

Hence, every function $f \in L^2(\mathbb{R})$ has the following (unique) series expansion:

$$f(x) = \sum_{j,k=-\infty}^{\infty} \langle f, \psi_{j,k}\rangle\psi^{j,k}(x). \qquad (1.3.15)$$

However, although the coefficients are values of the integral wavelet transform of $f$ relative to $\psi$, the series (1.3.15) is *not* necessarily a *wavelet series*. To qualify as a wavelet series, there must exist some function $\widetilde{\psi} \in L^2(\mathbb{R})$, such that the dual basis $\{\psi^{j,k}\}$ in the series (1.3.15) is obtained from $\widetilde{\psi}$ by

$$\psi^{j,k}(x) = \widetilde{\psi}_{j,k}(x) \tag{1.3.16}$$

where, as usual, the notation

$$\widetilde{\psi}_{j,k}(x) := 2^{j/2}\widetilde{\psi}(2^j x - k) \tag{1.3.17}$$

is used. If $\{\psi_{j,k}\}$ is an o.n. basis of $L^2(\mathbb{R})$, as already discussed in (1.1.14), (1.1.15), and (1.1.17), then it is clear that (1.3.14) holds with $\psi^{j,k} = \psi_{j,k}$, or $\widetilde{\psi} \equiv \psi$. In general, however, as we will see in the next section, $\widetilde{\psi}$ does not exist. If $\psi$ is so chosen that $\widetilde{\psi}$ exists, then the pair $(\psi, \widetilde{\psi})$ is very useful for displaying values of integral wavelet transforms of $f \in L^2(\mathbb{R})$ at the dyadic positions and different binary scale levels (or octaves) and for recovering $f$ from these values of its integral wavelet transforms. Precisely, we have

$$f(x) = \sum_{j,k=-\infty}^{\infty} \langle f, \psi_{j,k} \rangle \widetilde{\psi}_{j,k}(x) \tag{1.3.18}$$

$$= \sum_{j,k=-\infty}^{\infty} \langle f, \widetilde{\psi}_{j,k} \rangle \psi_{j,k}(x).$$

## 1.4. Classification of wavelets

Let $\psi \in L^2(\mathbb{R})$ be an $\mathcal{R}$-function; that is, $\{\psi_{j,k}\}$, as defined in (1.3.12), is a Riesz basis of $L^2(\mathbb{R})$. The first question we are faced with is if the dual basis $\{\psi^{j,k}\}$, relative to $\{\psi_{j,k}\}$, as defined in (1.3.14), is derived from some function $\widetilde{\psi} \in L^2(\mathbb{R})$ as in (1.3.16)-(1.3.17). Somewhat surprisingly, the answer is negative in general.

For instance, let $\eta \in L^2(\mathbb{R})$ be any orthogonal wavelet as introduced in Definition 1.1. For each complex number $z$ with $|z| < 1$, consider the function

$$\psi(x) := \psi_z(x) := \eta(x) - \bar{z}\sqrt{2}\eta(2x). \tag{1.4.1}$$

Then it is clear that the family $\{\psi_{j,k}\}$, as defined in (1.3.12), is a Riesz basis of $L^2(\mathbb{R})$. Now, let us consider the dual basis $\{\psi^{j,k}\}$ relative to $\{\psi_{j,k}\}$. It is easy to verify, in particular, that

$$\begin{cases} \psi^{0,0}(x) = \displaystyle\sum_{\ell=0}^{\infty} \eta_{-\ell,0}(x)z^{\ell}; \\ \\ \psi^{0,1}(x) = \eta_{0,1}(x). \end{cases} \tag{1.4.2}$$

If some function $\widetilde{\psi} = \widetilde{\psi}_z \in L^2(\mathbb{R})$ could be found such that (1.3.16)-(1.3.17) hold, then we have

$$\begin{aligned}
\eta(x) &= \eta_{0,1}(x+1) = \psi^{0,1}(x+1) \\
&= \widetilde{\psi}_{0,1}(x+1) = \widetilde{\psi}_{0,0}(x) \\
&= \psi^{0,0}(x) = \sum_{\ell=0}^{\infty} \eta_{-\ell,0}(x)z^\ell,
\end{aligned}$$

or

$$\sum_{\ell=1}^{\infty} \eta_{-\ell,0}(x)z^\ell = 0.$$

Since this is manifestly absurd, with the exception of at most a finite number of values of $z$ in $|z| \leq r$, where $0 < r < 1$ is arbitrary, we conclude that $\widetilde{\psi} = \widetilde{\psi}_z$ does not exist in general.

The above discussion motivates the following definition of "wavelets".

**Definition 1.5.** *An $\mathcal{R}$-function $\psi \in L^2(\mathbb{R})$ is called an $\mathcal{R}$-wavelet (or wavelet), if there exists a function $\widetilde{\psi} \in L^2(\mathbb{R})$, such that $\{\psi_{j,k}\}$ and $\{\widetilde{\psi}_{j,k}\}$, as defined in (1.3.12) and (1.3.17), are dual bases of $L^2(\mathbb{R})$. If $\psi$ is an $\mathcal{R}$-wavelet, then $\widetilde{\psi}$ is called a dual wavelet corresponding to $\psi$.*

It is clear that a dual wavelet $\widetilde{\psi}$ is unique and is itself an $\mathcal{R}$-wavelet. More precisely, the pair $(\psi, \widetilde{\psi})$ is symmetric in the sense that $\psi$ is the dual wavelet of $\widetilde{\psi}$ also. For convenience, we will simply call $\psi$ a "wavelet" and $\widetilde{\psi}$ the "dual" of $\psi$. As we already remarked in Section 1.3, if $\psi$ is an orthogonal wavelet, then it is self-dual in the sense of $\widetilde{\psi} \equiv \psi$.

It is important to emphasize once more that every wavelet $\psi$, orthogonal or not, generates a "wavelet series" representation of any $f \in L^2(\mathbb{R})$, namely:

$$f(x) = \sum_{j,k=-\infty}^{\infty} c_{j,k}\psi_{j,k}(x),$$

where each $c_{j,k}$ is the integral wavelet transform of $f$ relative to the dual $\widetilde{\psi}$ of $\psi$, evaluated at the time-scale coordinate

$$(b,a) = \left( \frac{k}{2^j}, \frac{1}{2^j} \right).$$

Let $\psi$ be any wavelet and consider the Riesz basis $\{\psi_{j,k}\}$ it generates. For each $j \in \mathbb{Z}$, let $W_j$ denote the closure of the linear span of $\{\psi_{j,k}: k \in \mathbb{Z}\}$, namely:

$$W_j := \text{clos}_{L^2(\mathbb{R})}\langle \psi_{j,k}: k \in \mathbb{Z}\rangle. \tag{1.4.3}$$

Then it is clear that $L^2(\mathbb{R})$ can be decomposed as a *direct sum* of the spaces $W_j$:

$$L^2(\mathbb{R}) = \overset{\bullet}{\sum_{j \in \mathbb{Z}}} W_j := \cdots \dotplus W_{-1} \dotplus W_0 \dotplus W_1 \dotplus \cdots, \qquad (1.4.4)$$

in the sense that every function $f \in L^2(\mathbb{R})$ has a unique decomposition:

$$f(x) = \cdots + g_{-1}(x) + g_0(x) + g_1(x) + \cdots, \qquad (1.4.5)$$

where $g_j \in W_j$ for all $j \in \mathbb{Z}$. The dots above the summation and plus signs in (1.4.4) indicate "direct sums".

If $\psi$ is an orthogonal wavelet, then the subspaces $W_j$ of $L^2(\mathbb{R})$ are mutually orthogonal, meaning:

$$\langle g_j, g_\ell \rangle = 0, \quad j \neq \ell, \quad \text{where} \quad g_j \in W_j \quad \text{and} \quad g_\ell \in W_\ell. \qquad (1.4.6)$$

In this case, we will use the notation

$$W_j \perp W_\ell, \qquad j \neq \ell. \qquad (1.4.7)$$

Consequently, the direct sum in (1.4.4) becomes an *orthogonal sum*:

$$L^2(\mathbb{R}) = \bigoplus_{j \in \mathbb{Z}} W_j := \cdots \oplus W_{-1} \oplus W_0 \oplus W_1 \oplus \cdots, \qquad (1.4.8)$$

where the circles around the plus signs in (1.4.8) indicate "orthogonal sums". The decomposition (1.4.8) is usually called an *orthogonal decomposition* of $L^2(\mathbb{R})$. This means that the decomposition (1.4.5) of any $f \in L^2(\mathbb{R})$ as the (infinite) sum of functions $g_j \in W_j$ is not only unique, but these components of $f$ are also mutually orthogonal, as described by (1.4.6).

So, an orthogonal wavelet $\psi$ generates an orthogonal decomposition of $L^2(\mathbb{R})$. However, we have not used all the orthogonality properties of $\{\psi_{j,k}\}$, namely: for each $j$, the orthogonality condition $\langle \psi_{j,k}, \psi_{j,\ell} \rangle = \delta_{k,\ell}$ is not reflected in (1.4.8). This means that there is a larger class of wavelets that can be used to generate orthogonal decompositions of $L^2(\mathbb{R})$. The available flexibility is important for constructing wavelets with certain desirable properties. The most important property that can be achieved for compactly supported wavelets $\psi$ with this flexibility is "symmetry" or "antisymmetry". Details will be studied in Chapters 5 and 6.

**Definition 1.6.** *A wavelet $\psi$ in $L^2(\mathbb{R})$ is called a semi-orthogonal wavelet (or s.o. wavelet) if the Riesz basis $\{\psi_{j,k}\}$ it generates satisfies*

$$\langle \psi_{j,k}, \psi_{\ell,m} \rangle = 0, \quad j \neq \ell; \qquad j, k, \ell, m \in \mathbb{Z}. \qquad (1.4.9)$$

Obviously, every s.o. wavelet generates an orthogonal decomposition (1.4.8) of $L^2(\mathbb{R})$, and every o.n. wavelet is also an s.o. wavelet. A wavelet (or more precisely, an $\mathcal{R}$-wavelet) $\psi$ is called a nonorthogonal (or n.o.) wavelet if it is not an s.o. wavelet. However, being an $\mathcal{R}$-wavelet, it has a dual $\tilde{\psi}$, and the pair $(\psi, \tilde{\psi})$ satisfies the *bi-orthogonality* property:

$$\langle \psi_{j,k}, \tilde{\psi}_{\ell,m} \rangle = \delta_{j,\ell} \cdot \delta_{k,m}, \qquad j, k, \ell, m \in \mathbb{Z}. \tag{1.4.10}$$

## 1.5. Multiresolution analysis, splines, and wavelets

Any wavelet, semi-orthogonal or not, generates a direct sum decomposition (1.4.4) of $L^2(\mathbb{R})$. For each $j \in \mathbb{Z}$, let us consider the closed subspaces

$$V_j = \cdots \dot{+} W_{j-2} \dot{+} W_{j-1}, \qquad j \in \mathbb{Z}, \tag{1.5.1}$$

of $L^2(\mathbb{R})$. These subspaces clearly have the following properties:

(1°) $\cdots \subset V_{-1} \subset V_0 \subset V_1 \subset \cdots$ ;

(2°) $\mathrm{clos}_{L^2} \left( \bigcup_{j \in \mathbb{Z}} V_j \right) = L^2(\mathbb{R})$;

(3°) $\bigcap_{j \in \mathbb{Z}} V_j = \{0\}$;

(4°) $V_{j+1} = V_j \dot{+} W_j, \quad j \in \mathbb{Z}$; and

(5°) $f(x) \in V_j \Leftrightarrow f(2x) \in V_{j+1}, j \in \mathbb{Z}$.

Hence, in contrast to the subspaces $W_j$ which satisfy

$$W_j \cap W_\ell = \{0\}, \qquad j \neq \ell,$$

the sequence of subspaces $V_j$ is nested, as described by (1°), and has the property that every function $f$ in $L^2(\mathbb{R})$ can be approximated as closely as desirable by its projections $P_j f$ in $V_j$, as described by (2°). But on the other hand, by decreasing $j$, the projections $P_j f$ could have arbitrarily small energy, as guaranteed by (3°). What is not described by (1°)–(3°) is the *most important* intrinsic property of these spaces which is that more and more "variations" of $P_j f$ are removed as $j \to -\infty$. In fact, these variations are peeled off, level by level in decreasing order of the "rate of variations" (better known as "*frequency bands*") and stored in the complementary subspaces $W_j$ as in (4°). This process can be made very efficient by an application of the property (5°).

In fact, if the reference subspace $V_0$, say, is generated by a single function $\phi \in L^2(\mathbb{R})$ in the sense that

$$V_0 = \mathrm{clos}_{L^2(\mathbb{R})} \langle \phi_{0,k} \colon k \in \mathbb{Z} \rangle, \tag{1.5.2}$$

where

$$\phi_{j,k}(x) := 2^{j/2} \phi(2^j x - k), \tag{1.5.3}$$

then all the subspaces $V_j$ are also generated by the same $\phi$ (just as the subspaces $W_j$ are generated by $\psi$ as in (1.4.3)), namely:

$$V_j = \mathrm{clos}_{L^2(\mathbb{R})} \langle \phi_{j,k} \colon k \in \mathbb{Z} \rangle, \qquad j \in \mathbb{Z}. \tag{1.5.4}$$

Hence, the "peeling-off" process from $V_j$ to $W_{j-1}, W_{j-2}, \ldots, W_{j-\ell}$ can be accomplished efficiently. We will return to this topic in the next section.

**Definition 1.7.** *A function $\phi \in L^2(\mathbb{R})$ is said to generate a multiresolution analysis (MRA) if it generates a nested sequence of closed subspaces $V_j$ that satisfy $(1^\circ), (2^\circ), (3^\circ)$, and $(5^\circ)$ in the sense of (1.5.4), such that $\{\phi_{0,k}\}$ forms a Riesz basis of $V_0$. Here, analogous to Definition 1.4, for $\{\phi_{0,k}\}$ to be a Riesz basis of $V_0$, there must exist two constants $A$ and $B$, with $0 < A \leq B < \infty$, such that*

$$A\|\{c_k\}\|_{\ell^2}^2 \leq \left\| \sum_{k=-\infty}^{\infty} c_k \phi_{0,k} \right\|_2^2 \leq B\|\{c_k\}\|_{\ell^2}^2, \tag{1.5.5}$$

*for all bi-infinite square summable sequences $\{c_k\}$; that is,*

$$\|\{c_k\}\|_{\ell^2}^2 = \sum_{k=-\infty}^{\infty} |c_k|^2 < \infty. \tag{1.5.6}$$

*If $\phi$ generates an MRA, then $\phi$ is called a "scaling function".*

An exact formulation of an MRA will be given in Section 5.1. Typical examples of scaling functions $\phi$ are the $m^{\text{th}}$ order cardinal $B$-splines $N_m$, where $m$ is an arbitrary positive integer. More precisely, the first order cardinal $B$-spline $N_1$ is the characteristic function of the unit interval $[0,1)$, and for $m \geq 2$, $N_m$ is defined recursively by (integral) convolution:

$$N_m(x) := \int_{-\infty}^{\infty} N_{m-1}(x-t)N_1(t)dt \tag{1.5.7}$$

$$= \int_0^1 N_{m-1}(x-t)dt.$$

To describe the space $V_0$ that $N_m$ generates, we need the following notations:

$$\left\{ \begin{array}{l} \pi_n \text{ denotes the collection of all polynomials of degree at most } n. \\ C^n \text{ denotes the collection of all functions } f \text{ such that} \\ f, f', \ldots, f^{(n)} \text{ are continuous everywhere. Also, let } C = C^0. \end{array} \right. \tag{1.5.8}$$

The subspace $V_0$ generated by $N_m$ consists of all functions $f \in C^{m-2} \cap L^2(\mathbb{R})$ such that the restriction of each function $f$ to any interval $[k, k+1)$, $k \in \mathbb{Z}$, is in $\pi_{m-1}$; that is,

$$f|_{[k,k+1)} \in \pi_{m-1}, \quad k \in \mathbb{Z}.$$

From the property $(5^\circ)$ of an MRA, we can now identify all the other subspaces $V_j$, namely:

$$V_j = \{f \in C^{m-2} \cap L^2(\mathbb{R}): f|_{[\frac{k}{2^j}, \frac{k+1}{2^j})} \in \pi_{m-1}, \quad k \in \mathbb{Z}\}.$$

Since splines are only piecewise polynomial functions, they are very easy to implement in the computer. In fact, algorithms for graphically displaying spline curves and for computing the polynomial pieces exactly in terms of $B$-nets (or Bernstein-Bézier coefficients) are extremely efficient. In addition, since $B$-splines have the smallest possible supports, local interpolation schemes for approximating functions in $C \cap L^2(\mathbb{R})$ from any desirable spline sub-space $V_j$ are also available. All the algorithms mentioned above can be implemented in real-time. Details will be studied in Chapter 4.

From the nested sequence of spline subspaces $V_j$, we have the orthogonal complementary subspaces $W_j$, namely:

$$V_{j+1} = V_j \oplus W_j, \qquad j \in \mathbb{Z}. \tag{1.5.9}$$

These subspaces $W_j$ are mutually orthogonal and are orthogonal summands of $L^2(\mathbb{R})$ as described by (1.4.7) and (1.4.8). Just as the $B$-spline $N_m$ is the minimally supported generator of $\{V_j\}$, we are interested in finding the minimally supported $\psi_m \in W_0$ that generates the mutually orthogonal subspaces $W_j$, in the sense of (1.4.3) with $\psi_{m;j,k}$ in place of $\psi_{j,k}$, where

$$\psi_{m,j,k}(x) = 2^{j/2} \psi_m(2^j x - k), \qquad j, k \in \mathbb{Z}. \tag{1.5.10}$$

These compactly supported functions $\psi_m$ will be called "*B-wavelets*" of order $m$. In Chapter 6, explicit formulas for all $\psi_m$ and their duals $\widetilde{\psi}_m$, $m = 1, 2, \ldots$, will be derived. It is perhaps interesting to compare the "*supports*" of the $B$-splines and the $B$-wavelets. By the support of a continuous function $f$, which vanishes outside some bounded interval, we mean the smallest closed set outside which $f$ vanishes identically. The standard notation is supp $f$. We will see that

$$\begin{cases} \text{supp } N_m = [0, m]; \\ \text{supp } \psi_m = [0, 2m - 1], \end{cases} \tag{1.5.11}$$

for all $m = 1, 2, \ldots$ . In addition to having minimum supports, the $B$-wavelets $\psi_m$ enjoy many other important properties. We only mention three of them here. Firstly, it is clear from (1.5.9) that each $\psi_m$ is an s.o. wavelet. Secondly, efficient algorithms for computing $\psi_m$ and all its derivatives are available. Finally, the $B$-wavelets $\psi_m$ are symmetric for even $m$ and antisymmetric for odd $m$, meaning:

$$\begin{cases} \psi_m(x) = \psi_m(2m - 1 - x), & \text{for even } m; \\ \psi_m(x) = -\psi_m(2m - 1 - x), & \text{for odd } m. \end{cases} \tag{1.5.12}$$

In applications to signal analysis, symmetry and antisymmetry of the wavelet functions are very important. For instance, they are essential to avoid distortion in reconstruction of compressed data. This will be discussed in Chapter 5. Other interesting properties of $\psi_m$ will be studied in Chapter 6.

## 1.6. Wavelet decompositions and reconstructions

Let us return to considering the general structure of multiresolution analysis and wavelets as discussed in (1.5.1), where $\{V_j\}$ is generated by some scaling function $\phi \in L^2(\mathbb{R})$ and $\{W_j\}$ is generated by some wavelet $\psi \in L^2(\mathbb{R})$. In this case, by the property (2°), every function $f$ in $L^2(\mathbb{R})$ can be approximated as closely as is desired by an $f_N \in V_N$, for some $N \in \mathbb{Z}$. Since $V_j = V_{j-1} \dotplus W_{j-1}$ for any $j \in \mathbb{Z}$, $f_N$ has a unique decomposition:

$$f_N = f_{N-1} + g_{N-1},$$

where $f_{N-1} \in V_{N-1}$ and $g_{N-1} \in W_{N-1}$. By repeating this process, we have

$$f_N = g_{N-1} + g_{N-2} + \cdots + g_{N-M} + f_{N-M} \tag{1.6.1}$$

where $f_j \in V_j$ and $g_j \in W_j$ for any $j$, and $M$ is so chosen that $f_{N-M}$ is sufficiently "blurred". The "decomposition" in (1.6.1), which is unique, is called "*wavelet decomposition*"; and the "blur" is measured in terms of the "variation" (or more precisely, frequency or number of cycles per unit length) of $f_{N-M}$. A less efficient "stopping criterion" is to require $\|f_{N-M}\|$ to be smaller than some threshold. In the following, we will discuss an algorithmic approach for expressing $f_N$ as a direct sum of its components $g_{N-1}, \dots, g_{N-M}$, and $f_{N-M}$, and recovering $f_N$ from these components.

Since both the scaling function $\phi \in V_0$ and the wavelet $\psi \in W_0$ are in $V_1$, and since $V_1$ is generated by $\phi_{1,k}(x) = 2^{1/2}\phi(2x - k)$, $k \in \mathbb{Z}$, there exist two sequences $\{p_k\}$ and $\{q_k\} \in \ell^2$ such that

$$\phi(x) = \sum_k p_k \phi(2x - k); \tag{1.6.2}$$

$$\psi(x) = \sum_k q_k \phi(2x - k), \tag{1.6.3}$$

for all $x \in \mathbb{R}$. The formulas (1.6.2) and (1.6.3) are called the "*two-scale relations*" of the scaling function and wavelet, respectively. On the other hand, since both $\phi(2x)$ and $\phi(2x - 1)$ are in $V_1$ and $V_1 = V_0 \dotplus W_0$, there are four $\ell^2$ sequences which we denote by $\{a_{-2k}\}$, $\{b_{-2k}\}$, $\{a_{1-2k}\}$, and $\{b_{1-2k}\}$, $k \in \mathbb{Z}$, such that

$$\phi(2x) = \sum_k [a_{-2k}\phi(x - k) + b_{-2k}\psi(x - k)]; \tag{1.6.4}$$

$$\phi(2x - 1) = \sum_k [a_{1-2k}\phi(x - k) + b_{1-2k}\psi(x - k)], \tag{1.6.5}$$

for all $x \in \mathbb{R}$. The two formulas (1.6.4) and (1.6.5) can be combined into a single formula:

$$\phi(2x - \ell) = \sum_k [a_{\ell-2k}\phi(x - k) + b_{\ell-2k}\psi(x - k)], \qquad \ell \in \mathbb{Z}, \tag{1.6.6}$$

which is called the "*decomposition relation*" of $\phi$ and $\psi$. Now, we have two pairs of sequences $(\{p_k\}, \{q_k\})$ and $(\{a_k\}, \{b_k\})$, all of which are unique due to the direct sum relationship $V_1 = V_0 \dot{+} W_0$. These sequences are used to formulate the following reconstruction and decomposition algorithms. Hence, $\{p_k\}$ and $\{q_k\}$ are called reconstruction sequences, while $\{a_k\}$ and $\{b_k\}$ are called decomposition sequences.

To describe these algorithms, let us first recall that both $f_j \in V_j$ and $g_j \in W_j$ have unique series representations:

$$\begin{cases} f_j(x) = \sum_k c_k^j \phi(2^j x - k), \\ \text{with } \mathbf{c}^j = \{c_k^j\} \in \ell^2; \end{cases} \tag{1.6.7}$$

and

$$\begin{cases} g_j(x) = \sum_k d_k^j \psi(2^j x - k), \\ \text{with } \mathbf{d}^j = \{d_k^j\} \in \ell^2, \end{cases} \tag{1.6.8}$$

where we have intentionally suppressed the normalization coefficient $2^{j/2}$, by writing out $\phi(2^j x - k)$ and $\psi(2^j x - k)$ instead of using $\phi_{j,k}$ and $\psi_{j,k}$, in order to drop the unnecessary multiple of $\sqrt{2}$ in the algorithms. In the following decomposition and reconstruction algorithms, the functions $f_j$ and $g_j$ are represented by the sequences $\mathbf{c}^j$ and $\mathbf{d}^j$ as defined in (1.6.7) and (1.6.8).

(i) *Decomposition algorithm*

By applying (1.6.6)-(1.6.8), we have:

$$\begin{cases} c_k^{j-1} = \sum_\ell a_{\ell-2k} c_\ell^j; \\ d_k^{j-1} = \sum_\ell b_{\ell-2k} c_\ell^j. \end{cases} \tag{1.6.9}$$

$$\mathbf{c}^N \xrightarrow{\quad} \mathbf{c}^{N-1} \xrightarrow{\quad} \mathbf{c}^{N-2} \xrightarrow{\quad} \cdots \xrightarrow{\quad} \mathbf{c}^{N-M}$$

with $\mathbf{d}^{N-1}$, $\mathbf{d}^{N-2}$, $\mathbf{d}^{N-M}$ branching upward.

**Figure 1.6.1.** Wavelet decomposition.

Observe that both $\mathbf{c}^{j-1}$ and $\mathbf{d}^{j-1}$ are obtained from $\mathbf{c}^j$ by "*moving average*" schemes, using the decomposition sequences as "weights", with the exception that these moving averages are sampled only at the even integers. This is called downsampling. Therefore, each of the arrows in Figure 1.6.1 indicates a moving average followed by a downsampling at the even indices.

(ii) *Reconstruction algorithm*

By applying (1.6.2), (1.6.3), (1.6.7), and (1.6.8), we have:

$$c_k^j = \sum_\ell [p_{k-2\ell} c_\ell^{j-1} + q_{k-2\ell} d_\ell^{j-1}] \qquad (1.6.10)$$

$$
\begin{array}{ccccccc}
\mathbf{d}^{N-M} & & \mathbf{d}^{N-M+1} & & & \mathbf{d}^{N-1} & \\
& \searrow & & \searrow & & & \searrow \\
\mathbf{c}^{N-M} & \longrightarrow & \mathbf{c}^{N-M+1} & \longrightarrow & \cdots \quad \mathbf{c}^{N-1} & \longrightarrow & \mathbf{c}^{N}
\end{array}
$$

**Figure 1.6.2.** Wavelet reconstruction.

Here, $\mathbf{c}^j$ is obtained from $\mathbf{c}^{j-1}$ and $\mathbf{d}^{j-1}$ by two moving averages, using the reconstruction sequences as "weights", with the exception that an upsampling is required before the moving averages are performed. More precisely, the samples $c_\ell^{j-1}$ and $d_\ell^{j-1}$ are used at the even indices $m = 2\ell$ and zeros are used at the odd indices $m = 2\ell + 1$, when the (discrete) convolutions are taken with respect to $\{p_m\}$ and $\{q_m\}$.

We end this section with some remarks on the two algorithms given above. Firstly, if a weight sequence $\{a_k\}$, $\{b_k\}$, $\{p_k\}$, or $\{q_k\}$, is finite, then the moving average is a very simple FIR (finite impulse response) filter. If, however, the weight sequence is infinite, then the moving average is an IIR (infinite impulse response) filter. As is well known, IIR filters can be implemented as ARMA (autoregressive-moving average) filters, provided that the symbol (or "$z$-transform") of the weight sequence is a rational function. We will call such weight sequences "ARMA sequences". Otherwise, the infinite weight sequence has to be truncated to give an FIR filter. Secondly, if the weight sequence consists of terms with irrationals or long decimal representations, rounding-off (or "quantization") of these numbers is necessary. Of course, both truncation and quantization induce errors which should be estimated a priori. Finally, since the scaling function and wavelet pair $(\phi, \psi)$ are used as "mirror filters", symmetry (or at least antisymmetry) is important in many applications in signal analysis. In reconstructing compressed images, for instance, non-symmetry and non-antisymmetry induce distortion. As will be seen in Chapter 5 the symmetric properties of $(\phi, \psi)$ are reflected by the symmetry of the decomposition and reconstruction sequences. A brief discussion of signal and image processing will be given in Chapter 3.

In Chapter 6, we will see that when spline-wavelets $\psi_m$ (with minimal supports) are used as $\psi$, the reconstruction sequences are finite and the decomposition sequences are ARMA. All these sequences are symmetric for even order $m$ and antisymmetric for odd order $m$. In addition, modulo a common factor of the reciprocal of an integer, all of these sequences consist of integer terms only.

On the other hand, when compactly supported orthogonal wavelets $\psi$ are considered, both the reconstruction and decomposition sequences are finite.

However, for continuous $\psi$, neither symmetry nor antisymmetry is possible, and the corresponding reconstruction and decomposition sequences have to be quantized. A detailed account of the structural analysis and constructive schemes of scaling functions and wavelets will be given in Chapter 5. In particular, the relationship between linear-phase filtering and symmetric scaling functions and wavelets will be studied there. The final two chapters will be devoted to semi-orthogonal and orthogonal wavelets, respectively. More precisely, a fairly complete analysis of cardinal spline-wavelets will be given in Chapter 6, and the topic of orthogonal wavelets, with emphasis on those with compact supports, will be presented in Chapter 7. Also included in this chapter will be a brief discussion of orthogonal wave packets which are introduced for better time-frequency localization.

# 2 Fourier Analysis

The subject of Fourier analysis is one of the oldest subjects in mathematical analysis and is of great importance to mathematicians and engineers alike. From a practical point of view, when one thinks of Fourier analysis, one usually refers to (integral) Fourier transforms and Fourier series. A Fourier transform is the Fourier integral of some function $f$ defined on the real line $\mathbb{R}$. When $f$ is thought of as an analog signal, then its domain of definition $\mathbb{R}$ is called the continuous time-domain. In this case, the Fourier transform $\hat{f}$ of $f$ describes the spectral behaviour of the signal $f$. Since the spectral information is given in terms of frequency, the domain of definition of the Fourier transform $\hat{f}$, which is again $\mathbb{R}$, is called the frequency domain. On the other hand, a Fourier series is a transformation of bi-infinite sequences to periodic functions. Hence, when a bi-infinite sequence is thought of as a digital signal, then its domain of definition, which is the set $\mathbb{Z}$ of integers, is called the discrete time-domain. In this case, its Fourier series again describes the spectral behaviour of the digital signal, and the domain of definition of a Fourier series is again the real line $\mathbb{R}$ which is the frequency domain. However, since Fourier series are $2\pi$-periodic, the frequency domain $\mathbb{R}$ in this situation is usually identified with the unit circle. To a mathematician, this identification is more satisfactory, since the "dual group" of $\mathbb{Z}$ is the "circle group".

The importance of both the Fourier transform and the Fourier series stems not only from the significance of their physical interpretations, such as time-frequency analysis of signals, but also from the fact that Fourier analytic techniques are extremely powerful. For instance, in the study of wavelet analysis, the Poisson summation formula, Parseval's identities for both series and integrals, Fourier transforms of the Gaussian, convolution of functions, and the delta distribution, etc., are often encountered. Since this monograph is intended to be self-contained, preliminary materials on the basic knowledge of Fourier analysis such as the above mentioned topics will be discussed in this chapter.

## 2.1. Fourier and inverse Fourier transforms

Throughout this text, all functions $f$ defined on the real line $\mathbb{R}$ are assumed to be measurable. For the reader who is not familiar with the basic Lebesgue theory, but is willing to believe some of the standard theorems, the sacrifice is very small in assuming that $f$ is piecewise continuous, and by this we mean the existence of points $\{x_j\}$ in $\mathbb{R}$ with no finite accumulation points, such that $x_j < x_{j+1}$ for all $j$, and that $f$ is continuous on each of the open intervals $(x_j, x_{j+1})$ as well as the unbounded intervals $(-\infty, \min x_j)$ and $(\max x_j, \infty)$, if $\min x_j$ or $\max x_j$ exist. For each $p$, $1 \leq p < \infty$, let $L^p(\mathbb{R})$ denote the class of measurable functions $f$ on $\mathbb{R}$ such that the (Lebesgue) integral

$$\int_{-\infty}^{\infty} |f(x)|^p dx$$

is finite. Also, let $L^\infty(\mathbb{R})$ be the collection of almost everywhere (a.e.) bounded functions; that is, functions bounded everywhere except on sets of (Lebesgue) measure zero. Hence, endowed with the "norm"

$$\|f\|_p := \begin{cases} \left\{ \int_{-\infty}^{\infty} |f(x)|^p dx \right\}^{\frac{1}{p}} & \text{for} \quad 1 \leq p < \infty; \\ \underset{-\infty < x < \infty}{\text{ess sup}} \ |f(x)| & \text{for} \quad p = \infty, \end{cases}$$

each $L^p(\mathbb{R})$, $1 \leq p \leq \infty$, is a Banach space. Since we do not require any knowledge of the Banach space structure in understanding wavelet and time-frequency analyses in this introductory monograph, the reader only has to know a few elementary properties of the $L^p(\mathbb{R})$ norms, such as the Minkowski Inequality:

$$\|f + g\|_p \leq \|f\|_p + \|g\|_p, \tag{2.1.1}$$

and the Hölder Inequality:

$$\|fg\|_1 \leq \|f\|_p \|g\|_{p(p-1)^{-1}}, \tag{2.1.2}$$

where $p(p-1)^{-1}$ should be replaced by 1 when $p = \infty$. A consequence of (2.1.2) is the Schwarz Inequality:

$$\|fg\|_1 \leq \|f\|_2 \|g\|_2. \tag{2.1.3}$$

Hence, in view of (2.1.3), we may define the "inner product"

$$\langle f, g \rangle = \int_{-\infty}^{\infty} f(x)\overline{g(x)}dx, \qquad f, g \in L^2(\mathbb{R}). \tag{2.1.4}$$

Endowed with this inner product, the Banach space $L^2(\mathbb{R})$ becomes a Hilbert space. Of course, it is clear that

$$\langle f, f \rangle = \|f\|_2^2, \qquad f \in L^2(\mathbb{R}). \tag{2.1.5}$$

In the following, we first concentrate our attention on functions in $L^1(\mathbb{R})$. As usual (to a mathematician), the imaginary unit will be denoted by $i$. The electrical engineer might want to replace $i$ by $j$ throughout the entire text.

**Definition 2.1.** *The Fourier transform of a function* $f \in L^1(\mathbb{R})$ *is defined by*

$$\hat{f}(\omega) = (\mathcal{F}f)(\omega) := \int_{-\infty}^{\infty} e^{-i\omega x} f(x)dx. \tag{2.1.6}$$

Some of the basic properties of $\hat{f}(\omega)$, for every $f \in L^1(\mathbb{R})$, are summarized in the following.

**Theorem 2.2.** *Let* $f \in L^1(\mathbb{R})$. *Then its Fourier transform* $\hat{f}$ *satisfies:*
(i) $\hat{f} \in L^{\infty}(\mathbb{R})$ *with* $\|\hat{f}\|_{\infty} \leq \|f\|_1$;
(ii) $\hat{f}$ *is uniformly continuous on* $\mathbb{R}$ ;
(iii) *if the derivative* $f'$ *of* $f$ *also exists and is in* $L^1(\mathbb{R})$, *then*

$$\widehat{f'}(\omega) = i\omega\hat{f}(\omega); \quad \text{and} \tag{2.1.7}$$

(iv) $\hat{f}(\omega) \to 0$, *as* $\omega \to \infty$ *or* $-\infty$.

**Proof.** Assertion (i) is obvious. To prove (ii), let $\delta$ be chosen arbitrarily and consider

$$\sup_{\omega}|\hat{f}(\omega + \delta) - \hat{f}(\omega)| = \sup_{\omega}\left|\int_{-\infty}^{\infty} e^{-i\omega x}(e^{-i\delta x} - 1)f(x)dx\right|$$

$$\leq \int_{-\infty}^{\infty}|e^{-i\delta x} - 1||f(x)|dx.$$

Now, since $|e^{-i\delta x}-1||f(x)| \leq 2|f(x)| \in L^1(\mathbb{R})$ and $|e^{-i\delta x}-1| \to 0$ as $\delta \to 0$, the Lebesgue Dominated Convergence Theorem implies that the quantity above tends to zero as $\delta \to 0$.

To establish (iii), we simply apply another standard theorem in Lebesgue integration theory to integrate (2.1.6) by parts, using the fact that $f(x) \to 0$ as $x \to \pm\infty$.

Finally, the statement in (iv) is usually called the *"Riemann-Lebesgue Lemma"*. To prove it, we first observe that if $f'$ exists and is in $L^1(\mathbb{R})$, then by (iii) and (i), we have, indeed,

$$|\hat{f}(\omega)| = \frac{1}{|\omega|}|\widehat{f'}(\omega)| \leq \frac{1}{|\omega|}\|f'\|_1 \to 0,$$

as $\omega \to \pm\infty$. In general, for any given $\varepsilon > 0$, we can find a function $g$ such that $g, g' \in L^1$ and $\|f - g\|_1 < \varepsilon$. Then by (i), we have

$$|\hat{f}(\omega)| \leq |\hat{f}(\omega) - \hat{g}(\omega)| + |\hat{g}(\omega)|$$

$$\leq \|f - g\|_1 + |\hat{g}(\omega)| < \varepsilon + |\hat{g}(\omega)|,$$

completing the proof of (iv).    ■

Although $\hat{f}(\omega) \to 0$, as $\omega \to \pm\infty$, for every $f \in L^1(\mathbb{R})$, it does not mean that $\hat{f}$ is necessarily in $L^1(\mathbb{R})$. To demonstrate this remark with a counterexample, we need the notion of the so-called "Heaviside unit step" function:

$$u_a(x) = \begin{cases} 1 & \text{for} \quad x \geq a; \\ 0 & \text{for} \quad x < a. \end{cases} \tag{2.1.8}$$

where $a \in \mathbb{R}$.

**Example 2.3.** The function

$$f(x) = e^{-x} u_0(x)$$

is in $L^1(\mathbb{R})$, but its Fourier transform, which is

$$\hat{f}(\omega) = \frac{1}{1 - i\omega}$$

is not in $L^1(\mathbb{R})$.

**Proof.** From $e^{-i\omega x} = \cos \omega x - i \sin \omega x$, we have

$$\hat{f}(\omega) = \int_0^\infty e^{-x} \cos \omega x \, dx - i \int_0^\infty e^{-x} \sin \omega x \, dx$$
$$= \frac{1}{1 + \omega^2} - \frac{i\omega}{1 + \omega^2} = \frac{1}{1 - i\omega},$$

which behaves like $O(|\omega|^{-1})$ at $\infty$, and hence, is not in $L^1(\mathbb{R})$. ∎

If it happens that $\hat{f}$ is in $L^1(\mathbb{R})$, then we can usually "recover" $f$ from $\hat{f}$, by using the "inverse Fourier transform" defined as follows.

**Definition 2.4.** *Let $\hat{f} \in L^1(\mathbb{R})$ be the Fourier transform of some function $f \in L^1(\mathbb{R})$. Then the inverse Fourier transform of $\hat{f}$ is defined by*

$$(\mathcal{F}^{-1}\hat{f})(x) := \frac{1}{2\pi} \int_{-\infty}^\infty e^{ix\omega} \hat{f}(\omega) d\omega. \tag{2.1.9}$$

So, the important question is: When can $f$ be recovered from $\hat{f}$ by using the operator $\mathcal{F}^{-1}$, or when is $(\mathcal{F}^{-1}\hat{f})(x) = f(x)$? The answer is: At every point $x$ where $f$ is continuous. That is, we have the following.

**Theorem 2.5.** *Let $f \in L^1(\mathbb{R})$ such that its Fourier transform $\hat{f}$ is also in $L^1(\mathbb{R})$. Then*

$$f(x) = (\mathcal{F}^{-1}\hat{f})(x) \tag{2.1.10}$$

*at every point $x$ where $f$ is continuous.*

We will delay the proof of this theorem to the next section. Instead, we end this section by deriving the Fourier transform of the so-called "Gaussian function".

**Example 2.6.** Let $a > 0$. Then

$$\int_{-\infty}^{\infty} e^{-i\omega x} e^{-ax^2} dx = \sqrt{\frac{\pi}{a}} e^{-\frac{\omega^2}{4a}}. \qquad (2.1.11)$$

In particular, the Fourier transform of the Gaussian function $e^{-x^2}$ is $\sqrt{\pi} e^{-\omega^2/4}$.

**Proof.** Consider the function

$$f(y) := \int_{-\infty}^{\infty} e^{-ax^2 + xy} dx, \qquad y \in \mathbb{R}. \qquad (2.1.12)$$

By completing squares, we have

$$f(y) = \int_{-\infty}^{\infty} e^{-a(x - \frac{y}{2a})^2 + \frac{y^2}{4a}} dx \qquad (2.1.13)$$

$$= \frac{1}{\sqrt{a}} e^{y^2/4a} \int_{-\infty}^{\infty} e^{-x^2} dx$$

$$= \sqrt{\frac{\pi}{a}} e^{y^2/4a}.$$

Now, since both $f(y)$ as defined in (2.1.12) and the function

$$g(y) := \sqrt{\frac{\pi}{a}} e^{y^2/4a}$$

can be extended to be entire (analytic) functions, and since they agree on $\mathbb{R}$ as shown in (2.1.13), they must agree on the complex plane $\mathbb{C}$. In particular, by setting $y$ to be $-i\omega$, we have

$$\int_{-\infty}^{\infty} e^{-i\omega x} e^{-ax^2} dx = \sqrt{\frac{\pi}{a}} e^{-\frac{\omega^2}{4a}}. \qquad \blacksquare$$

## 2.2. Continuous-time convolution and the delta function

Let $f$ and $g$ be functions in $L^1(\mathbb{R})$. Then the (continuous-time) *convolution* of $f$ and $g$ is also an $L^1(\mathbb{R})$ function $h$ defined by

$$h(x) = (f * g)(x) := \int_{-\infty}^{\infty} f(x - y)g(y)dy. \qquad (2.2.1)$$

It is clear that $h \in L^1(\mathbb{R})$, and in fact,

$$\|h\|_1 \le \|f\|_1 \|g\|_1, \qquad (2.2.2)$$

since

$$\int_{-\infty}^{\infty} |h(x)|dx \le \int_{-\infty}^{\infty} \int_{-\infty}^{\infty} |f(x - y)||g(y)|dydx$$

$$= \int_{-\infty}^{\infty} |g(y)| \left[ \int_{-\infty}^{\infty} |f(x - y)|dx \right] dy$$

$$= \int_{-\infty}^{\infty} |g(y)| \left[ \int_{-\infty}^{\infty} |f(x)|dx \right] dy.$$

A change of the variable of integration in (2.2.1) yields:

$$f * g = g * f, \qquad f, g \in L^1(\mathbb{R}). \qquad (2.2.3)$$

That is, the convolution operation is "*commutative*". Since $f * g$ is in $L^1(\mathbb{R})$, we can again convolve $f * g$ with another function $u \in L^1(\mathbb{R})$; that is, we may consider $(f * g) * u$. It is easy to see that

$$(f * g) * u = f * (g * u), \qquad f, g, u \in L^1(\mathbb{R}). \qquad (2.2.4)$$

Hence, the convolution operation is "*associative*".

Now, the question is: Does there exist some function, say $d \in L^1(\mathbb{R})$, such that

$$f * d = f, \qquad f \in L^1(\mathbb{R})? \qquad (2.2.5)$$

The answer is negative, and this can be shown by using a Fourier transform argument. First, let us record the following important property of the Fourier transform operator.

**Theorem 2.7.** *Let $f$ and $g$ be in $L^1(\mathbb{R})$. Then*

$$(f * g)^\wedge(\omega) = \hat{f}(\omega)\hat{g}(\omega). \qquad (2.2.6)$$

Since the proof is a trivial application of the Fubini Theorem, it is omitted here. ∎

Now, if a function $d \in L^1(\mathbb{R})$ exists such that (2.2.5) holds, then by applying Theorem 2.7, we have

$$\hat{f}(\omega)\hat{d}(\omega) = \hat{f}(\omega), \qquad f \in L^1(\mathbb{R}).$$

That is, we must have $\hat{d}(\omega) = 1$, and this violates the Riemann-Lebesgue Lemma as stated in Theorem 2.2 (iv).

However, we still wish to "approximate" $d$ in (2.2.5), since even an "*approximation of the convolution identity*" (or simply "*approximate identity*") is a very important tool in Fourier analysis.

From the preceding discussion, we see that the first requirement for such a family $\{d_\alpha\} \subset L^1(\mathbb{R})$ that seeks to approximate the identity is

$$\hat{d}_\alpha(\omega) \approx 1, \qquad \omega \in \mathbb{R}, \qquad (2.2.7)$$

as $\alpha \to 0$, say. In particular, we may use the normalization $\hat{d}_\alpha(0) = 1$, or equivalently,

$$\int_{-\infty}^{\infty} d_\alpha(x)dx = 1. \qquad (2.2.8)$$

An excellent candidate is the family of Gaussian functions

$$g_\alpha(x) := \frac{1}{2\sqrt{\pi\alpha}}e^{-\frac{x^2}{4\alpha}}, \qquad \alpha > 0. \qquad (2.2.9)$$

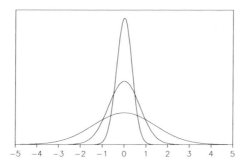

**Figure 2.2.1.** Gaussian functions $g_\alpha, \alpha = 1, \frac{1}{4}, \frac{1}{16}$.

Indeed, by applying (2.1.11) in Example 2.6 with $\alpha = 1/4a$, we have

$$\hat{g}_\alpha(\omega) = e^{-\alpha\omega^2}, \tag{2.2.10}$$

which clearly satisfies (2.2.7) and (2.2.8). The graphs of $g_\alpha$ for a sequence of decreasing values of $\alpha > 0$ is shown in Figure 2.2.1.

Observe that if $g_\alpha$ is used as the "weight" function in taking the mean of a continuous function $f$ in $L^1(\mathbb{R})$, then the concentration of the weight approaches the origin as $\alpha \to 0^+$; that is,

$$\int_{-\infty}^{\infty} f(x-y)g_\alpha(y)dy \sim f(x-0) = f(x), \quad \alpha \to 0^+;$$

which is the same as

$$(f * g_\alpha)(x) \sim f(x), \qquad \alpha \to 0^+.$$

More precisely, we have the following.

**Theorem 2.8.** *Let $f \in L^1(\mathbb{R})$. Then*

$$\lim_{\alpha \to 0^+} (f * g_\alpha)(x) = f(x) \tag{2.2.11}$$

*at every point $x$ where $f$ is continuous.*

**Proof.** Let $f$ be continuous at $x$ and $\varepsilon > 0$ be arbitrarily given. We select $\eta > 0$ such that

$$|f(x-y) - f(x)| < \varepsilon,$$

for all $y \in \mathbb{R}$ with $|y| < \eta$. Then in view of (2.2.8) with $d_\alpha = g_\alpha$, we have

$$|(f * g_\alpha)(x) - f(x)| = \left| \int_{-\infty}^{\infty} [f(x-y) - f(x)] g_\alpha(y) dy \right|$$

$$\leq \int_{-\eta}^{\eta} |f(x-y) - f(x)| g_\alpha(y) dy$$

$$+ \int_{|y| \geq \eta} |f(x-y) - f(x)| g_\alpha(y) dy$$

$$\leq \epsilon \int_{-\eta}^{\eta} g_\alpha(y) dy + \|f\|_1 \max_{|y| \geq \eta} g_\alpha(y) + |f(x)| \int_{|y| \geq \eta} g_\alpha(y) dy$$

$$\leq \epsilon \int_{-\infty}^{\infty} g_\alpha(y) dy + \|f\|_1 g_\alpha(\eta) + |f(x)| \int_{|y| \geq \eta/\sqrt{\alpha}} g_1(y) dy$$

$$= \epsilon + \|f\|_1 g_\alpha(\eta) + |f(x)| \int_{|y| \geq \eta/\sqrt{\alpha}} g_1(y) dy.$$

Since both $g_\alpha(\eta)$ and the last term obviously converge to zero as $\alpha \to 0^+$, this completes the proof of the theorem. ■

Now, let us consider the collection $\mathcal{C}$ of continuous functions in $L^1(\mathbb{R})$. Then for any fixed $x \in \mathbb{R}$, each $g_\alpha$ may be considered a *"linear functional"* on $\mathcal{C}$, defined by

$$g_\alpha \circ f(x - \cdot) := (f * g_\alpha)(x).$$

Similarly, we consider the linear functional $\delta$ on $\mathcal{C}$, defined by

$$\delta \circ f(x - \cdot) := f(x). \tag{2.2.12}$$

Then (2.2.11) in Theorem 2.8 indeed says that

$$g_\alpha \to \delta \quad \text{on} \quad \mathcal{C}, \qquad \alpha \to 0^+. \tag{2.2.13}$$

Since $\delta * f = f$, $\delta$ is the "convolution identity"; and hence, $\{g_\alpha\}$ is an approximation of the convolution identity. Recall that $\delta$ is not an $L^1(\mathbb{R})$ function. In fact, it is not a function at all, since it satisfies

$$\begin{cases} \delta(x) = 0 \quad \text{for all} \quad x \neq 0 \\ \int_{-\infty}^{\infty} \delta(x) dx = 1. \end{cases} \tag{2.2.14}$$

Although this linear functional $\delta$ is usually called the "delta function", it is really a "generalized function" or "distribution". As we remarked earlier, since $\delta * f = f$ for all $f \in \mathcal{C}$, we could assign the Fourier transform of $\delta$ to be the constant 1, namely:

$$\hat{\delta}(\omega) = 1. \tag{2.2.15}$$

To end this section, we demonstrate the power of the approximate identity $\{g_\alpha\}$ by returning to establish Theorem 2.5 in the previous section. Let us first introduce the identity:

$$\int_{-\infty}^{\infty} f(x)\hat{g}(x)dx = \int_{-\infty}^{\infty} \hat{f}(x)g(x)dx, \qquad f, g \in L^1(\mathbb{R}). \qquad (2.2.16)$$

Note that since $\hat{f}$ and $\hat{g}$ are in $L^\infty(\mathbb{R})$ as shown in Theorem 2.2 (i), both integrals in (2.2.16) are finite in view of the Hölder inequality (2.1.2) with $p = 1$. A trivial application of the Fubini Theorem yields (2.2.16).

**Proof of Theorem 2.5.** Let $x \in \mathbb{R}$ be fixed and set

$$g(y) := \frac{1}{2\pi} e^{iyx} e^{-\alpha y^2}. \qquad (2.2.17)$$

Then by applying (2.1.11) in Example 2.6, we have

$$\begin{aligned}
\hat{g}(y) &= \frac{1}{2\pi} \int_{-\infty}^{\infty} e^{-iyt} e^{itx} e^{-\alpha t^2} dt \\
&= \frac{1}{2\pi} \int_{-\infty}^{\infty} e^{-i(y-x)t} e^{-\alpha t^2} dt \\
&= \frac{1}{2\pi} \sqrt{\frac{\pi}{\alpha}} e^{-\frac{(y-x)^2}{4\alpha}} = g_\alpha(x-y),
\end{aligned}$$

where $g$ was defined in (2.2.9). Hence, it follows from (2.2.16) and (2.2.17) that

$$\begin{aligned}
(f * g_\alpha)(x) &= \int_{-\infty}^{\infty} f(y)g_\alpha(x-y)dy \qquad (2.2.18) \\
&= \int_{-\infty}^{\infty} f(y)\hat{g}(y)dy \\
&= \int_{-\infty}^{\infty} \hat{f}(y)g(y)dy \\
&= \frac{1}{2\pi} \int_{-\infty}^{\infty} e^{iyx} \hat{f}(y)e^{-\alpha y^2} dy.
\end{aligned}$$

Now, if $f$ is continuous at $x$, then by Theorem 2.8, the left-hand side of (2.2.18) converges to $f(x)$ as $\alpha \to 0^+$. Therefore, since the right-hand side of (2.2.18) tends to $(\mathcal{F}^{-1}\hat{f})(x)$, we have

$$f(x) = (\mathcal{F}^{-1}\hat{f})(x). \qquad \blacksquare$$

### 2.3. Fourier transform of square-integrable functions

In this section, we introduce the definition of the Fourier transform of functions in $L^2(\mathbb{R})$. To do so, the notion of "autocorrelation" is needed.

**Definition 2.9.** *The autocorrelation function of an $f \in L^2(\mathbb{R})$ is defined by*

$$F(x) = \int_{-\infty}^{\infty} f(x+y)\overline{f(y)}dy. \tag{2.3.1}$$

Note that in view of the Schwarz Inequality (2.1.3), the integrand in (2.3.1) is in $L^1(\mathbb{R})$, so that $F(x)$ is a finite value for every $x \in \mathbb{R}$. In fact, we can say a little more as in the following.

**Lemma 2.10.** *Let $F$ be the autocorrelation function of $f \in L^2(\mathbb{R})$. Then*
*(i) $|F(x)| \le \|f\|_2^2$, all $x \in \mathbb{R}$; and*
*(ii) $F$ is uniformly continuous on $\mathbb{R}$.*

**Proof.** As mentioned above, (i) is a consequence of the Schwarz Inequality

$$\begin{aligned}|F(x)| &\le \int_{-\infty}^{\infty} |f(x+y)||\overline{f(y)}|dy \\ &\le \left\{\int_{-\infty}^{\infty} |f(x+y)|^2 dy\right\}^{1/2} \|f\|_2 \\ &= \|f\|_2^2.\end{aligned}$$

To prove (ii), we consider an arbitrary real number $\eta$ and again apply the Schwarz Inequality to obtain

$$\begin{aligned}|F(x+\eta) - F(x)| &= \left|\int_{-\infty}^{\infty} \{f(x+\eta+y) - f(x+y)\}\overline{f(y)}dy\right| \\ &\le \left\{\int_{-\infty}^{\infty} |f(x+\eta+y) - f(x+y)|^2 dy\right\}^{1/2} \|f\|_2 \\ &= \left\{\int_{-\infty}^{\infty} |f(y+\eta) - f(y)|^2 dy\right\}^{1/2} \|f\|_2.\end{aligned}$$

Since $f \in L^2(\mathbb{R})$, by a basic property in Lebesgue integration theory, the integral inside the braces, which is independent of $x$, tends to zero as $\eta \to 0$. ∎

The following result is instrumental in extending the notion of Fourier transform to include $L^2(\mathbb{R})$ functions.

**Theorem 2.11.** *Let $f \in L^1(\mathbb{R}) \cap L^2(\mathbb{R})$. Then the Fourier transform $\hat{f}$ of $f$ is in $L^2(\mathbb{R})$, and satisfies the following "Parseval Identity":*

$$\|\hat{f}\|_2^2 = 2\pi\|f\|_2^2. \tag{2.3.2}$$

**Proof.** Since $\hat{f}$ is continuous and tends to zero at infinity as assured by Theorem 2.2, the family $\{\hat{g}_\alpha\}$ introduced in (2.2.10) can be used as weight functions, so that $\hat{g}_\alpha |\hat{f}|^2 \in L^1(\mathbb{R})$. Observe that

$$\int_{-\infty}^{\infty} \hat{g}_\alpha(x)|\hat{f}(x)|^2 dx = \int_{-\infty}^{\infty} \hat{g}_\alpha(x)\hat{f}(x)\overline{\hat{f}(x)}dx$$

$$= \int_{-\infty}^{\infty} f(y) \int_{-\infty}^{\infty} \overline{f(u)} \left[ \int_{-\infty}^{\infty} e^{ix(y-u)}\hat{g}_\alpha(x)dx \right] du\,dy,$$

where, with the exception of a multiple of $(2\pi)^{-1}$, the term inside the brackets is the inverse Fourier transform of $\hat{g}_\alpha$. Hence, by Theorem 2.5, we have

$$\int_{-\infty}^{\infty} \hat{g}_\alpha(x)|\hat{f}(x)|^2 dx = 2\pi \int_{-\infty}^{\infty} f(y) \int_{-\infty}^{\infty} \overline{f(u)} g_\alpha(y-u)dy\,du,$$

where $g_\alpha$ is given in (2.2.9). So, by using the notion of autocorrelation introduced in (2.3.1), the identity (2.3.2) becomes

$$\int_{-\infty}^{\infty} \hat{g}_\alpha(x)|\hat{f}(x)|^2 dx = 2\pi \int_{-\infty}^{\infty} F(x)g_\alpha(x)dx.$$

Since $F$ is continuous and $\{g_\alpha\}$ is an approximation of the $\delta$ distribution, we have

$$\lim_{\alpha\to 0+} \int_{-\infty}^{\infty} \hat{g}_\alpha(x)|\hat{f}(x)|^2 dx = 2\pi F(0). \tag{2.3.3}$$

Now, by Fatou's lemma, we have $\hat{f} \in L^2(\mathbb{R})$; and because of $0 \le \hat{g}_\alpha|\hat{f}|^2 \le |\hat{f}|^2$, the Lebesgue Dominated Convergence Theorem allows us to interchange limit and integration in (2.3.3), yielding

$$\int_{-\infty}^{\infty} |\hat{f}(x)|^2 dx = 2\pi F(0) = 2\pi\|f\|_2^2.$$

This completes the proof of Theorem 2.11. ∎

As a consequence of Theorem 2.11, we observe that the Fourier transform $\mathcal{F}$ may be considered as a "*bounded linear operator*" on $L^1(\mathbb{R}) \cap L^2(\mathbb{R})$ with range in $L^2(\mathbb{R})$; that is,

$$\mathcal{F}\colon L^1(\mathbb{R}) \cap L^2(\mathbb{R}) \to L^2(\mathbb{R}),$$

such that $\|\mathcal{F}\| = \sqrt{2\pi}$. Since $L^1(\mathbb{R}) \cap L^2(\mathbb{R})$ is dense in $L^2(\mathbb{R})$, $\mathcal{F}$ has a norm-preserving extension to all of $L^2(\mathbb{R})$. More precisely, if $f \in L^2(\mathbb{R})$, then its truncations:

$$f_N(x) := \begin{cases} f(x) & \text{for } |x| \le N; \\ 0 & \text{otherwise,} \end{cases} \tag{2.3.4}$$

where $N = 1, 2, \ldots$, are in $L^1(\mathbb{R}) \cap L^2(\mathbb{R})$, so that $\hat{f}_N \in L^2(\mathbb{R})$. In fact, it is easy to see that $\{\hat{f}_N\}$ is a Cauchy sequence in $L^2(\mathbb{R})$, and by the completeness of $L^2(\mathbb{R})$, there is a function $\hat{f}_\infty \in L^2(\mathbb{R})$, such that

$$\lim_{N \to \infty} \|\hat{f}_N - \hat{f}_\infty\|_2 = 0.$$

**Definition 2.12.** *The Fourier transform $\hat{f}$ of a function $f \in L^2(\mathbb{R})$ is defined to be the Cauchy limit $\hat{f}_\infty$ of $\{\hat{f}_N\}$, and the notation*

$$\hat{f}(\omega) = \underset{N \to \infty}{\text{l.i.m.o.t.}}\ \hat{f}_N(\omega) \qquad (2.3.5)$$

$$= \underset{N \to \infty}{\text{l.i.m.o.t.}} \int_{-N}^{N} e^{-i\omega x} f(x)\,dx,$$

*which stands for "limit in the mean of order two", will be used.*

Of course, the definition of $\hat{f}$, for $f \in L^2(\mathbb{R})$, should be independent of the choice of $f_N \in L^1(\mathbb{R}) \cap L^2(\mathbb{R})$. In other words, any other Cauchy sequence from $L^1(\mathbb{R}) \cap L^2(\mathbb{R})$ that approximates $f$ in $L^2(\mathbb{R})$ can be used to define $\hat{f}$. But in view of their simplicity, the truncations of $f$ as in (2.3.4) are often chosen, particularly in signal analysis. We also remark that the extension of $\mathcal{F}$ from $L^1(\mathbb{R}) \cap L^2(\mathbb{R})$ to $L^2(\mathbb{R})$ is consistent with $\mathcal{F}$ originally defined on $L^1(\mathbb{R})$. This can be easily verified by using basic Lebesgue theory. Finally, we must emphasize that the Parseval Identity (2.3.2) extends to all of $L^2(\mathbb{R})$. In fact, a little more can also be said, as follows.

**Theorem 2.13.** *For all $f, g \in L^2(\mathbb{R})$, the following relation holds:*

$$\langle f, g \rangle = \frac{1}{2\pi} \langle \hat{f}, \hat{g} \rangle. \qquad (2.3.6)$$

*In particular, $\|f\|_2 = (2\pi)^{-\frac{1}{2}} \|\hat{f}\|_2$.*

The relation in (2.3.6) is also called the Parseval Identity.

**Proof.** It is clear from the foregoing discussions that

$$\|\hat{h}\|_2^2 = 2\pi \|h\|_2^2, \qquad h \in L^2(\mathbb{R}).$$

Hence, (2.3.6) follows by setting $h$ to be each of the four functions

$$f + g, f - g, f - ig, f + ig$$

in the inner product identity

$$\langle f, g \rangle = \frac{\|f + g\|_2^2 - \|f - g\|_2^2}{4} + \frac{\|f - ig\|_2^2 - \|f + ig\|_2^2}{4i}. \qquad \blacksquare \qquad (2.3.7)$$

Recall that when the inverse Fourier transform $\mathcal{F}^{-1}$ was introduced in Definition 2.4, we had to restrict $\mathcal{F}^{-1}$ to the intersection of $L^1(\mathbb{R})$ with the image of $\mathcal{F}$ because $\mathcal{F}$ does not map $L^1(\mathbb{R})$ into $L^1(\mathbb{R})$. In addition, we could not even write

$$f(x) = (\mathcal{F}^{-1}\hat{f})(x),$$

unless $f$ was continuous at $x$. The $L^2(\mathbb{R})$ theory, on the other hand, is much more elegant. We have seen that $\mathcal{F}$ maps $L^2(\mathbb{R})$ into itself. In the following, we will show that this map is actually *one-one* and *onto*, so that the inverse Fourier transform $\mathcal{F}^{-1}$ can be easily formulated.

Before we proceed any further, we need the following preliminary lemma and some notation.

**Lemma 2.14.** *Let* $f, g \in L^2(\mathbb{R})$. *Then*

$$\int_{-\infty}^{\infty} f(x)\hat{g}(x)dx = \int_{-\infty}^{\infty} \hat{f}(x)g(x)dx. \tag{2.3.8}$$

**Proof.** Since (2.3.8) holds for $f, g \in L^1(\mathbb{R})$ as shown in (2.2.16), and $L^1(\mathbb{R}) \cap L^2(\mathbb{R})$ is dense in $L^2(\mathbb{R})$, it is easy to see that (2.3.8) also holds for $f, g \in L^2(\mathbb{R})$. ∎

**Definition 2.15.** *For every* $f$ *defined on* $\mathbb{R}$, *the function* $f^-$ *is defined as follows:*

$$f^-(x) := f(-x). \tag{2.3.9}$$

*We call* $f^-$ *the "reflection" of* $f$ *(relative to the origin).*

The following observation is trivial.

**Lemma 2.16.** *Let* $f \in L^2(\mathbb{R})$. *Then*

$$\overline{\hat{f}}(x) = (\widehat{\overline{f}^-})(x); \quad (\widehat{f^-})(x) = (\hat{f})^-(x). \tag{2.3.10}$$

We are now ready to establish the invertibility of the Fourier transform operator on $L^2(\mathbb{R})$.

**Theorem 2.17.** *The Fourier transform* $\mathcal{F}$ *is a one-one map of* $L^2(\mathbb{R})$ *onto itself. In other words, to every* $g \in L^2(\mathbb{R})$, *there corresponds one and only one* $f \in L^2(\mathbb{R})$ *such that* $\hat{f} = g$; *that is,*

$$f(x) := (\mathcal{F}^{-1}g)(x) =: \check{g}(x) \tag{2.3.11}$$

*is the inverse Fourier transform of* $g$.

**Proof.** Let $g \in L^2(\mathbb{R})$. Then its reflection $g^-$ as defined in (2.3.9) is also in $L^2(\mathbb{R})$. We will first show that the $L^2(\mathbb{R})$ function

$$f(x) := \frac{1}{2\pi}\widehat{(g^-)}(x) \tag{2.3.12}$$

satisfies $\hat{f} = g$ a.e.

Indeed, by applying (2.3.10), Lemma 2.14, (2.3.12), (2.3.10) again, and the Parseval Identity, consecutively, we have

$$
\begin{aligned}
\|g - \hat{f}\|_2^2 &= \|g\|_2^2 - 2Re\langle g, \hat{f}\rangle + \|\hat{f}\|_2^2 \\
&= \|g\|_2^2 - 2Re\langle g, (\widehat{\overline{f^-}})\rangle + \|\hat{f}\|_2^2 \\
&= \|g\|_2^2 - 2Re\langle \hat{g}, \overline{(\hat{f})^-}\rangle + \|\hat{f}\|_2^2 \\
&= \|g\|_2^2 - 2Re\langle \hat{g}, f^-\rangle + \|\hat{f}\|_2^2 \\
&= \|g\|_2^2 - 2Re\left\langle \hat{g}, \frac{1}{2\pi}\hat{g}\right\rangle + \|\hat{f}\|_2^2 \\
&= \frac{1}{2\pi}\|\hat{g}\|_2^2 - \frac{2}{2\pi}\|\hat{g}\|_2^2 + 2\pi\|f\|_2^2 \\
&= -\frac{1}{2\pi}\|\hat{g}\|_2^2 + \frac{1}{2\pi}\|\widehat{g^-}\|_2^2 = 0,
\end{aligned}
$$

so that $\hat{f} = g$, a.e.

Showing that $f$, as defined in (2.3.12), is the only function in $L^2(\mathbb{R})$ that satisfies $\hat{f} = g$, is equivalent to showing that $\hat{f} = 0$ implies $f = 0$, a.e. and this is an immediate consequence of the Parseval Identity in Theorem 2.13. ∎

The $L^2(\mathbb{R})$ theory of Fourier transform as discussed above is usually known as the Plancherel theory.

## 2.4. Fourier series

We now turn to the study of $2\pi$-periodic functions. For each $p$, $1 \le p \le \infty$, the following notation will be used:

$$
\|f\|_{L^p(0,2\pi)} := \begin{cases} \left[\frac{1}{2\pi}\int_0^{2\pi}|f(x)|^p dx\right]^{1/p}, & \text{for } 1 \le p < \infty; \\ \operatorname*{ess\,sup}_{0 \le x \le 2\pi}|f(x)|, & \text{for } p = \infty. \end{cases} \tag{2.4.1}
$$

For each $p$, $L^p(0, 2\pi)$ denotes the Banach space of functions $f$ satisfying $f(x + 2\pi) = f(x)$ a.e. in $\mathbb{R}$, and $\|f\|_{L^p(0,2\pi)} < \infty$. Sometimes the subspace $C^*[0, 2\pi]$ of $L^\infty(0, 2\pi)$ consisting of only continuous functions is more useful than the whole space $L^\infty(0, 2\pi)$. Here, the asterisk $*$ is used to remind us that $f(0) = f(2\pi)$ for $f \in C^*[0, 2\pi]$.

The inequalities of Minkowski, Hölder, and Schwarz for $L^p(\mathbb{R})$ in (2.1.1), (2.1.2), and (2.1.3) are also valid for $L^p(0, 2\pi)$. In particular, for $p = 2$, we can again define the "inner product"

$$
\langle f, g\rangle^* = \frac{1}{2\pi}\int_0^{2\pi} f(x)\overline{g(x)}dx, \qquad f, g \in L^2(0, 2\pi), \tag{2.4.2}
$$

where the asterisk is used to distinguish this inner product from that in $L^2(\mathbb{R})$. Later, we will also need the following generalized Minkowski Inequality:

$$\left\{\frac{1}{2\pi}\int_0^{2\pi}\left|\int_a^b g(t,x)dt\right|^p dx\right\}^{\frac{1}{p}} \tag{2.4.3}$$

$$\leq \int_a^b\left\{\frac{1}{2\pi}\int_0^{2\pi}|g(t,x)|^p dx\right\}^{\frac{1}{p}} dt,$$

where the generalization is simply replacing a finite sum by a definite (Lebesgue) integral. Also, note that, in contrast to the spaces $L^p(\mathbb{R})$ which are not nested, we have

$$L^p(0,2\pi) \subseteq L^q(0,2\pi), \qquad p \geq q.$$

This can be easily verified by applying the Hölder Inequality.

The companions of the spaces $L^p(0,2\pi)$ are the (sequence) spaces $\ell^p = \ell^p(\mathbb{Z})$ of bi-infinite sequences $\{a_k\}$, $k \in \mathbb{Z}$, that satisfy: $\|\{a_k\}\|_{\ell^p} < \infty$, where

$$\|\{a_k\}\|_{\ell^p} := \begin{cases} \left\{\sum_{k\in\mathbb{Z}}|a_k|^p\right\}^{\frac{1}{p}}, & \text{for } 1 \leq p < \infty; \\ \sup_k |a_k|, & \text{for } p = \infty. \end{cases} \tag{2.4.4}$$

Again, the inequalities of Minkowski, Hölder, and Schwarz remain valid for the sequence spaces. Analogous to the Hilbert spaces $L^2(\mathbb{R})$ and $L^2(0,2\pi)$, the space $\ell^2 = \ell^2(\mathbb{Z})$ is also a Hilbert space with the inner product:

$$\langle\{a_k\},\{b_k\}\rangle_{\ell^2} := \sum_{k\in\mathbb{Z}} a_k\bar{b}_k. \tag{2.4.5}$$

Recall that the (integral) Fourier transform is used to describe the spectral behavior of an analog signal $f$ with finite energy (i.e., $f \in L^2(\mathbb{R})$). Here, we introduce the *"discrete Fourier transform"* $\mathcal{F}^*$ of a *"digital signal"* $\{c_k\} \in \ell^p$ to describe its spectral behavior, as follows:

$$(\mathcal{F}^*\{c_k\})(x) := \sum_{k\in\mathbb{Z}} c_k e^{ikx}. \tag{2.4.6}$$

That is, the discrete Fourier transform of $\{c_k\}$ is the "Fourier series" with "Fourier coefficients" given by $\{c_k\}$. We have not discussed the convergence of the Fourier series in (2.4.6) yet; but for $\{c_k\} \in \ell^1$, it is clear that the series converges absolutely and uniformly for all $x \in \mathbb{R}$. In general, the formal series (2.4.6) may be viewed simply as the "symbol" of the sequence $\{c_k\}$.

Since $e^{ix} = \cos x + i\sin x$, the Fourier series in (2.4.6) can also be written as

$$f(x) := \sum_{k=-\infty}^{\infty} c_k e^{ikx} = \frac{a_0}{2} + \sum_{k=1}^{\infty} a_k \cos kx + \sum_{k=1}^{\infty} b_k \sin kx \tag{2.4.7}$$

with

$$\begin{cases} a_k = c_k + c_{-k}; \\ b_k = i(c_k - c_{-k}). \end{cases} \tag{2.4.8}$$

The formulas in (2.4.8) can be easily derived by using the following identities:

$$\begin{cases} \sin x = \dfrac{e^{ix} - e^{-ix}}{2i}; \\ \cos x = \dfrac{e^{ix} + e^{-ix}}{2}. \end{cases}$$

The function notation $f(x)$ in (2.4.7) is only used as a notation for the Fourier series. It may not even be a function. In any case, we can always consider the trigonometric polynomials

$$(S_N f)(x) := \sum_{k=-N}^{N} c_k e^{ikx} = \frac{a_0}{2} + \sum_{k=1}^{N} \{a_k \cos kx + b_k \sin kx\}, \tag{2.4.9}$$

where $N$ is a nonnegative integer. These are called "*partial sums*" of the Fourier series $f$.

The $N^{\text{th}}$ degree trigonometric polynomial

$$D_N(x) := \frac{1}{2} + \sum_{k=1}^{N} \cos kx = \frac{\sin\left(N + \frac{1}{2}\right)x}{2\sin\frac{x}{2}} \tag{2.4.10}$$

is of special importance. It is called the "*Dirichlet kernel*" of degree $N$. Observe that, at least *formally*, the $N^{\text{th}}$ partial sum $S_N f$ of a Fourier series $f$ can be obtained by the "convolution" of $f$ with the Dirichlet kernel of degree $N$, namely:

$$(S_N f)(x) = \frac{1}{\pi} \int_0^{2\pi} f(x - t) D_N(t) dt \tag{2.4.11}$$

$$= \frac{1}{\pi} \int_0^{2\pi} f(t) D_N(x - t) dt.$$

The integration in (2.4.11) is certainly meaningful if $f \in L^1(0, 2\pi)$.

On the other hand, if $f$ is any function in $L^p(0, 2\pi)$, $1 \le p \le \infty$, then we can define the "inverse discrete Fourier transform" $\mathcal{F}^{*-1}$ of $f$ by:

$$(\mathcal{F}^{*-1} f)(k) = c_k(f) := \frac{1}{2\pi} \int_0^{2\pi} f(x) e^{-ikx} dx. \tag{2.4.12}$$

That is, $\mathcal{F}^{*-1}$ takes $f \in L^p(0, 2\pi)$ to a bi-infinite sequence $\{c_k(f)\}$, $k \in \mathbb{Z}$. This sequence, of course, defines a Fourier series

$$\sum_{k \in \mathbb{Z}} c_k(f) e^{ikx} \tag{2.4.13}$$

and is called the sequence of "Fourier coefficients" of the Fourier series. A fundamental question is if this series "converges" to the original function $f$. A discussion of this topic is delayed to the next section. In the following, we will only study the $L^2(0, 2\pi)$ theory.

**Theorem 2.18.** *Let $f \in L^2(0, 2\pi)$. Then the sequence $\{c_k(f)\}$ of Fourier coefficients of $f$ is in $\ell^2$ and satisfies the Bessel Inequality:*

$$\sum_{k=-\infty}^{\infty} |c_k(f)|^2 \leq \|f\|_{L^2(0,2\pi)}^2. \tag{2.4.14}$$

**Proof.** Let $S_N(f)$ denote the $N^{\text{th}}$ partial sum of the Fourier series (2.4.13). Then we have

$$0 \leq \|f - S_N(f)\|_{L^2(0,2\pi)}^2 = \|f\|_{L^2(0,2\pi)}^2 - 2Re\langle f, S_N(f)\rangle^* + \|S_N(f)\|_{L^2(0,2\pi)}^2, \tag{2.4.15}$$

where it is easy to verify that

$$\langle f, S_N(f)\rangle^* = \sum_{k=-N}^{N} |c_k(f)|^2 \tag{2.4.16}$$

and

$$\|S_N(f)\|_{L^2(0,2\pi)}^2 = \sum_{k=-N}^{N} |c_k(f)|^2. \tag{2.4.17}$$

Hence, putting (2.4.16) and (2.4.17) into (2.4.15), we have

$$\sum_{k=-N}^{N} |c_k(f)|^2 \leq \|f\|_{L^2(0,2\pi)}^2.$$

Since this inequality holds for any $N$, we have established (2.4.14). ■

The converse of Theorem 2.18 is the following so-called Riesz-Fischer Theorem.

**Theorem 2.19.** *Let $\{c_k\} \in \ell^2$. Then there exists some $f \in L^2(0, 2\pi)$ such that $c_k$ is the $k^{\text{th}}$ Fourier coefficient of $f$. Furthermore,*

$$\sum_{k=-\infty}^{\infty} |c_k|^2 = \|f\|_{L^2(0,2\pi)}^2. \tag{2.4.18}$$

This theorem asserts that the discrete Fourier transform $\mathcal{F}^*$ maps $\ell^2$ into $L^2(0, 2\pi)$ and the identity (2.4.18) holds for all $f$ in the image of $\ell^2$ under $\mathcal{F}^*$.

**Proof.** For each positive integer $N$, consider the trigonometric polynomial

$$S_N(x) = \sum_{k=-N}^{N} c_k e^{ikx}. \tag{2.4.19}$$

Since $\{c_k\}$ is in $\ell^2$, the sequence

$$\sum_{k=-N}^{N} |c_k|^2$$

is a Cauchy sequence of real numbers. Hence, by considering an identity similar to (2.4.17) for $\|S_N - S_M\|^2_{L^2(0,2\pi)}$, it is clear that $\{S_N\}$ is a Cauchy sequence in $L^2(0, 2\pi)$. Let $f \in L^2(0, 2\pi)$ be the limit of this sequence. Then by the Bessel Inequality (2.4.14), the Fourier coefficients $c_k(f)$ of $f$ satisfy the estimate

$$\sum_{k=-N}^{N} |c_k(f) - c_k|^2 \leq \|f - S_N\|^2_{L^2(0,2\pi)}.$$

Hence, taking $N \to \infty$, we obtain

$$c_k(f) = c_k, \qquad k \in \mathbb{Z}.$$

Furthermore, in view of (2.4.16) and (2.4.17), we have

$$\|f - S_N\|^2_{L^2(0,2\pi)} = \|f\|^2_{L^2(0,2\pi)} - 2Re\langle f, S_N \rangle^* + \|S_N\|^2_{L^2(0,2\pi)}$$
$$= \|f\|^2_{L^2(0,2\pi)} - \sum_{k=-N}^{N} |c_k|^2,$$

which, as $N \to \infty$, yields (2.4.18). ∎

We emphasize again that Theorem 2.19 only asserts that the identity (2.4.18) is valid for all functions $f$ in the image of the space $\ell^2$ under the discrete Fourier transform operation. That (2.4.18) can indeed be extended to all of $L^2(0, 2\pi)$ is a consequence of the Weierstraß Theorem, which says that the set of all trigonometric polynomials is dense in $L^2(0, 2\pi)$. The identity (2.4.18), so extended to all of $L^2(0, 2\pi)$ is called the *"Parseval Identity"* for $L^2(0, 2\pi)$. A simple way to establish the Weierstraß Theorem is to consider the Cesàro means of the sequences of partial sums of the Fourier series of $f \in L^2(0, 2\pi)$.

Let $f \in L^2(0, 2\pi)$ and denote by $S_n f$ the $n^{\text{th}}$ partial sum of the Fourier series (2.4.7) as defined in (2.4.9). Then the $N^{\text{th}}$ Cesàro means of $\{S_n f\}$ is given by

$$\sigma_N f := \frac{S_0 f + \cdots + S_N f}{N + 1}. \tag{2.4.20}$$

Since $S_n f$ is the convolution of $f$ with the Dirichlet kernel $D_n$ as defined in (2.4.11) (for $f \in L^2(0, 2\pi) \subset L^1(0, 2\pi)$), it follows that $\sigma_N f$ is the convolution of $f$ with the so-called *"Fejér kernel"*, defined by

$$K_N(x) := \frac{D_0(x) + \cdots + D_N(x)}{N + 1} = \frac{1}{N + 1} \frac{\sin^2\left(\frac{N+1}{2}x\right)}{2\sin^2\left(\frac{x}{2}\right)}, \tag{2.4.21}$$

namely:

$$(\sigma_N f)(x) = \frac{1}{\pi} \int_0^{2\pi} f(x-t)K_N(t)dt. \qquad (2.4.22)$$

Observe that the trigonometric polynomial $K_N$ differs from $D_N(x)$ in that $K_N(x) \geq 0$ for all $x$. This property is crucial in establishing the following polynomial density result.

**Theorem 2.20.** *Let* $f \in L^2(0, 2\pi)$. *Then*

$$\lim_{N \to \infty} \|f - \sigma_N f\|_{L^2(0,2\pi)} = 0.$$

Before we prove this result, we find it convenient to introduce the notation of "$L^p(0, 2\pi)$ modulus of continuity":

$$\begin{cases} \omega_p(f;\eta) := \sup_{0 < h \leq \eta} \left\{ \frac{1}{2\pi} \int_0^{2\pi} |f(x+h) - f(x)|^p dx \right\}^{\frac{1}{p}}, \text{ for } f \in L^p(0, 2\pi) \\ \quad \text{where } 1 \leq p < \infty; \text{ and if } p = \infty, \text{ then} \\ \omega(f;\eta) := \omega_\infty(f;\eta) := \sup_{0 < h \leq \eta} \max_x |f(x+h) - f(x)|, \text{ for } f \in C^*[0, 2\pi]. \end{cases} \qquad (2.4.23)$$

Note that $L^\infty(0, 2\pi)$ is replaced by its subspace $C^*[0, 2\pi]$. We also remark that $\omega_p(f; \eta)$ and $\omega(f; \eta)$ are nondecreasing functions of $\eta$ and that

$$\begin{cases} \omega_p(f;\eta) \to 0 \text{ as } \eta \to 0^+ \quad \text{for} \quad f \in L^p(0, 2\pi); \\ \omega(f;\eta) \to 0 \text{ as } \eta \to 0^+ \quad \text{for} \quad f \in C^*[0, 2\pi]. \end{cases} \qquad (2.4.24)$$

Let us now turn to the proof of the Theorem 2.20.

**Proof.** Since

$$\frac{1}{\pi} \int_0^{2\pi} F_N(x)dx = 1,$$

we have, by the generalized Minkowski Inequality in (2.4.3) and the definition of $\omega_2(f; |t|)$, consecutively, that

$$\|f - \sigma_N f\|_{L^2(0,2\pi)} = \left\{ \frac{1}{2\pi} \int_0^{2\pi} \left| \frac{1}{2\pi(N+1)} \int_0^{2\pi} \{f(x) - f(x-t)\} \right. \right.$$
$$\left. \left. \times \left( \frac{\sin \frac{(N+1)t}{2}}{\sin \frac{t}{2}} \right)^2 dt \right|^2 dx \right\}^{1/2}$$

$$\leq \frac{1}{2\pi(N+1)} \int_0^{2\pi} \left\{ \left( \frac{\sin \frac{(N+1)t}{2}}{\sin \frac{t}{2}} \right)^2 \right.$$

$$\times \left. \left[ \frac{1}{2\pi} \int_0^{2\pi} |f(x) - f(x-t)|^2 dx \right]^{1/2} \right\} dt$$

$$\leq \frac{1}{2\pi(N+1)} \int_{-\pi}^{\pi} \left( \frac{\sin \frac{(N+1)t}{2}}{\sin \frac{t}{2}} \right)^2 \omega_2(f; |t|) dt$$

$$\leq \frac{\pi}{N+1} \int_0^{\pi} \frac{\sin^2 \frac{(N+1)t}{2}}{t^2} \omega_2(f; t) dt$$

$$= \frac{\pi}{2} \int_0^{(N+1)\pi/2} \left( \frac{\sin u}{u} \right)^2 \omega_2 \left( f; \frac{2u}{N+1} \right) du.$$

Let $\varepsilon > 0$ be arbitrarily given. Choose $M > 0$ so that

$$\pi \|f\|_{L^2(0,2\pi)} \int_M^{\infty} \frac{du}{u^2} < \varepsilon.$$

Since $\omega_2(f; \cdot) \leq 2\|f\|_{L^2(0,2\pi)}$ and $\omega_2(f; \cdot)$ is a nondecreasing function, it follows that, for $(N+1)\pi \geq 2M$,

$$\|f - \sigma_N f\|_{L^2(0,2\pi)} < \varepsilon + \frac{\pi}{2} \int_0^M \left( \frac{\sin u}{u} \right)^2 \omega_2 \left( f; \frac{2u}{N+1} \right) du$$

$$< \varepsilon + \frac{M\pi}{2} \omega_2 \left( f; \frac{2M}{N+1} \right) \to \varepsilon + 0,$$

as $N \to \infty$. This completes the proof of Theorem 2.20. ∎

By using the preceding result on the density of trigonometric polynomials in $L^2(0, 2\pi)$, we can now establish the following main result of this section.

**Theorem 2.21.** *The discrete Fourier transform $\mathcal{F}^*$ defined in (2.4.6) is an isometric isomorphism of $\ell^2$ onto $L^2(0, 2\pi)$. In other words, $\mathcal{F}^*$ maps $\ell^2$ one-one onto $L^2(0, 2\pi)$, such that the Parseval Identity*

$$\sum_{k=-\infty}^{\infty} |c_k|^2 = \frac{1}{2\pi} \int_0^{2\pi} |f(x)|^2 dx, \qquad f \in L^2(0, 2\pi), \qquad (2.4.25)$$

*holds, where $c_k = c_k(f)$ is the $k^{\text{th}}$ Fourier coefficient of $f$.*

**Proof.** Theorem 2.19 already says that $\mathcal{F}^*$ maps $\ell^2$ into $L^2(0, 2\pi)$. To prove that the map is onto, let $f \in L^2(0, 2\pi)$ be arbitrarily chosen and let $\{c_k\}$ denote the sequence of Fourier coefficients of $f$. Then by the Bessel Inequality in Theorem 2.18, we have

$$\sum_{k=-\infty}^{\infty} |c_k|^2 \leq \|f\|_{L^2(0,2\pi)}^2.$$

On the other hand, by the definition of $\sigma_N f$ in (2.4.20), and referring to $S_n f$ in (2.4.9), it is clear that

$$(\sigma_N f)(x) = \sum_{k=-N}^{N} \left(1 - \frac{|k|}{N+1}\right) c_k e^{ikx},$$

so that

$$\|\sigma_N f\|^2_{L^2(0,2\pi)} = \sum_{k=-N}^{N} \left(1 - \frac{|k|}{N+1}\right)^2 |c_k|^2$$

$$\leq \sum_{k=-N}^{N} |c_k|^2 \leq \|f\|^2_{L^2(0,2\pi)}.$$

That is, we have

$$\|f\|_{L^2(0,2\pi)} \geq \left\{ \sum_{k=-N}^{N} |c_k|^2 \right\}^{\frac{1}{2}} \geq \|\sigma_N f\|_{L^2(0,2\pi)}$$

$$\geq \|f\|_{L^2(0,2\pi)} - \|f - \sigma_N f\|_{L^2(0,2\pi)}.$$

Hence, by applying Theorem 2.20, the Parseval Identity (2.4.45) is established. This identity of course guarantees that $\mathcal{F}^*$ is one-one, since if all the Fourier coefficients of $f$ are zero, then $\|f\|_{L^2(0,2\pi)} = 0$, or $f = 0$ a.e.. ∎

## 2.5. Basic convergence theory and Poisson's summation formula

Although the theory of convergence of Fourier series is a very fascinating subject, a detailed study is beyond the scope of this book. We will only discuss two basic tests of convergence and omit their derivations.

First, let us mention that there exists a $2\pi$-periodic continuous function whose Fourier series diverges at every rational number. In addition, there even exists a function in $L^1(0, 2\pi)$ whose Fourier series diverges everywhere. Hence, certain conditions must be imposed to guarantee convergence. The convergence result that requires the weakest assumption is a very deep result which says that the Fourier series of every function $f$ in $L^p(0, 2\pi)$, where $1 < p \leq \infty$, converges to $f$ almost everywhere. In the following, we are interested in uniform convergence, or at least, convergence at certain specific points.

The following result is called the *Dini-Lipschitz* Test of convergence. The notation $\omega(f; \eta)$ for uniform modulus of continuity introduced in (2.4.23) will be used here.

**Theorem 2.22.** Let $f \in C^*[0, 2\pi]$ such that

$$\int_0^a \frac{\omega(f; t)}{t} dt < \infty \tag{2.5.1}$$

*for some $a > 0$. Then the Fourier series of $f$ converges uniformly to $f$; that is,*

$$\lim_{N \to \infty} \|f - S_N f\|_{L^\infty[0,2\pi]} = 0.$$

For instance, if $\omega(f; \eta) = O(\eta^\alpha)$ for some $\alpha > 0$, then the condition in (2.5.1) certainly holds.

The second convergence test to be stated below is called the *Dirichlet-Jordan Test*. It is valid for functions which do not oscillate too drastically. Such functions are said to be of "*bounded variation*". It is well known (and not too difficult to derive from the definition) that every function of bounded variation on an interval $[a, b]$ can be written as the difference of two non-decreasing functions. Hence, if $f$ is of bounded variation on $[a, b]$, then both of the one-sided limits

$$\begin{cases} f(x^+) := \lim_{h \to 0+} f(x + h); \\ f(x^-) := \lim_{h \to 0+} f(x - h) \end{cases} \tag{2.5.2}$$

exist at every $x$, $a < x < b$.

**Theorem 2.23.** *Let $f$ be a $2\pi$-periodic function of bounded variation on $[0, 2\pi]$. Then the Fourier series of $f$ converges to $(f(x^+) + f(x^-))/2$ everywhere; that is,*

$$\lim_{N \to \infty} (S_N f)(x) = \frac{f(x^+) + f(x^-)}{2} \tag{2.5.3}$$

*for every $x \in \mathbb{R}$. Furthermore, if $f$ is also continuous on any compact interval $[a, b]$, then the Fourier series of $f$ converges uniformly to $f$ on $[a, b]$.*

While the (integral) Fourier transform studied in the first three sections of this chapter is defined on $L^p(\mathbb{R})$, the Fourier series represents only periodic functions. To periodize a function $f \in L^p(\mathbb{R})$, the simplest way is to consider

$$\Phi_f(x) := \sum_{k=-\infty}^{\infty} f(x + 2\pi k). \tag{2.5.4}$$

The first question that arises is whether or not $\Phi_f$ a function. The answer is positive for $p = 1$, as shown in the following.

**Lemma 2.24.** *Let $f \in L^1(\mathbb{R})$. Then the series as defined in (2.5.4) converges a.e. to some $2\pi$-periodic function $\Phi_f$. Furthermore, the a.e. convergence is absolute, and $\Phi_f \in L^1(0, 2\pi)$ with*

$$\|\Phi_f\|_{L^1(0,2\pi)} \le \frac{1}{2\pi} \|f\|_1. \tag{2.5.5}$$

**Proof.** The almost everywhere absolute convergence will be established, once we have

$$\sum_{k=-\infty}^{\infty} \int_0^{2\pi} |f(x + 2\pi k)| dx < \infty.$$

But this, along with (2.5.5), follows immediately from the simple observation that

$$\int_0^{2\pi} |\Phi_f(x)| dx \leq \sum_{k=-\infty}^{\infty} \int_0^{2\pi} |f(x+2\pi k)| dx$$

$$= \sum_{k=-\infty}^{\infty} \int_{2\pi k}^{2\pi(k+1)} |f(x)| dx = \int_{-\infty}^{\infty} |f(x)| dx < \infty. \quad \blacksquare$$

In view of this lemma, we may consider the Fourier series of $\Phi_f$, namely:

$$\Phi_f(x) = \sum_{k=-\infty}^{\infty} c_k(\Phi_f) e^{ikx},$$

where

$$c_k(\Phi_f) = \frac{1}{2\pi} \int_0^{2\pi} e^{-ikx} \Phi_f(x) dx$$

$$= \frac{1}{2\pi} \sum_{j=-\infty}^{\infty} \int_0^{2\pi} e^{-ikx} f(x+2\pi j) dx$$

$$= \frac{1}{2\pi} \sum_{j=-\infty}^{\infty} \int_{2\pi j}^{2\pi(j+1)} e^{-ikx} f(x) dx = \frac{1}{2\pi} \hat{f}(k).$$

Hence, if the Fourier series of $\Phi_f$ converges to $\Phi_f$, then the two quantities

$$\sum_{k=-\infty}^{\infty} f(x+2\pi k) \tag{2.5.6}$$

and

$$\frac{1}{2\pi} \sum_{k=-\infty}^{\infty} \hat{f}(k) e^{ikx} \tag{2.5.7}$$

can be equated. Unfortunately, since $\Phi_f$ is only in $L^1(0, 2\pi)$, its Fourier series may even diverge everywhere. So, some conditions must be imposed on $\Phi_f$ or $f$ in order to be able to insure that (2.5.6) and (2.5.7) are the same. We first settle for a very general statement.

**Theorem 2.25.** *Let $f \in L^1(\mathbb{R})$ satisfy the following two conditions:*
*(i) the series (2.5.6) converges everywhere to some continuous function, and*
*(ii) the Fourier series (2.5.7) converges everywhere.*

*Then the following "Poisson Summation Formula" holds:*

$$\sum_{k=-\infty}^{\infty} f(x+2\pi k) = \frac{1}{2\pi} \sum_{k=-\infty}^{\infty} \hat{f}(k) e^{ikx}, \qquad x \in \mathbb{R}. \tag{2.5.8}$$

In particular,

$$\sum_{k=-\infty}^{\infty} f(2\pi k) = \frac{1}{2\pi} \sum_{k=-\infty}^{\infty} \hat{f}(k). \qquad (2.5.9)$$

Before we give some sufficient conditions on $f$ to guarantee both (i) and (ii), let us remark that Poisson's summation formula (2.5.8) or (2.5.9) can be formulated a little differently. To do so, we simply observe that if $f_a(x) := f(ax)$ where $a > 0$, then $\hat{f}_a(x) = a^{-1}\hat{f}(\frac{x}{a})$. Hence, (2.5.8) and (2.5.9) become:

$$\begin{cases} \displaystyle\sum_{k=-\infty}^{\infty} f(x+2\pi ak) = \frac{1}{2\pi a} \sum_{k=-\infty}^{\infty} \hat{f}\left(\frac{k}{a}\right) e^{i\frac{k}{a}x}; \\ \displaystyle\sum_{k=-\infty}^{\infty} f(2\pi ak) = \frac{1}{2\pi a} \sum_{k=-\infty}^{\infty} \hat{f}\left(\frac{k}{a}\right). \end{cases} \qquad (2.5.10)$$

In particular, by choosing $a = (2\pi)^{-1}$, we have

$$\begin{cases} \displaystyle\sum_{k=-\infty}^{\infty} f(x+k) = \sum_{k=-\infty}^{\infty} \hat{f}(2\pi k) e^{i2\pi kx}; \\ \displaystyle\sum_{k=-\infty}^{\infty} f(k) = \sum_{k=-\infty}^{\infty} \hat{f}(2\pi k). \end{cases} \qquad (2.5.11)$$

Now, we list some conditions under which both (i) and (ii) in Theorem 2.25 hold.

**Corollary 2.26.** *Let $f$ be a measurable function satisfying:*

$$f(x), \hat{f}(x) = O\left(\frac{1}{1+|x|^\alpha}\right) \qquad (2.5.12)$$

*for some $\alpha > 1$. Then Poisson's summation formula (2.5.8) holds for all $x \in \mathbb{R}$.*

Observe that since $\hat{f}$ satisfies (2.5.12), $f$ is necessarily continuous, and it is clear that both (i) and (ii) are valid.

**Corollary 2.27.** *Let $f \in L^1(\mathbb{R})$ and suppose that the series in (2.5.4) converges everywhere to a continuous function of bounded variation on $[0, 2\pi]$. Then Poisson's summation formula (2.5.8) holds for every $x \in \mathbb{R}$.*

If $\Phi_f$ is a continuous function of bounded variation on $[0, 2\pi]$, then by Theorem 2.23, its Fourier series, which is (2.5.7), converges everywhere to $\Phi_f$. That is, (2.5.6) and (2.5.7) are identical.

The most important example is any compactly supported continuous function $f$ of bounded variation. For such an $f$, the series (2.5.4) is only a finite sum, and hence, $\Phi_f$ is also a continuous function and is of bounded variation on $[0, 2\pi]$. All $B$-splines of order at least 2, to be studied in detail in Chapter 4, are typical examples.

We end this chapter by applying Poisson's summation formula to study the Fourier transform of the autocorrelation function $F$ of a function $f \in L^2(\mathbb{R})$ as defined in (2.3.1). This will better prepare us to study the construction of semi-orthogonal wavelets in Chapter 5. By using the notation $f^-$ for the reflection of $f$ as introduced in (2.3.9), we may reformulate $F$ as

$$F(x) = (f * (\bar{f})^-)(x).$$

Hence, by applying Lemma 2.16, we obtain

$$\widehat{F}(x) = |\hat{f}(x)|^2. \tag{2.5.13}$$

Now, since $f$ is in $L^2(\mathbb{R})$, so is $\hat{f}$ by the Parseval Identity. Hence, $\widehat{F} \in L^1(\mathbb{R})$ and by Lemma 2.24,

$$\Phi_{\widehat{F}}(x) = \sum_{k=-\infty}^{\infty} \widehat{F}(x + 2\pi k) = \sum_{k=-\infty}^{\infty} |\hat{f}(x + 2\pi k)|^2 \tag{2.5.14}$$

converges a.e. and $\Phi_{\widehat{F}} \in L^1(0, 2\pi)$.

To study the Fourier series of $\Phi_{\widehat{F}}$, let us impose the extra condition, namely: $f \in L^1(\mathbb{R})$. Then $(\bar{f})^-$ is also in $L^1(\mathbb{R})$, and so is the convolution $F = f * (\bar{f})^-$. Therefore, by Theorem 2.5, we have

$$F(x) = (\mathcal{F}^{-1}\widehat{F})(x) = \frac{1}{2\pi} \int_{-\infty}^{\infty} e^{ixy} \widehat{F}(y) dy \tag{2.5.15}$$

$$= \left(\frac{1}{2\pi}\widehat{F}\right)^{\wedge}(-x);$$

and hence, the Fourier coefficients of $\Phi_{\widehat{F}}$ are given by

$$c_k = c_k(\Phi_{\widehat{F}}) = \left(\frac{1}{2\pi}\widehat{F}\right)^{\wedge}(k) = F(-k).$$

That is, the Fourier series of $\Phi_{\widehat{F}}$ can be written as:

$$\Phi_{\widehat{F}}(x) = \sum_{k=-\infty}^{\infty} F(-k)e^{ikx} = \sum_{k=-\infty}^{\infty} F(k)e^{-ikx}. \tag{2.5.16}$$

So, we have:

$$\sum_{k=-\infty}^{\infty} |\hat{f}(x + 2\pi k)|^2 = \sum_{k=-\infty}^{\infty} F(k)z^k, \quad \text{a.e.,} \tag{2.5.17}$$

where $z = e^{-ix}$, and the right-hand side of (2.5.17) is called the "*symbol*" of the sequence $\{F(k)\}$. If $f$ happens to have compact support, then so does its autocorrelation function $F$; and the symbol of $\{F(k)\}$ is a "*Laurent polynomial*". This Laurent polynomial is also called the "*Euler-Frobenius polynomial*" generated by $f$. Hence, (2.5.17) gives a very important relation between the Euler-Frobenius polynomial generated by a compactly supported function $f$ and a nonnegative function $\Phi_{\hat{F}}$ in (2.5.14), which is instrumental to the study of unconditional basis, orthogonalization, and duality. Details on these topics will be studied in later chapters, particularly in Chapter 5.

Returning to (2.5.17) without assuming that $f$ is of compact support, we only have equality a.e. for $f \in L^1(\mathbb{R}) \cap L^2(\mathbb{R})$. In the following, we give three different sets of conditions that ensure the validity of (2.5.17) for all $x \in \mathbb{R}$.

**Theorem 2.28.** *Let $f \in L^2(\mathbb{R})$ satisfy any one of the following three conditions:*

(i) $f(x) = O(|x|^{-\beta})$, $\beta > 1$; and $\hat{f}(x) = O(|x|^{-\alpha})$, $\alpha > \frac{1}{2}$, as $|x| \to \infty$.

(ii) $\hat{f}$ *is of compact support and belongs to class* Lip$(\gamma)$ *for some* $\gamma > 0$, *meaning:*

$$\sup_{x} \sup_{0 < t \le h} |\hat{f}(x+t) - \hat{f}(x)| = O(h^{\gamma}), \text{ as } h \to 0^{+}. \tag{2.5.18}$$

(iii) $\hat{f}$ *is a continuous function of compact support, and is of bounded variation in its support.*

*Then it follows that*

$$\sum_{k=-\infty}^{\infty} |\hat{f}(x + 2\pi k)|^2 = \sum_{k=-\infty}^{\infty} \left\{ \int_{-\infty}^{\infty} f(y+k)\overline{f(y)}dy \right\} e^{-ikx} \tag{2.5.19}$$

*for all $x \in \mathbb{R}$.*

We remark that each of the conditions in (i)-(iii) already implies that the left-hand side of (2.5.19) is a $2\pi$-periodic continuous function whose Fourier series is given by the right-hand side. This is obvious for (i), but requires a little more work for (ii) and (iii). The convergence of the Fourier series at every $x \in \mathbb{R}$ for each of (i), (ii), and (iii) follows from Corollary 2.26, Theorem 2.22, and Theorem 2.23, respectively. ∎

# 3  Wavelet Transforms and Time-Frequency Analysis

In order to study the spectral behavior of an analog signal from its Fourier transform, full knowledge of the signal in the time-domain must be acquired. This even includes future information. In addition, if a signal is altered in a small neighborhood of some time instant, then the entire spectrum is affected. Indeed, in the extreme case, the Fourier transform of the delta distribution $\delta(t - t_0)$, with support at a single point $t_0$, is $e^{-it_0\omega}$, which certainly covers the whole frequency domain. Hence, in many applications such as analysis of non-stationary signals and real-time signal processing, the formula of Fourier transform alone is quite inadequate.

The deficiency of the formula of Fourier transform in time-frequency analysis was already observed by D. Gabor, who, in his 1946 paper, introduced a time-localization "window function" $g(t - b)$, where the parameter $b$ is used to translate the window in order to cover the whole time-domain, for extracting local information of the Fourier transform of the signal. In fact, Gabor used a Gaussian function for the window function $g$. Since the Fourier transform of a Gaussian function is again a Gaussian, the inverse Fourier transform is localized simultaneously.

The first section in this chapter is devoted to a study of the Gabor transform. A discussion of this so-called "short-time Fourier transform" (STFT) in general and the Uncertainty Principle that governs the size of the window will be the content of the second section. In particular, it will be observed there that the time-frequency window of any STFT is rigid, and hence, is not very effective for detecting signals with high frequencies and investigating signals with low frequencies. This motivates the introduction of the integral wavelet transform (IWT) in Section 3. Instead of windowing the Fourier and inverse Fourier transforms as the STFT does, the IWT windows the function (or signal) and its Fourier transform directly. This allows room for a dilation (or scale) parameter that narrows and widens the time-frequency window according to high and low frequencies. Inverting the IWT is required for reconstructing the signal from its decomposed local spectral information. Information on both continuous and discrete time observations will be considered. This leads to the study of frames and wavelet series in the last two sections of the chapter.

49

### 3.1. The Gabor transform

A function $f$ in $L^2(\mathbb{R})$ is used to represent an analog signal with finite energy, and its Fourier transform

$$\hat{f}(\omega) = \int_{-\infty}^{\infty} e^{-i\omega t} f(t)dt \qquad (3.1.1)$$

reveals the spectral information of the signal. Here and throughout this chapter, $t$ and $\omega$ will be reserved for the time and frequency variables, respectively. Unfortunately, the formula (3.1.1) alone is not very useful for extracting information of the spectrum $\hat{f}$ from local observation of the signal $f$. What is needed is a "good" time-window.

The "optimal" window for time-localization is achieved by using any Gaussian function

$$g_\alpha(t) := \frac{1}{2\sqrt{\pi\alpha}} e^{-\frac{t^2}{4\alpha}}, \qquad (3.1.2)$$

where $\alpha > 0$ is fixed, as the window function (see Figure. 2.2.1). Here, optimality is characterized by the Uncertainty Principle to be discussed in the next section. For any fixed value of $\alpha > 0$, the "*Gabor transform*" of an $f \in L^2(\mathbb{R})$ is defined by

$$(\mathcal{G}_b^\alpha f)(\omega) = \int_{-\infty}^{\infty} (e^{-i\omega t} f(t)) g_\alpha(t - b)dt; \qquad (3.1.3)$$

that is, $(\mathcal{G}_b^\alpha f)(\omega)$ localizes the Fourier transform of $f$ around $t = b$. The "width" of the window is determined by the (fixed) positive constant $\alpha$ to be discussed below. Observe that from (2.1.11) in Example 2.6 with $\omega = 0$ and $a = (4\alpha)^{-1}$, we have

$$\int_{-\infty}^{\infty} g_\alpha(t - b)db = \int_{-\infty}^{\infty} g_\alpha(x)dx = 1, \qquad (3.1.4)$$

so that

$$\int_{-\infty}^{\infty} (\mathcal{G}_b^\alpha f)(\omega)db = \hat{f}(\omega), \quad \omega \in \mathbb{R}.$$

That is, the set

$$\{\mathcal{G}_b^\alpha f \colon b \in \mathbb{R}\}$$

of Gabor transforms of $f$ decomposes the Fourier transform $\hat{f}$ of $f$ exactly, to give its local spectral information. To select a measurement of the width of the window function, we employ the notion of standard deviation, or root mean square (RMS) duration, defined by

$$\Delta_{g_\alpha} := \frac{1}{\|g_\alpha\|_2} \left\{ \int_{-\infty}^{\infty} x^2 g_\alpha^2(x)dx \right\}^{1/2}. \qquad (3.1.5)$$

Note that since $g_\alpha$ is an even function, its center, defined by (1.25), is 0, and hence, $\Delta_{g_\alpha}$ agrees with the general notion of "radius" introduced in Definition 1.2. In particular, the width of the window function $g_\alpha$ is $2\Delta_{g_\alpha}$.

**Theorem 3.1.** *For each $\alpha > 0$,*

$$\Delta_{g_\alpha} = \sqrt{\alpha}. \tag{3.1.6}$$

*That is, the width of the window function $g_\alpha$ is $2\sqrt{\alpha}$.*

**Proof.** By setting $\omega = 0$ in (2.1.11), we have

$$\int_{-\infty}^{\infty} e^{-ax^2}\, dx = \sqrt{\pi}\, a^{-\frac{1}{2}}; \tag{3.1.7}$$

and differentiating both sides with respect to the parameter $a$ yields

$$\int_{-\infty}^{\infty} x^2 e^{-ax^2}\, dx = \frac{\sqrt{\pi}}{2} a^{-3/2}. \tag{3.1.8}$$

Hence, by setting $a = (2\alpha)^{-1}$ in (3.1.7) and (3.1.8), it follows that

$$\|g_\alpha\|_2 = (8\pi\alpha)^{-1/4}, \tag{3.1.9}$$

and consequently,

$$\Delta_{g_\alpha} = (8\pi\alpha)^{1/4} \left\{ \frac{1}{4\pi\alpha} \cdot \frac{\sqrt{\pi}}{2}(2\alpha)^{3/2} \right\}^{1/2} = \sqrt{\alpha}. \quad \blacksquare$$

We may interpret the Gabor transform $\mathcal{G}_b^\alpha f$ in (3.1.3) somewhat differently, namely: by setting

$$G_{b,\omega}^\alpha(t) := e^{i\omega t} g_\alpha(t - b), \tag{3.1.10}$$

we have

$$(\mathcal{G}_b^\alpha f)(\omega) = \langle f, G_{b,\omega}^\alpha \rangle = \int_{-\infty}^{\infty} f(t)\overline{G_{b,\omega}^\alpha(t)}\, dt. \tag{3.1.11}$$

In other words, instead of considering $\mathcal{G}_b^\alpha f$ as localization of the Fourier transform of $f$, we may interpret it as windowing the function (or signal) $f$ by using the window function $G_{b,\omega}^\alpha$ in (3.1.10). We will follow this point of view in comparing it with the "integral wavelet transform" later. Graphs of the real and imaginary parts of $G_{b,\omega}^\alpha$ for $b = 0$, $\omega = 2\pi$, and $\alpha = 1, \frac{1}{4}, \frac{1}{16}$ are shown in Figures 3.1.1-3.1.2.

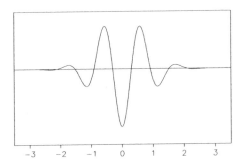

**Figure 3.1.1.** $Re\, G_{0,2\pi}^\alpha, \alpha = 0.2925$.

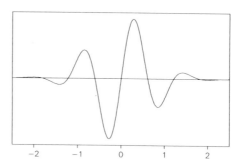

**Figure 3.1.2.** $Im\, G_{0,2\pi}^{\alpha}, \alpha = 0.2300.$

One advantage of the formulation (3.1.11) is that the Parseval Identity in (2.3.6) can be applied to relate the Gabor transform of $f$ with the Gabor transform of $\hat{f}$. In fact, since

$$\widehat{G}_{b,\omega}^{\alpha}(\eta) = e^{-ib(\eta-\omega)}e^{-\alpha(\eta-\omega)^2}, \tag{3.1.12}$$

which follows from (2.2.10), we have

$$(\mathcal{G}_b^{\alpha}f)(\omega) = \langle f, G_{b,\omega}^{\alpha}\rangle = \frac{1}{2\pi}\langle \hat{f}, \widehat{G}_{b,\omega}^{\alpha}\rangle \tag{3.1.13}$$

$$= \frac{1}{2\pi}\int_{-\infty}^{\infty} \hat{f}(\eta)e^{ib(\eta-\omega)}e^{-\alpha(\eta-\omega)^2}\,d\eta$$

$$= \frac{e^{-ib\omega}}{2\sqrt{\pi\alpha}}\int_{-\infty}^{\infty} (e^{ib\eta}\hat{f}(\eta))g_{1/4\alpha}(\eta-\omega)d\eta$$

$$= \frac{e^{-ib\omega}}{2\sqrt{\pi\alpha}}(\mathcal{G}_{\omega}^{1/4\alpha}\hat{f})(-b).$$

Let us interpret (3.1.13) from two different points of view. First, we consider

$$\int_{-\infty}^{\infty} (e^{-i\omega t}f(t))g_{\alpha}(t-b)dt \tag{3.1.14}$$

$$= \left(\sqrt{\frac{\pi}{\alpha}}e^{-ib\omega}\right)\cdot\frac{1}{2\pi}\int_{-\infty}^{\infty} (e^{ib\eta}\hat{f}(\eta))g_{1/4\alpha}(\eta-\omega)d\eta,$$

which says that, with the exception of the multiplicative term $\sqrt{\frac{\pi}{\alpha}}e^{-ib\omega}$, the *"window Fourier transform"* of $f$ with window function $g_{\alpha}$ at $t = b$ agrees with the *"window inverse Fourier transform"* of $\hat{f}$ with window function $g_{1/4\alpha}$ at $\eta = \omega$. By Theorem 3.1, the product of the widths of these two windows is

$$(2\Delta_{g_{\alpha}})(2\Delta_{g_{1/4\alpha}}) = 2. \tag{3.1.15}$$

On the other hand, by considering

$$H_{b,\omega}^\alpha(\eta) := \frac{1}{2\pi}\widehat{G_{b,\omega}^\alpha}(\eta) = \left(\frac{e^{ib\omega}}{2\sqrt{\pi\alpha}}\right) e^{-ib\eta}g_{1/4\alpha}(\eta - \omega), \tag{3.1.16}$$

we have

$$\langle f, G_{b,\omega}^\alpha \rangle = \langle \hat{f}, H_{b,\omega}^\alpha \rangle. \tag{3.1.17}$$

This identity says that the information obtained by investigating an analog signal $f(t)$ at $t = b$ by using the window function $G_{b,\omega}^\alpha$ as defined in (3.1.10) can also be obtained by observing the spectrum $\hat{f}(\eta)$ of the signal in a neighborhood of the frequency $\eta = \omega$ by using the window function $H_{b,\omega}^\alpha$ as defined in (3.1.16). Again the product of the width of the time-window $G_{b,\omega}^\alpha$ and that of the frequency-window $H_{b,\omega}^\alpha$ is

$$\left(2\Delta_{G_{b,\omega}^\alpha}\right)\left(2\Delta_{H_{b,\omega}^\alpha}\right) = (2\Delta_{g_\alpha})(2\Delta_{g_{1/4\alpha}}) = 2. \tag{3.1.18}$$

The Cartesian product

$$[b - \sqrt{\alpha},\, b + \sqrt{\alpha}] \times \left[\omega - \frac{1}{2\sqrt{\alpha}},\, \omega + \frac{1}{2\sqrt{\alpha}}\right] \tag{3.1.19}$$

of these two windows is called a rectangular time-frequency window. It is usually plotted in the time-frequency domain to show how a signal is localized. The width $2\sqrt{\alpha}$ of the time-window is called the "*width of the time-frequency window*", and the width $\frac{1}{\sqrt{\alpha}}$ of the frequency window is called the "*height of the time-frequency window*". A plot of this window is shown in Figure 3.1.3. Observe that the width of the time-frequency window is unchanged for observing the spectrum at all frequencies. That this restricts the application of the Gabor transform to study signals with unusually high and low frequencies will be discussed in Section 3.3.

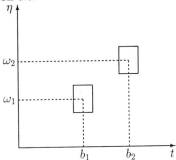

**Figure 3.1.3.** Gabor window.

## 3.2. Short-time Fourier transforms and the Uncertainty Principle

The Gabor transform is a window Fourier transform with any Gaussian function $g_\alpha$ as the window function. For various reasons such as computational efficiency or convenience in implementation, other functions may also be used as window functions instead. For a non-trivial function $w \in L^2(\mathbb{R})$ to qualify as a window function, it must satisfy the requirement that

$$tw(t) \in L^2(\mathbb{R}), \tag{3.2.1}$$

so that $|t|^{1/2}w(t) \in L^2(\mathbb{R})$ also. From (3.2.1) and by an application of the Schwarz Inequality to the product of $(1+|t|)^{-1}$ and $(1+|t|)w(t)$, it is clear that $w \in L^1(\mathbb{R})$ also. Hence, by Theorem 2.2, its Fourier transform $\widehat{w}$ is continuous. However, although it follows from the Parseval Identity that $\widehat{w}$ is also in $L^2(\mathbb{R})$, it does not necessarily satisfy (3.2.1), and hence, may not be a (frequency) window function. Recall from the previous section that the importance of a Gaussian function $g_\alpha$ is that its Fourier transform is also a Gaussian function, so that $g_\alpha$ and $\hat{g}_\alpha$ can be used for time-frequency localization.

**Example 3.2.** Both the first order $B$-spline

$$N_1(t) := \begin{cases} 1 & \text{for } 0 \le t < 1; \\ 0 & \text{otherwise,} \end{cases} \tag{3.2.2}$$

and the Haar function

$$\psi_1(t) = \psi_H(t) := \begin{cases} 1 & \text{for } 0 \le t < \frac{1}{2}; \\ -1 & \text{for } \frac{1}{2} \le t < 1; \\ 0 & \text{otherwise,} \end{cases} \tag{3.2.3}$$

as already defined in (1.5.7) and (1.1.16), are window functions; but their Fourier transforms $\widehat{N}_1$ and $\widehat{\psi}_1$ do not satisfy (3.2.1), and hence $N_1$ and $\psi_1$ cannot be used for time-frequency localization.

**Proof.** Since both $N_1$ and $\psi_1$ have compact support, they certainly satisfy (3.2.1). On the other hand, in view of Theorem 2.5 and Theorem 2.2, (ii), since $N_1$ and $\psi_1$ are not continuous, $\widehat{N}_1$ and $\widehat{\psi}_1$ cannot be in $L^1(\mathbb{R})$. Consequently, they do not satisfy (3.2.1). ■

In general, for any $w \in L^2(\mathbb{R})$ that satisfies (3.2.1), we define the center and radius of $w$, as in Definition 1.2, by

$$x^* := \frac{1}{\|w\|_2^2} \int_{-\infty}^{\infty} t|w(t)|^2 dt \tag{3.2.4}$$

and

$$\Delta_w := \frac{1}{\|w\|_2} \left\{ \int_{-\infty}^{\infty} (t - x^*)^2 |w(t)|^2 dt \right\}^{1/2}. \tag{3.2.5}$$

We also use the value $2\Delta_w$ to measure the width of the window function $w$. In signal analysis, if $w$ is considered as an analog signal itself, then $\Delta_w$ is called the root mean square (RMS) duration of the analog signal, and $\Delta_{\widehat{w}}$ is called its RMS bandwidth, provided that $\widehat{w}$ also satisfies (3.2.1). The Gabor transform (3.1.3), can be generalized to any "window Fourier transform" of an $f \in L^2(\mathbb{R})$, by using a function $w$ that satisfies (3.2.1) as the window function, as follows:

$$(\widetilde{\mathcal{G}}_b f)(\omega) := \int_{-\infty}^{\infty} (e^{-i\omega t} f(t)) \overline{w(t-b)} \, dt. \qquad (3.2.6)$$

Hence, by setting

$$W_{b,\omega}(t) := e^{i\omega t} w(t-b), \qquad (3.2.7)$$

we have

$$(\widetilde{\mathcal{G}}_b f)(\omega) = \langle f, W_{b,\omega} \rangle = \int_{-\infty}^{\infty} f(t) \overline{W_{b,\omega}(t)} \, dt, \qquad (3.2.8)$$

so that $(\widetilde{\mathcal{G}}_b f)(\omega)$ gives local information of $f$ in the time-window

$$[x^* + b - \Delta_w, \, x^* + b + \Delta_w]. \qquad (3.2.9)$$

Now, suppose that the Fourier transform $\widehat{w}$ of $w$ also satisfies (3.2.1). Then we can determine the center $\omega^*$ and radius $\Delta_{\widehat{w}}$, of the window function $\widehat{w}$, by using formulas analogous to (3.2.4) and (3.2.5). By setting

$$V_{b,\omega}(\eta) := \frac{1}{2\pi} \widehat{W}_{b,\omega}(\eta) \qquad (3.2.10)$$

$$= \left( \frac{e^{ib\omega}}{2\pi} \right) e^{-ib\eta} \widehat{w}(\eta - \omega),$$

which is also a window function with center at $\omega^* + \omega$ and radius equal to $\Delta_{\widehat{w}}$, we have, by the Parseval Identity,

$$(\widetilde{\mathcal{G}}_b f)(\omega) = \langle f, W_{b,\omega} \rangle = \langle \widehat{f}, V_{b,\omega} \rangle. \qquad (3.2.11)$$

Hence, $(\widetilde{\mathcal{G}}_b f)(\omega)$ also gives local spectral information of $f$ in the frequency-window:

$$[\omega^* + \omega - \Delta_{\widehat{w}}, \, \omega^* + \omega + \Delta_{\widehat{w}}]. \qquad (3.2.12)$$

In summary, by choosing any $w \in L^2(\mathbb{R})$ such that both $w$ and $\widehat{w}$ satisfy (3.2.1) to define the window Fourier transform in (3.2.6), we have a time-frequency window

$$[x^* + b - \Delta_w, \, x^* + b + \Delta_w] \times [\omega^* + \omega - \Delta_{\widehat{w}}, \, \omega^* + \omega + \Delta_{\widehat{w}}] \qquad (3.2.13)$$

with width $2\Delta_w$ (as determined by the width of the time-window) and constant window area

$$4\Delta_w \Delta_{\widehat{w}}. \qquad (3.2.14)$$

Again, the width of the time-frequency window remains unchanged for localizing signals with both high and low frequencies.

**Definition 3.3.** *If $w \in L^2(\mathbb{R})$ is so chosen that both $w$ and its Fourier transform $\widehat{w}$ satisfy (3.2.1), then the window Fourier transform introduced in (3.2.6), by using $w$ as the window function, is called a "short-time Fourier transform" (STFT).*

As observed earlier, since both $w$ and $\widehat{w}$ satisfy (3.2.1), they must be continuous functions. In addition to the Gaussian functions, every $B$-spline of order higher than one can be used to define an STFT.

**Example 3.4.** The $m^{\text{th}}$ order cardinal $B$-spline

$$N_m(t) := \int_0^1 N_{m-1}(t-x)dx, \qquad (3.2.15)$$

where $m \geq 2$, as defined recursively in (1.5.7) with $N_1$ given in (3.2.2), is a window function that defines an STFT. Furthermore,

$$\widehat{N}_m(\omega) = \left(\frac{1-e^{-i\omega}}{i\omega}\right)^m = e^{-im\omega/2}\left(\frac{\sin\omega/2}{\omega/2}\right)^m. \qquad (3.2.16)$$

**Proof.** Since $N_m$ is the $m$-fold convolution of $N_1$ and $\widehat{N}_1(\omega) = (1-e^{-i\omega})/i\omega$, the result in (3.2.16) follows from an application of Theorem 2.7. Hence, it is clear that $\widehat{N}_m$ satisfies (3.2.1). That $N_m$ itself satisfies (3.2.1) is trivial because it has compact support. ∎

For accurate time-frequency localization, one chooses a window function $w$ such that the time-frequency window has sufficiently small area $4\Delta_w\Delta_{\widehat{w}}$. We have already seen in (3.1.18) that if $w$ is any Gaussian function $g_\alpha$, $\alpha > 0$, then the window area is 2. So, the first question to be answered is whether a smaller area can be achieved. In the following theorem, known as the "Uncertainty Principle", we will see that it is not possible to find a window with size smaller than or equal to that of the Gaussian functions.

**Theorem 3.5.** *Let $w \in L^2(\mathbb{R})$ be chosen such that both $w$ and its Fourier transform $\widehat{w}$ satisfy (3.2.1). Then*

$$\Delta_w\Delta_{\widehat{w}} \geq \frac{1}{2}. \qquad (3.2.17)$$

*Furthermore, equality is attained if and only if*

$$w(t) = ce^{iat}g_\alpha(t-b),$$

*where $c \neq 0$, $\alpha > 0$, and $a, b \in \mathbb{R}$.*

**Remark.** In the engineering literature, if $w$ is considered as an analog signal with $t$ as the variable in the time-domain, then the domain of definition of its spectrum $\hat{w}$ is expressed in terms of the frequency (variable) $f = \omega/2\pi$ (in Hertz). Hence, if we replace $\omega$ by $2\pi f$ in the definition of $\Delta_{\hat{w}}$, then a factor of $2\pi$ is introduced. More precisely, by setting

$$\begin{cases} \Delta_t = \Delta_w; \\ \Delta_f = \dfrac{1}{2\pi}\Delta_{\hat{w}} \end{cases} \tag{3.2.18}$$

as usually found in the engineering literature, the Uncertainty Principle (3.2.17) now states that

$$\Delta_t \Delta_f \geq \frac{1}{4\pi}, \tag{3.2.19}$$

where equality holds if and only if the signal $w$ is a Gaussian.

To facilitate our proof of Theorem 3.5, we need the following result.

**Lemma 3.6.** *Let $f$ be a nontrivial almost everywhere differentiable function such that both $(1 + |x|)f(x)$ and $f'(x)$ belong to $L^2(\mathbb{R})$. Then*

$$\left| Re \int_{-\infty}^{\infty} x f(x)\overline{f'}(x)dx \right|^2 \leq \left\{ \int_{-\infty}^{\infty} |xf(x)|^2 dx \right\} \left\{ \int_{-\infty}^{\infty} |f'(x)|^2 dx \right\}. \tag{3.2.20}$$

*Furthermore, equality holds if and only if $f(x)$ is a Gaussian function.*

**Proof.** The inequality in (3.2.20) is a trivial application of the Schwarz Inequality. Now, if equality in (3.2.20) holds, then it follows that

$$\begin{cases} -Re\, xf(x)\overline{f'}(x) = |xf(x)f'(x)|, \text{ and} \\ |xf(x)| = 2\alpha|f'(x)| \end{cases} \tag{3.2.21}$$

for some positive constant $\alpha$. (Here, as we will see later, the other possibility $Re\, xf(x)\overline{f'}(x) = |xf(x)f'(x)|$ is ruled out, since $f$ must be in $L^2(\mathbb{R})$. Also, that $\alpha \neq 0$ follows from the assumption that $f$ is a nontrivial function in $L^2(\mathbb{R})$.) From the second identity in (3.2.21), we have

$$xf(x) = 2\alpha f'(x)e^{i\theta(x)}$$

for some real-valued function $\theta(x)$. The first identity in (3.2.21) then implies that

$$-xf(x)\overline{f'}(x) \geq 0,$$

so that

$$-2\alpha|f'(x)|^2 e^{i\theta(x)} \geq 0,$$

which in turn implies that $e^{i\theta(\alpha)} = -1$. Hence, we conclude that

$$xf(x) = -2\alpha f'(x),$$

so that $f'$ is continuous, and

$$f(x) = ce^{-x^2/4\alpha}$$

for some constant $c \neq 0$. That is, $f$ is a nonconstant multiple of the Gaussian function $g_\alpha$ defined in (2.2.9).   ∎

Before we turn to the proof of Theorem 3.5, we first observe that

$$Re\, xf(x)\overline{f'}(x) = \frac{1}{2}x\frac{d}{dx}|f(x)|^2 \tag{3.2.22}$$

and that if each of the functions $f(x)$, $f'(x)$, and $xf(x)$, is in $L^2(\mathbb{R})$, then

$$\lim_{|x|\to\infty} x|f(x)|^2 = 0. \tag{3.2.23}$$

The reason for (3.2.23) to hold is that under our assumptions, we have

$$\frac{d}{dx}(x|f(x)|^2) \in L^1(\mathbb{R}).$$

**Proof of Theorem 3.5.**

Let us first assume that the centers of $w$ and $\widehat{w}$ are at the origin. Then by applying Theorem 2.2 (iii), the Parseval Identity, (3.2.20), (3.2.22), and (3.2.23), consecutively, we have

$$
\begin{aligned}
(\Delta_w\Delta_{\widehat{w}})^2 &= \frac{\left(\int_{-\infty}^{\infty} t^2|w(t)|^2dt\right)\left(\int_{-\infty}^{\infty} \omega^2|\widehat{w}(\omega)|^2d\omega\right)}{\|w\|_2^2\|\widehat{w}\|_2^2} \tag{3.2.24}\\
&= \frac{\left(\int_{-\infty}^{\infty} t^2|w(t)|^2dt\right)\left(\int_{-\infty}^{\infty} |\widehat{w}'(\omega)|^2d\omega\right)}{\|w\|_2^2\|\widehat{w}\|_2^2}\\
&= \frac{2\pi\|xw(x)\|_2^2\|w'\|_2^2}{2\pi\|w\|_2^4}\\
&\geq \frac{1}{\|w\|_2^4}\left|Re\int_{-\infty}^{\infty} xw(x)\overline{w'}(x)dx\right|^2\\
&= \frac{1}{\|w\|_2^4}\left|\frac{1}{2}\int_{-\infty}^{\infty} x\frac{d}{dx}|w(x)|^2dx\right|^2\\
&= \frac{1}{\|w\|_2^4}\left(\frac{1}{2}\int_{-\infty}^{\infty} |w(x)|^2dx\right)^2 = \frac{1}{4}.
\end{aligned}
$$

Furthermore, by Lemma 3.6, the only inequality in the preceding derivation becomes equality, if and only if $w$ is a Gaussian function.

In general, if the centers of $w$ and $\widehat{w}$ are at $t = b$ and $\omega = a$, respectively, then by a simple change of variables, the above derivation shows that $\Delta_w\Delta_{\widehat{w}} = \frac{1}{2}$ only if

$$w(t) = ce^{iat}g_\alpha(t - b),$$

where $\alpha > 0$ and $c \neq 0$. ∎

Hence, the Gabor transform introduced in the previous section is the STFT with the smallest time-frequency window. In some applications, a larger window must be chosen in order to achieve other desirable properties. For instance, a second or higher order $B$-spline introduced in Example 3.4 facilitates computational and implementational effectiveness. The most important property not possessed by the Gabor transform is the additional condition:

$$\int_{-\infty}^{\infty} \psi(x)dx = 0, \tag{3.2.25}$$

where $\psi$ is the window function. This property gives us an extra degree of freedom for introducing a dilation (or scale) parameter in order to make the time-frequency window flexible. With this dilation parameter, the time-localization integral transform to be discussed in the next section will be called an *"integral wavelet transform"* (IWT), and any window function for defining the IWT will be called a "basic wavelet".

Before we end this section, let us derive a formula for recovering any finite-energy signal from its STFT values.

**Theorem 3.7.** *Let $w \in L^2(\mathbb{R})$ be so chosen that $\|w\|_2 = 1$ and both $w$ and $\hat{w}$ satisfy (3.2.1). Also, let $W_{b,\omega}(t)$ be as defined in (3.2.7). Then*

$$\int_{-\infty}^{\infty} \int_{-\infty}^{\infty} \langle f, W_{b,\omega} \rangle \overline{\langle g, W_{b,\omega} \rangle} db d\omega = 2\pi \langle f, g \rangle, \tag{3.2.26}$$

*for any $f, g \in L^2(\mathbb{R})$.*

**Proof.** For any $f \in L^2(\mathbb{R})$, let $\overset{\vee}{f}$ denote the inverse Fourier transform of $f$; that is, $\overset{\vee}{f}(x) = \frac{1}{2\pi}\hat{f}(-x)$. Then by the Parseval Identity and (3.2.6), we have

$$\int_{-\infty}^{\infty} (\widetilde{\mathcal{G}}_b f)(\omega) \overline{(\widetilde{\mathcal{G}}_b g)(\omega)} d\omega = 2\pi \int_{-\infty}^{\infty} (\widetilde{\mathcal{G}}_b f)^{\vee}(x) \overline{(\widetilde{\mathcal{G}}_b g)^{\vee}(x)} \, dx$$

$$= 2\pi \int_{-\infty}^{\infty} f(t) \overline{w(t-b)} \, \overline{g(t)} \, w(t-b) dt$$

$$= 2\pi \int_{-\infty}^{\infty} f(t) \overline{g(t)} \, |w(t-b)|^2 dt.$$

Therefore, it follows from the assumption $\|w\|_2 = 1$ that

$$\int_{-\infty}^{\infty} \int_{-\infty}^{\infty} \langle f, W_{b,\omega} \rangle \overline{\langle g, W_{b,\omega} \rangle} \, db d\omega$$

$$= \int_{-\infty}^{\infty} \int_{-\infty}^{\infty} (\widetilde{\mathcal{G}}_b f)(\omega) \overline{(\widetilde{\mathcal{G}} g)(\omega)} \, db d\omega$$

$$= \int_{-\infty}^{\infty} 2\pi \int_{-\infty}^{\infty} f(t) \overline{g(t)} \, |w(t-b)|^2 db dt$$

$$= 2\pi \langle f, g \rangle. \quad ∎$$

By selecting $g$ to be the Gaussian function $g_\alpha(\cdot - x)$ and letting $\alpha \to 0^+$, we arrive at the following result.

**Corollary 3.8.** *Let $w$ satisfy the hypotheses stated in Theorem 3.7 and let $f \in L^2(\mathbb{R})$. Then at every point $x$ where $f$ is continuous,*

$$f(x) = \frac{1}{2\pi} \int_{-\infty}^{\infty} \int_{-\infty}^{\infty} [e^{i\omega x}(\widetilde{\mathcal{G}}_b f)(\omega)] w(x - b) d\omega db. \tag{3.2.27}$$

## 3.3. The integral wavelet transform

We have seen that in analyzing a signal with any STFT, the time-frequency window is rigid, in the sense that its width is unchanged in observing any frequency band (or octave)

$$[\omega^* + \omega - \Delta_{\widehat{w}}, \, \omega^* + \omega + \Delta_{\widehat{w}}]$$

with center frequency $\omega^* + \omega$. Since frequency is directly proportional to the number of cycles per unit time, it takes a narrow time-window to locate high-frequency phenomena more precisely and a wide time-window to analyze low-frequency behaviors more thoroughly. Hence, the STFT is not suitable for analyzing signals with both very high and very low frequencies. On the other hand, the integral wavelet transform (IWT) relative to some basic wavelet, to be defined below, provides a flexible time-frequency window which automatically narrows when observing high-frequency phenomena and widens when studying low-frequency environments.

**Definition 3.9.** *If $\psi \in L^2(\mathbb{R})$ satisfies the "admissibility" condition:*

$$C_\psi := \int_{-\infty}^{\infty} \frac{|\widehat{\psi}(\omega)|^2}{|\omega|} d\omega < \infty, \tag{3.3.1}$$

*then $\psi$ is called a "basic wavelet". Relative to every basic wavelet $\psi$, the integral wavelet transform (IWT) on $L^2(\mathbb{R})$ is defined by*

$$(W_\psi f)(b, a) := |a|^{-\frac{1}{2}} \int_{-\infty}^{\infty} f(t) \overline{\psi\left(\frac{t-b}{a}\right)} dt, \qquad f \in L^2(\mathbb{R}), \tag{3.3.2}$$

*where $a, b \in \mathbb{R}$ with $a \neq 0$.*

**Remark.** If, in addition, both $\psi$ and $\widehat{\psi}$ satisfy (3.2.1), then the basic wavelet $\psi$ provides a time-frequency window with finite area given by $4\Delta_\psi \cdot \Delta_{\widehat{\psi}}$. In addition, under this additional assumption, it follows that $\widehat{\psi}$ is a continuous function, so that the finiteness of $C_\psi$ in (3.3.1) implies $\widehat{\psi}(0) = 0$, or equivalently,

$$\int_{-\infty}^{\infty} \psi(t) dt = 0. \tag{3.3.3}$$

This is the reason that $\psi$ is called a "wavelet". We will see later in this section that the admissibility condition (3.3.1) is needed in obtaining the inverse of the IWT.

By setting

$$\psi_{b;a}(t) := |a|^{-\frac{1}{2}}\psi\left(\frac{t-b}{a}\right), \tag{3.3.4}$$

the IWT defined in (3.3.2) can be written as

$$(W_\psi f)(b,a) = \langle f, \psi_{b;a}\rangle. \tag{3.3.5}$$

In the following discussion, let us assume that both $\psi$ and $\widehat{\psi}$ satisfy (3.2.1). Then if the center and radius of the window function $\psi$ are given by $t^*$ and $\Delta_\psi$, respectively, the function $\psi_{b;a}$ is a window function with center at $b + at^*$ and radius equal to $a\Delta_\psi$. Hence, the IWT, as formulated in (3.3.5), gives local information of an analog signal $f$ with a time-window

$$[b + at^* - a\Delta_\psi, \, b + at^* + a\Delta_\psi]. \tag{3.3.6}$$

This window narrows for small values of $a$ and widens for allowing $a$ to be large.

Next, consider

$$\frac{1}{2\pi}\widehat{\psi}_{b;a}(\omega) = \frac{|a|^{-\frac{1}{2}}}{2\pi}\int_{-\infty}^{\infty} e^{-i\omega t}\psi\left(\frac{t-b}{a}\right)dt \tag{3.3.7}$$

$$= \frac{a|a|^{-\frac{1}{2}}}{2\pi}e^{-ib\omega}\widehat{\psi}(a\omega),$$

and suppose that the center and radius of the window function $\widehat{\psi}$ are given by $\omega^*$ and $\Delta_{\widehat{\psi}}$, respectively. Then by setting

$$\eta(\omega) := \widehat{\psi}(\omega + \omega^*), \tag{3.3.8}$$

we have a window function $\eta$ with center at the origin and radius equal to $\Delta_{\widehat{\psi}}$. Now from (3.3.5) and (3.3.7), and applying the Parseval Identity, we have

$$(W_\psi f)(b,a) = \frac{a|a|^{-\frac{1}{2}}}{2\pi}\int_{-\infty}^{\infty} \hat{f}(\omega)e^{ib\omega}\overline{\eta\left(a\left(\omega - \frac{\omega^*}{a}\right)\right)}\,d\omega. \tag{3.3.9}$$

Since it is clear that the window function $\eta\left(a\left(\omega - \frac{\omega^*}{a}\right)\right) = \eta(a\omega - \omega^*) = \widehat{\psi}(a\omega)$ has radius given by $\frac{1}{a}\Delta_{\widehat{\psi}}$, the expression in (3.3.9) says that, with the exception of a multiple of $a|a|^{-1/2}/2\pi$ and a linear phase-shift of $e^{ib\omega}$, the IWT $W_\psi f$ also gives local information of $\hat{f}$ with a frequency-window

$$\left[\frac{\omega^*}{a} - \frac{1}{a}\Delta_{\widehat{\psi}}, \, \frac{\omega^*}{a} + \frac{1}{a}\Delta_{\widehat{\psi}}\right]. \tag{3.3.10}$$

In the following discussion, the center $\omega^*$ of $\hat{\psi}$ is assumed to be positive. In doing so, we may think of this window as a frequency band (or octave) with center-frequency $\omega^*/a$ and bandwidth $2\Delta_{\hat{\psi}}/a$. The importance of this identification is that the ratio

$$\frac{\text{center frequency}}{\text{bandwidth}} = \frac{\omega^*/a}{2\Delta_{\hat{\psi}}/a} = \frac{\omega^*}{2\Delta_{\hat{\psi}}} \qquad (3.3.11)$$

is independent of the scaling $a$. Hence, if the frequency variable is identified as a constant multiple of $a^{-1}$, then an adaptive bandpass filter, with pass-band given by (3.3.10), has the property that the center-frequency to bandwidth ratio is independent of the location of the center-frequency. This is called "constant-$Q$ filtering".

Now, if $\omega^*/a$ is considered to be the frequency variable $\omega$, then we may consider the $t$-$\omega$ plane as the time-frequency plane. Hence, with the time-window in (3.3.6) and the frequency-window in (3.3.10), we have a rectangular time-frequency window

$$[b + at^* - a\Delta_\psi,\ b + at^* + a\Delta_\psi] \times \left[\frac{\omega^*}{a} - \frac{1}{a}\Delta_{\hat{\psi}},\ \frac{\omega^*}{a} + \frac{1}{a}\Delta_{\hat{\psi}}\right] \qquad (3.3.12)$$

in the $t$-$\omega$ plane, with width $2a\Delta_\psi$ (determined by the width of the time-window). Hence, this window automatically narrows for detecting high-frequency phenomena (i.e., small $a > 0$), and widens for investigating low-frequency behavior (i.e. large $a > 0$). (See Figure 1.2.1.)

We next derive a formula for reconstructing any finite-energy signal from its IWT values. For completeness, we first allow the scaling $a$ to be negative, and later restrict our attention to positive values of $a$ in order to apply the IWT to time-frequency analysis.

**Theorem 3.10.** *Let $\psi$ be a basic wavelet which defines an IWT $W_\psi$. Then*

$$\int_{-\infty}^{\infty} \int_{-\infty}^{\infty} \left[(W_\psi f)(b, a)\overline{(W_\psi g)(b, a)}\right] \frac{da}{a^2}\, db = C_\psi \langle f, g\rangle, \qquad (3.3.13)$$

*for all $f, g \in L^2(\mathbb{R})$. Furthermore, for any $f \in L^2(\mathbb{R})$ and $x \in \mathbb{R}$ at which $f$ is continuous,*

$$f(x) = \frac{1}{C_\psi} \int_{-\infty}^{\infty} \int_{-\infty}^{\infty} [(W_\psi f)(b, a)]\psi_{b;a}(x) \frac{da}{a^2}\, db, \qquad (3.3.14)$$

*where $\psi_{b;a}$ is defined in (3.3.4).*

**Proof.** By applying the Parseval Identity and (3.3.7), and using the notation

$$\begin{cases} F(x) := \hat{f}(x)\overline{\hat{\psi}(ax)}; \\ G(x) := \hat{g}(x)\overline{\hat{\psi}(ax)}, \end{cases} \qquad (3.3.15)$$

we have

$$\int_{-\infty}^{\infty} \left[ (W_\psi f)(b,a)\overline{(W_\psi g)(b,a)} \right] db$$

$$= \frac{1}{|a|} \int_{-\infty}^{\infty} \left\{ \int_{-\infty}^{\infty} f(t)\overline{\psi\left(\frac{t-b}{a}\right)} dt \int_{-\infty}^{\infty} \overline{g(s)}\, \psi\left(\frac{s-b}{a}\right) ds \right\} db$$

$$= \frac{a^2}{|a|} \int_{-\infty}^{\infty} \left\{ \frac{1}{2\pi} \overline{\int_{-\infty}^{\infty} F(x)e^{-ibx}dx} \right\} \left\{ \frac{1}{2\pi} \int_{-\infty}^{\infty} \overline{G(y)}e^{-iby}dy \right\} db$$

$$= \frac{a^2}{2\pi|a|} \left\{ \frac{1}{2\pi} \int_{-\infty}^{\infty} \widehat{\overline{G}}(b)\overline{\widehat{\overline{F}}(b)}db \right\}$$

$$= \frac{a^2}{2\pi|a|} \int_{-\infty}^{\infty} \overline{G}(x)F(x)dx,$$

where the Parseval Identity is applied again to arrive at the last equality. Hence, by substituting (3.3.15) into the above expression, integrating with respect to $da/a^2$ on $(-\infty, \infty)$, and recalling the definition of $C_\psi$ from (3.3.1), we obtain

$$\int_{-\infty}^{\infty} \left\{ \int_{-\infty}^{\infty} \left[ (W_\psi f)(b,a)\overline{(W_\psi g)(b,a)} \right] db \right\} \frac{da}{a^2} \qquad (3.3.16)$$

$$= \frac{1}{2\pi} \int_{-\infty}^{\infty} \left\{ \hat{f}(x)\overline{\hat{g}(x)} \int_{-\infty}^{\infty} \frac{|\widehat{\psi}(ax)|^2}{|a|}da \right\} dx$$

$$= \frac{1}{2\pi} \int_{-\infty}^{\infty} \left\{ \hat{f}(x)\overline{\hat{g}(x)} \int_{-\infty}^{\infty} \frac{|\widehat{\psi}(y)|^2}{|y|}dy \right\} dx$$

$$= C_\psi \frac{1}{2\pi} \langle \hat{f}, \hat{g} \rangle = C_\psi \langle f, g \rangle.$$

Furthermore, if $f$ is continuous at $x$, then using the Gaussian function $g_\alpha(\cdot - x)$ for the function $g$ and allowing $\alpha$ to tend to 0 from above, we arrive at

$$f(x) = \frac{1}{C_\psi} \lim_{\alpha \to 0^+} \int_{-\infty}^{\infty} \int_{-\infty}^{\infty} \left[ (W_\psi f)(b,a)\overline{\langle g_\alpha(\cdot - x), \psi_{b;a} \rangle} \right] \frac{da}{a^2} db$$

$$= \frac{1}{C_\psi} \int_{-\infty}^{\infty} \int_{-\infty}^{\infty} [(W_\psi f)(b,a)]\psi_{b;a}(x)\frac{da}{a^2} db.$$

This completes the proof of the theorem. ∎

In signal analysis, we only consider positive frequencies $\omega$. Hence, if the frequency variable $\omega$ is identified as a positive constant multiple of the reciprocal of the dilation parameter $a$, such as $\omega = \omega^*/a$ (where the center $\omega^*$ of $\widehat{\psi}$ is always assumed to be positive), then we must only consider positive values of $a$. Consequently, in reconstructing $f$ from the IWT of $f$, we are only allowed to use the values $(W_\psi f)(b, a)$, $a > 0$. As one might expect, the basic wavelet $\psi$ must be somewhat more restrictive for this to be possible. The extra condition on $\psi$ is:

$$\int_0^\infty \frac{|\widehat{\psi}(\omega)|^2}{\omega} d\omega = \int_0^\infty \frac{|\widehat{\psi}(-\omega)|^2}{\omega} d\omega = \frac{1}{2} C_\psi < \infty. \tag{3.3.17}$$

**Theorem 3.11.** Let $\psi$ be a basic wavelet that satisfies *(3.3.17)*. Then

$$\int_0^\infty \left[ \int_{-\infty}^\infty (W_\psi f)(b, a)\overline{(W_\psi g)(b, a)} \, db \right] \frac{da}{a^2} = \frac{1}{2} C_\psi \langle f, g \rangle, \tag{3.3.18}$$

for all $f, g \in L^2(\mathbb{R})$. Furthermore, for any $f \in L^2(\mathbb{R})$ and $x \in \mathbb{R}$ at which $f$ is continuous,

$$f(x) = \frac{2}{C_\psi} \int_0^\infty \left[ \int_{-\infty}^\infty (W_\psi f)(b, a)\psi_{b;a}(x) db \right] \frac{da}{a^2}, \tag{3.3.19}$$

where $\psi_{b;a}$ is defined in *(3.3.4)*.

We remark that for the left-hand side of (3.3.18) to be equal to $C\langle f, g \rangle$ for all $f, g \in L^2(\mathbb{R})$, the assumption (3.3.17) is necessary, and it is also necessary that $C = \frac{1}{2} C_\psi$.

**Proof of Theorem 3.11.**

Under the assumption (3.3.17), it is easy to verify that

$$\int_0^\infty \frac{|\widehat{\psi}(ax)|^2}{a} da = \int_0^\infty \frac{|\widehat{\psi}(y)|^2}{y} dy, \quad x \neq 0. \tag{3.3.20}$$

Hence, by following the same derivation as in (3.3.16) (with the only exception being that the integral with respect to $da/a^2$ is over $(0, \infty)$ instead of $(-\infty, \infty)$), we obtain (3.3.18). The proof of (3.3.19) is the same as that of (3.3.14) in Theorem 3.10.  ∎

### 3.4. Dyadic wavelets and inversions

In signal analysis, it is sometimes necessary to partition the (positive) frequency axis into disjoint frequency bands (or octaves). For computational efficiency and convenience in discussions, we will only consider "*binary partitions*", namely:

$$(0, \infty) = \bigcup_{j=-\infty}^\infty (2^j \Delta_{\widehat{\psi}}, 2^{j+1} \Delta_{\widehat{\psi}}], \tag{3.4.1}$$

where $\Delta_{\widehat{\psi}} > 0$ is the radius of the Fourier transform $\widehat{\psi}$ of a basic wavelet $\psi$. Here, we have again assumed that $\widehat{\psi}$ satisfies (3.2.1). Observe that for any basic wavelet $\psi$, a phase-shift of $\psi$ by $\alpha$ is equivalent to the corresponding forward-shift in frequency of $\hat{\psi}$ by the same $\alpha$; that is

$$\psi^\circ(t) = e^{i\alpha t}\psi(t) \Leftrightarrow \hat{\psi}^\circ(\omega) = \hat{\psi}(\omega - \alpha). \qquad (3.4.2)$$

Hence, since $\Delta_{\psi^\circ} = \Delta_\psi$ and $\Delta_{\hat{\psi}^\circ} = \Delta_{\hat{\psi}}$, we may always assume, without loss of generality, that the center of $\hat{\psi}$ is at $\omega^* = 3\Delta_{\hat{\psi}}$. In doing so, we have

$$\left(\frac{\omega^*}{a_j} - \frac{1}{a_j}\Delta_{\hat{\psi}}, \frac{\omega^*}{a_j} + \frac{1}{a_j}\Delta_{\hat{\psi}}\right] = (2^{j+1}\Delta_{\hat{\psi}}, 2^{j+2}\Delta_{\hat{\psi}}], \qquad (3.4.3)$$

provided that

$$a_j = \frac{1}{2^j}, \qquad j \in \mathbb{Z}. \qquad (3.4.4)$$

The center-frequency of the frequency band described in (3.4.3) is given by

$$\omega_j := \frac{\omega^*}{a_j} = \frac{3\Delta_{\hat{\psi}}}{a_j} = 3 \times 2^j \Delta_{\hat{\psi}}. \qquad (3.4.5)$$

So, by using $\omega^*/a$ to represent the frequency variable $\omega$, where $a > 0$ is the dilation (or scale) parameter, the disjoint union in (3.4.1) indeed gives a partition of the (positive) frequency domain $(0, \infty)$.

In this section, we study the problem of recovering any finite-energy signal $f$ (i.e. any $f \in L^2(\mathbb{R})$) from its integral wavelet transform $(W_\psi f)(b, a)$, only at the discrete set of frequencies:

$$\{\omega_j = 3 \times 2^j \Delta_{\hat{\psi}} : j \in \mathbb{Z}\},$$

(or using only the scale samples $a = a_j := \frac{1}{2^j}, j \in \mathbb{Z}$). For this problem to have a solution, one naturally expects the basic wavelets $\psi$ to be more restrictive than the admissibility condition in (3.3.1).

**Definition 3.12.** *A function $\psi \in L^2(\mathbb{R})$ is called a dyadic wavelet if there exist two positive constants $A$ and $B$, with $0 < A \leq B < \infty$, such that*

$$A \leq \sum_{j=-\infty}^{\infty} |\hat{\psi}(2^{-j}\omega)|^2 \leq B, \quad \text{a.e.} \qquad (3.4.6)$$

The condition in (3.4.6) is called the "stability" condition imposed on $\psi$. To account for this terminology, let us make use of the notation (2.3.9) of the reflection of a function to introduce the following "normalized" IWT:

$$(W_j^\psi f)(b) := 2^{j/2}(W_\psi f)\left(b, \frac{1}{2^j}\right) = 2^j(f * \overline{\psi^-(2^j \cdot)})(b). \qquad (3.4.7)$$

Then we see that (3.4.6) is equivalent to

$$A\|f\|_2^2 \le \sum_{j=-\infty}^{\infty} \|W_j^\psi f\|_2^2 \le B\|f\|_2^2, \qquad f \in L^2(\mathbb{R}), \qquad (3.4.8)$$

for the same constants $A$ and $B$. Indeed, by the Parseval Identity and the first identity in (2.3.10) of Lemma 2.16, the set of inequalities in (3.4.8) can be written as

$$A\|\hat{f}\|_2^2 \le \int_{-\infty}^{\infty} \sum_{j=-\infty}^{\infty} |\hat{f}(\omega)\overline{\hat{\psi}}(2^{-j}\omega)|^2 d\omega \le B\|\hat{f}\|_2^2$$

which is equivalent to

$$A \le \int_{-\infty}^{\infty} \sum_{j=-\infty}^{\infty} \left|\frac{g(x)}{\|g\|_2}\hat{\psi}(2^{-j}x)\right|^2 dx \le B, \qquad g \in L^2(\mathbb{R}). \qquad (3.4.9)$$

By choosing $g/\|g\|_2$ to be the Gaussian functions $g_\alpha(\cdot - \omega)$ and allowing $\alpha \to 0^+$, we see that (3.4.9) yields (3.4.6). Since (3.4.6) clearly implies (3.4.9), these two sets of inequalities are equivalent.

In the following, we will see that the stability condition of $\psi$ implies that any dyadic wavelet $\psi$ must be a basic wavelet.

**Theorem 3.13.** *Let $\psi$ satisfy the stability condition (3.4.6). Then $\psi$ is a basic wavelet satisfying*

$$A \ln 2 \le \int_0^\infty \frac{|\hat{\psi}(\omega)|^2}{\omega} d\omega, \int_0^\infty \frac{|\hat{\psi}(-\omega)|^2}{\omega} d\omega \le B \ln 2. \qquad (3.4.10)$$

*Furthermore, if $A$ and $B$ in (3.4.6) agree, then*

$$C_\psi := \int_{-\infty}^{\infty} \frac{|\hat{\psi}(\omega)|^2}{|\omega|} d\omega = 2A \ln 2. \qquad (3.4.11)$$

**Proof.**   We first note that

$$\int_1^2 \frac{|\hat{\psi}(2^{-j}\omega)|^2}{\omega} d\omega = \int_{2^{-j}}^{2^{-j+1}} \frac{|\hat{\psi}(x)|^2}{x} dx.$$

Hence, dividing each term in (3.4.6) by $\omega$ and integrating over the interval $(1,2)$, we have

$$A \ln 2 \le \int_0^\infty \frac{|\hat{\psi}(\omega)|^2}{\omega} d\omega \le B \ln 2.$$

Similarly, dividing by $-\omega$ and integrating over $(-2, -1)$ yields

$$A \ln 2 \le \int_0^\infty \frac{|\hat{\psi}(-\omega)|^2}{\omega} d\omega \le B \ln 2. \qquad \blacksquare$$

The stability condition is instrumental for recovering any $f \in L^2(\mathbb{R})$ from its IWT values $(W_\psi f)(b, 2^{-j})$, $j \in \mathbb{Z}$. The approach we take is to introduce another dyadic wavelet $\psi^*$ which we define by considering its Fourier transform:

$$\widehat{\psi}^*(\omega) := \frac{\widehat{\psi}(\omega)}{\sum\limits_{k=-\infty}^{\infty} |\widehat{\psi}(2^{-k}\omega)|^2}. \tag{3.4.12}$$

With this function $\psi^*$, it follows from (3.4.7) that for any $f \in L^2(\mathbb{R})$,

$$\sum_{j=-\infty}^{\infty} \int_{-\infty}^{\infty} (W_j^\psi f)(b)\{2^j \psi^*(2^j(x-b))\} db \tag{3.4.13}$$

$$= \sum_{j=-\infty}^{\infty} \frac{1}{2\pi} \int_{-\infty}^{\infty} (W_j^\psi f)^\wedge(\omega)\widehat{\psi}^*(2^{-j}\omega)e^{ix\omega} d\omega$$

$$= \sum_{j=-\infty}^{\infty} \frac{1}{2\pi} \int_{-\infty}^{\infty} \widehat{f}(\omega)\overline{\widehat{\psi}(2^{-j}\omega)}\widehat{\psi}^*(2^{-j}\omega)e^{ix\omega} d\omega$$

$$= \frac{1}{2\pi} \int_{-\infty}^{\infty} \widehat{f}(\omega)e^{ix\omega} d\omega = f(x),$$

where we have applied the formula $h_1 * h_2 = \mathcal{F}^{-1}(\widehat{h}_1 \widehat{h}_2)$, (3.4.7), Lemma 2.16, and (3.4.12), consecutively. This leads to the following notion of "*dyadic duals*".

**Definition 3.14.** *A function $\widetilde{\psi} \in L^2(\mathbb{R})$ is called a dyadic dual of a dyadic wavelet $\psi$ if every $f \in L^2(\mathbb{R})$ can be expressed as*

$$f(x) = \sum_{j=-\infty}^{\infty} \int_{-\infty}^{\infty} (W_j^\psi f)(b)\{2^j \widetilde{\psi}(2^j(x-b))\} db \tag{3.4.14}$$

$$= \sum_{j=-\infty}^{\infty} 2^{3j/2} \int_{-\infty}^{\infty} (W_\psi f)\left(b, \frac{1}{2^j}\right) \widetilde{\psi}(2^j(x-b)) db.$$

Hence, in (3.4.13) we have established the following result.

**Theorem 3.15.** *Let $\psi$ be a dyadic wavelet. Then the function $\psi^*$, whose Fourier transform is given by (3.4.12), is a dyadic dual of $\psi$. Furthermore, $\psi^*$ is also a dyadic wavelet with*

$$\frac{1}{B} \leq \sum_{j=-\infty}^{\infty} |\widehat{\psi}^*(2^{-j}\omega)|^2 \leq \frac{1}{A}, \quad \text{a.e.} \tag{3.4.15}$$

We remark that dyadic duals of a given dyadic wavelet $\psi$ may not be unique. A discussion of nonuniqueness is delayed to Section 3.6. It will depend on the following characterization result.

**Theorem 3.16.** *Let $\psi$ be a dyadic wavelet and $\tilde{\psi}$ be any function in $L^2(\mathbb{R})$ that satisfies*

$$\operatorname*{ess\,sup}_{-\infty < x < \infty} \sum_{j=-\infty}^{\infty} |\widehat{\tilde{\psi}}(2^{-j}x)|^2 < \infty. \tag{3.4.16}$$

*Then $\tilde{\psi}$ is a dyadic dual of $\psi$ if and only if the following identity is satisfied:*

$$\sum_{j=-\infty}^{\infty} \overline{\widehat{\tilde{\psi}}(2^{-j}\omega)} \, \widehat{\tilde{\psi}}(2^{-j}\omega) = 1, \quad a.e. \tag{3.4.17}$$

**Proof.** Following the same derivation as in (3.4.13), we observe that $\tilde{\psi}$ is a dyadic dual of $\psi$ if and only if for any $f \in L^2(\mathbb{R})$, we have

$$\hat{f}(\omega) = \sum_{j=-\infty}^{\infty} \hat{f}(\omega) \overline{\widehat{\tilde{\psi}}(2^{-j}\omega)} \, \widehat{\tilde{\psi}}(2^{-j}\omega), \quad a.e., \tag{3.4.18}$$

where the a.e. convergence of the infinite series is assured by the hypothesis (3.4.16) and the definition of a dyadic wavelet $\psi$. It is obvious that (3.4.17) and (3.4.18) are equivalent.  ∎

### 3.5. Frames

In the previous section, we partitioned the (positive) frequency axis into disjoint frequency bands $(2^j \Delta_{\widehat{\psi}}, 2^{j+1} \Delta_{\widehat{\psi}}]$, $j \in \mathbb{Z}$, by choosing the dilation parameter $a$ to be $a_j := 2^{-j}$, $j \in \mathbb{Z}$, while the translation parameter was allowed to vary over all of $\mathbb{R}$. In doing so, we considered semi-discrete information on the IWT of an $f \in L^2(\mathbb{R})$, namely:

$$(W_\psi f)\left(b, \frac{1}{2^j}\right), \qquad b \in \mathbb{R}, \quad j \in \mathbb{Z}.$$

For computational efficiency, let us also discretize the translation parameter $b$ by restricting $b$ to the discrete set of sampling points

$$b_{j,k} := \frac{k}{2^j} b_0, \qquad j, k \in \mathbb{Z}, \tag{3.5.1}$$

where $b_0 > 0$ is a fixed constant, called the "*sampling rate*". Hence, by introducing the notation

$$\psi_{b_0;j,k}(t) := \psi_{b_{j,k};a_j}(t) = 2^{\frac{j}{2}} \psi(2^j t - k b_0), \tag{3.5.2}$$

(see (3.3.4) for the definition of $\psi_{b;a}$), the values of the IWT of any $f \in L^2(\mathbb{R})$ we are going to consider are given by

$$(W_\psi f)(b_{j,k}, a_j) = \langle f, \psi_{b_0;j,k} \rangle, \qquad j, k \in \mathbb{Z}. \tag{3.5.3}$$

Analogous to the semi-discrete setting in the previous section, we are also interested in recovering any $f \in L^2(\mathbb{R})$ from the values of its IWT in (3.5.3). The "stability" condition for this reconstruction is the existence of positive constants $A$ and $B$, with $0 < A \leq B < \infty$, such that

$$A\|f\|_2^2 \leq \sum_{j,k \in \mathbb{Z}} |\langle f, \psi_{b_0;j,k} \rangle|^2 \leq B\|f\|_2^2, \qquad f \in L^2(\mathbb{R}). \tag{3.5.4}$$

In other words, the stability condition on the function $\psi$ requires that $\psi$ generates a "frame" of $L^2(\mathbb{R})$ with sampling rate $b_0$, as follows.

**Definition 3.17.** *A function $\psi \in L^2(\mathbb{R})$ is said to generate a frame $\{\psi_{b_0;j,k}\}$ of $L^2(\mathbb{R})$ with sampling rate $b_0 > 0$ if (3.5.4) holds for some positive constants $A$ and $B$, which are called frame bounds. If $A = B$, then the frame is called a tight frame.*

Under the stability condition (3.5.4); that is, under the condition that $\psi$ generates a frame, we are assured that any $f \in L^2(\mathbb{R})$ can be recovered from its IWT values in (3.5.3). To see this, let us consider the linear operator $T$ on $L^2(\mathbb{R})$, defined by

$$Tf := \sum_{j,k \in \mathbb{Z}} \langle f, \psi_{b_0;j,k} \rangle \psi_{b_0;j,k}, \qquad f \in L^2(\mathbb{R}). \tag{3.5.5}$$

From the stability condition in (3.5.4), it is clear that $T$ is a one-one bounded linear operator. In fact, because of the lower bound in (3.5.4), $T$ also maps $L^2(\mathbb{R})$ onto its range, and by the Interior Mapping Principle, its inverse $T^{-1}$ is bounded. In our setting, we can even include the following simple argument. For any $g = Tf$, where $f \in L^2(\mathbb{R})$, since

$$\langle Tf, f \rangle = \sum_{j,k \in \mathbb{Z}} |\langle f, \psi_{b_0;jk} \rangle|^2, \tag{3.5.6}$$

we have

$$A\|T^{-1}g\|_2^2 = A\|f\|_2^2 \leq \langle Tf, f \rangle$$
$$= \langle g, T^{-1}g \rangle \leq \|g\|_2 \|T^{-1}g\|_2,$$

so that

$$\|T^{-1}g\|_2 \leq \frac{1}{A}\|g\|_2,$$

or $\|T^{-1}\| \leq A^{-1}$. Hence, every $f \in L^2(\mathbb{R})$ can be reconstructed from its IWT values in (3.5.3) by applying the formula

$$f = T^{-1}Tf = \sum_{j,k \in \mathbb{Z}} \langle f, \psi_{b_0;j,k} \rangle T^{-1}\psi_{b_0;j,k}. \tag{3.5.7}$$

By setting

$$\psi_{b_0}^{j,k} := T^{-1}\psi_{b_0;j,k}, \qquad j, k \in \mathbb{Z}, \tag{3.5.8}$$

the reconstruction formula (3.5.7) may be written as

$$\begin{cases} \langle f, g \rangle = \sum_{j,k \in \mathbb{Z}} \langle f, \psi_{b_0;j,k} \rangle \langle \psi_{b_0}^{j,k}, g \rangle; \\[2ex] f = \sum_{j,k \in \mathbb{Z}} \langle f, \psi_{b_0;j,k} \rangle \psi_{b_0}^{j,k}, \end{cases} \tag{3.5.9}$$

for all $f, g \in L^2(\mathbb{R})$. We may call $\{\psi_{b_0}^{j,k}\}$ the "*dual*" of the frame $\{\psi_{b_0;j,k}\}$. However, the reconstruction formula (3.5.7), or (3.5.9), is not useful unless we have some knowledge of the dual. Unfortunately, in general, the dual $\{\psi_{b_0}^{j,k}\}$ may not be generated by some $\tilde{\psi} \in L^2(\mathbb{R})$, the same way as $\{\psi_{b_0;j,k}\}$ is generated by $\psi$. We will return to a discussion of this topic in the next section.

In the following, we see that a frame may not be a linearly independent family.

**Example 3.18.** Let $\psi_1$ be the Haar function defined in (3.2.3) and consider the sampling rate $b_0 = \frac{1}{3}$. Then the linearly dependent family $S := \{\psi_{1,\frac{1}{3};j,k} \colon j, k \in \mathbb{Z}\}$ is a frame of $L^2(\mathbb{R})$.

**Proof.** Let us use the notation

$$\gamma_j = \begin{cases} 1 & \text{for even } j \in \mathbb{Z} \\ 2 & \text{for odd } j \in \mathbb{Z}, \end{cases}$$

and decompose the family $S$ into a disjoint union of three subfamilies:

$$S_1 = \{\psi_{1;j,k}(x) = 2^{j/2}\psi_1(2^j x - k) \colon j, k \in \mathbb{Z}\},$$
$$S_2 = \left\{2^{j/2}\psi_1\left(2^j x - k + \frac{\gamma_j}{3}\right) \colon j, k \in \mathbb{Z}\right\},$$

and

$$S_3 = \left\{2^{j/2}\psi_1\left(2^j x - k - \frac{\gamma_j}{3}\right) \colon j, k \in \mathbb{Z}\right\}.$$

Since $\psi_1$ is the Haar function, $S_1$ is already an o.n. basis of $L^2(\mathbb{R})$ (see Section 1.5 and also Chapters 5 and 6 for more details). Hence, $S$ is a linearly dependent family. It is also easy to verify that both $S_2$ and $S_3$ are o.n. families, so that the (generalized) Bessel Inequality applies (see (2.4.14) for the Bessel Inequality for trigonometric polynomials). Consequently, we have

$$\|f\|_2^2 = \sum_{j,k \in \mathbb{Z}} |\langle f, \psi_{1;j,k} \rangle|^2$$
$$\leq \sum_{j,k \in \mathbb{Z}} |\langle f, \psi_{1,\frac{1}{3};j,k} \rangle|^2$$
$$= \|f\|_2^2 + \sum_{s \in S_2} |\langle f, s \rangle|^2 + \sum_{s \in S_3} |\langle f, s \rangle|^2$$
$$\leq 3\|f\|_2^2. \qquad \blacksquare$$

Hence, the stability condition (3.5.4) is weaker than the requirement that $\psi$ generates a Riesz basis, defined as follows.

**Definition 3.19.** *A function $\psi \in L^2(\mathbb{R})$ is said to generate a Riesz basis (or unconditional basis) $\{\psi_{b_0;j,k}\}$ with sampling rate $b_0$ if both of the following two properties are satisfied:*

(i) *the linear span*

$$\langle \psi_{b_0;j,k} \colon j, k \in \mathbb{Z} \rangle \tag{3.5.10}$$

*is dense in $L^2(\mathbb{R})$; and*

(ii) *there exist positive constants $A$ and $B$, with $0 < A \leq B < \infty$ such that*

$$A\|\{c_{j,k}\}\|_{\ell^2}^2 \leq \left\| \sum_{j,k \in \mathbb{Z}} c_{j,k}\psi_{b_0;j,k} \right\|_2^2 \leq B\|\{c_{j,k}\}\|_{\ell^2}^2, \tag{3.5.11}$$

*for all $\{c_{j,k}\} \in \ell^2(\mathbb{Z}^2)$. Here, $A$ and $B$ are called the Riesz bounds of $\{\psi_{b_0;j,k}\}$.*

If $\psi$ generates a Riesz basis with sampling rate $b_0 = 1$, then $\psi$ is called an $\mathcal{R}$-function (see Definition 1.4).

**Remark.** Throughout this book, we will always use the notation

$$\psi_{j,k}(x) := \psi_{1;j,k}(x) = 2^{j/2}\psi(2^j x - k). \tag{3.5.12}$$

This notation should not be confused with the notation $\psi_{b;a}$ introduced in (3.3.4).

The following result clarifies the difference between a frame and a Riesz basis.

**Theorem 3.20.** *Let $\psi \in L^2(\mathbb{R})$ and $b_0 > 0$. Then the following two statements are equivalent.*

(i) *$\{\psi_{b_0;j,k}\}$ is a Riesz basis of $L^2(\mathbb{R})$.*
(ii) *$\{\psi_{b_0;j,k}\}$ is a frame of $L^2(\mathbb{R})$, and is also an $\ell^2$-linearly independent family, in the sense that if $\sum c_{j,k}\psi_{b_0;j,k} = 0$ and $\{c_{j,k}\} \in \ell^2$ then $c_{j,k} = 0$. Furthermore, the Riesz bounds and frame bounds agree.*

**Proof.** It is clear from (3.5.11) that any Riesz basis is $\ell^2$-linearly independent. Let $\{\psi_{b_0;j,k}\}$ be a Riesz basis with Riesz bounds $A$ and $B$, and consider the "matrix operator"

$$M := [\gamma_{\ell,m;j,k}]_{(\ell,m),(j,k) \in \mathbb{Z}^2}$$

where the entries are defined by

$$\gamma_{\ell,m;j,k} := \langle \psi_{b_0;\ell,m}, \psi_{b_0;j,k} \rangle. \tag{3.5.13}$$

By (3.5.11), we have

$$A\|\{c_{j,k}\}\|_{\ell^2}^2 \leq \sum_{\ell,m,j,k} c_{\ell,m}\gamma_{\ell,m;j,k}\bar{c}_{j,k} \leq B\|\{c_{j,k}\}\|_{\ell^2}^2,$$

so that $M$ is positive definite. We denote the inverse of $M$ by

$$M^{-1} := [\mu_{\ell,m;j,k}]_{(\ell,m),(j,k)\in\mathbb{Z}^2},$$

which means that both

$$\sum_{r,s} \mu_{\ell,m;r,s}\gamma_{r,s;j,k} = \delta_{\ell,j}\delta_{m,k}, \quad \ell,m,j,k \in \mathbb{Z}, \qquad (3.5.14)$$

and

$$B^{-1}\|\{c_{j,k}\}\|_{\ell^2}^2 \leq \sum_{\ell,m,j,k} c_{\ell,m}\mu_{\ell,m;j,k}\bar{c}_{j,k} \leq A^{-1}\|\{c_{j,k}\}\|_{\ell^2}^2, \quad \{c_{j,k}\} \in \ell^2,$$

$$(3.5.15)$$

are satisfied. This allows us to introduce

$$\psi^{\ell,m}(x) := \sum_{j,k} \mu_{\ell,m;j,k}\psi_{b_0;j,k}(x). \qquad (3.5.16)$$

Clearly, $\psi^{\ell,m} \in L^2(\mathbb{R})$; and it follows from (3.5.13) and (3.5.14) that

$$\langle \psi^{\ell,m}, \psi_{b_0;j,k} \rangle = \delta_{\ell,j}\delta_{m,k}, \quad \ell,m,j,k \in \mathbb{Z},$$

which means that $\{\psi^{\ell,m}\}$ is the basis of $L^2(\mathbb{R})$ which is dual to $\{\psi_{b_0;j,k}\}$. Furthermore, from (3.5.14) and (3.5.15), we conclude that

$$\langle \psi^{\ell,m}, \psi^{j,k} \rangle = \mu_{\ell,m;j,k}$$

and that the Riesz bounds of $\{\psi^{\ell,m}\}$ are $B^{-1}$ and $A^{-1}$. In particular, for any $f \in L^2(\mathbb{R})$, we may write

$$f(x) = \sum_{j,k} \langle f, \psi_{b_0;j,k} \rangle \psi^{j,k}(x)$$

and

$$B^{-1}\sum_{j,k}|\langle f, \psi_{b_0;j,k}\rangle|^2 \leq \|f\|_2^2 \leq A^{-1}\sum_{j,k}|\langle f, \psi_{b_0;j,k}\rangle|^2. \qquad (3.5.17)$$

Since it is clear that (3.5.17) is equivalent to (3.5.4), we have established that statement (i) implies statement (ii).

To establish the converse, we have to rely on two basic results in functional analysis, namely: the Banach-Steinhaus Theorem and the Open Mapping Theorem, which are, unfortunately, beyond the scope of this book. We only give a very brief outline without going into any details. Recall from (3.5.5) that if $\{\psi_{b_0;j,k}\}$ is a frame of $L^2(\mathbb{R})$, then for any $g \in L^2(\mathbb{R})$ and $f = T^{-1}g$, we have

$$g(x) = \sum_{j,k} \langle f, \psi_{b_0;j,k} \rangle \psi_{b_0;j,k}(x).$$

Also, by the $\ell^2$-linear independence of $\{\psi_{b_0;j,k}\}$, this representation is unique. It can also be shown that in using this "basis" $\{\psi_{b_0;j,k}\}$ to represent functions in $L^2(\mathbb{R})$, a series

$$\sum_{j,k} c_{j,k} \psi_{b_0;j,k}(x)$$

converges in $L^2(\mathbb{R})$ if and only if the coefficient sequence $\{c_{j,k}\}$ is in $\ell^2$. Then, as mentioned above, the Banach-Steinhaus Theorem and the Open Mapping Theorem can be applied to conclude that $\{\psi_{b_0;j,k}\}$ is a Riesz basis of $L^2(\mathbb{R})$.∎

We end this section by showing that if $\psi \in L^2(\mathbb{R})$ generates a frame of $L^2(\mathbb{R})$, then it must be a dyadic wavelet.

**Theorem 3.21.** Let $\psi \in L^2(\mathbb{R})$ generate a frame $\{\psi_{b_0;j,k}\}$ of $L^2(\mathbb{R})$ with frame bounds $A$ and $B$, and sampling rate $b_0 > 0$. Then its Fourier transform $\hat{\psi}$ satisfies:

$$b_0 A \le \sum_{j=-\infty}^{\infty} |\hat{\psi}(2^{-j}\omega)|^2 \le b_0 B \quad \text{a.e.} \tag{3.5.18}$$

**Partial Proof.** Let $f \in L^2(\mathbb{R})$. By introducing the notation

$$\tau := \frac{2\pi}{b_0},$$

and applying the Parseval Identities both to $\langle f, \psi_{b_0;j,k} \rangle$ and in the circle setting, we have

$$\sum_{j,k \in \mathbb{Z}} |\langle f, \psi_{b_0;j,k} \rangle|^2 = \sum_{j=-\infty}^{\infty} \frac{2^j}{4\pi^2} \sum_{k=-\infty}^{\infty} \left| \int_{-\infty}^{\infty} \hat{f}(2^j\omega)\overline{\hat{\psi}(\omega)}e^{ikb_0\omega} d\omega \right|^2$$

$$= \sum_{j=-\infty}^{\infty} \frac{2^j \tau^2}{4\pi^2} \sum_{k=-\infty}^{\infty} \left| \frac{1}{\tau} \int_0^{\tau} \left[ \sum_{\ell=-\infty}^{\infty} \hat{f}(2^j(\omega + \ell\tau)) \right. \right.$$

$$\left. \left. \times \overline{\hat{\psi}(\omega + \ell\tau)} \right] e^{i\frac{k 2\pi\omega}{\tau}} d\omega \right|^2$$

$$= \sum_{j=-\infty}^{\infty} \frac{2^j}{2\pi b_0} \int_0^{\tau} \left| \sum_{\ell=-\infty}^{\infty} \hat{f}(2^j(\omega + \ell\tau))\overline{\hat{\psi}(\omega + \ell\tau)} \right|^2 d\omega.$$

Hence, the frame (or stability) condition (3.5.4) becomes:

$$A\|\hat{f}\|_2^2 \le \sum_{j=-\infty}^{\infty} \frac{2^j}{b_0} \int_0^\tau \left| \sum_{\ell=-\infty}^{\infty} \hat{f}(2^j(x+\ell\tau))\overline{\hat{\psi}(x+\ell\tau)} \right|^2 dx \le B\|\hat{f}\|_2^2.$$

Now, if the summation over $j$ were a finite sum, say $-M \le j \le M$, then for any $\omega \in \mathbb{R}$ and sufficiently small $\varepsilon > 0$, the choice of

$$\hat{f}(x) = \frac{1}{\sqrt{2\varepsilon}} \chi_{[\omega-\varepsilon,\omega+\varepsilon]}(x)$$

in the above inequalities readily yields:

$$A \le \sum_j \frac{1}{b_0} \cdot \frac{2^j}{2\varepsilon} \int_{2^{-j}(\omega-\varepsilon)}^{2^{-j}(\omega+\varepsilon)} |\hat{\psi}(x)|^2 dx \le B,$$

and consequently the inequalities in (3.5.18) may be obtained by taking $\varepsilon \to 0^+$. Unfortunately, the sum over $j$ is not finite. Nevertheless, since any finite truncation preserves the upper bound, the second inequality in (3.5.18) certainly holds. To derive the first inequality in (3.5.18), the "tails" of the sum over $j$ must be estimated very carefully before the preceding argument to a finite sum can be applied. We omit the technical details here. ∎

### 3.6. Wavelet series

We continue our discussion of time-frequency analysis by considering discrete time-scale samples of the IWT as in the previous section. To simplify our discussion, we only consider the sampling rate $b_0 = 1$ and use the notation $\psi_{j,k} = \psi_{1;j,k}$ as introduced in (3.5.12). We further restrict our attention to $\mathcal{R}$-functions $\psi$, in the sense that $\{\psi_{j,k}\}$ is a Riesz basis of $L^2(\mathbb{R})$, as in Definition 3.19. Hence, by Theorems 3.20 and 3.21, respectively, $\{\psi_{j,k}\}$ is a frame of $L^2(\mathbb{R})$ and $\psi$ is a dyadic wavelet. Let $\{\psi^{j,k}\}$ be the dual basis relative to the Riesz basis $\{\psi_{j,k}\}$ as defined in (3.5.13). When $\{\psi_{j,k}\}$ is considered as a frame of $L^2(\mathbb{R})$, we may consider $\psi^{j,k} = \psi_1^{j,k} = T^{-1}\psi_{1;j,k}$ to be the dual of this frame, as discussed in (3.5.7)-(3.5.8).

There are two very important subclasses of $\mathcal{R}$-functions that constitute the central theme of our study in this text. They are "semi-orthogonal wavelets" and, more restrictively, "orthogonal wavelets", to be defined below. For these two classes of functions, it is quite easy to characterize their "duals".

**Definition 3.22.** Let $\psi \in L^2(\mathbb{R})$ be an $\mathcal{R}$-function that generates $\{\psi_{j,k}\}$ as in (3.5.12). Then

(i) $\psi$ is called an orthogonal wavelet (or o.n. wavelet), if $\{\psi_{j,k}\}$ satisfies the orthonormality condition:

$$\langle \psi_{j,k}, \psi_{\ell,m} \rangle = \delta_{j,\ell}\delta_{k,m}, \qquad j,k,\ell,m \in \mathbb{Z}; \qquad (3.6.1)$$

(ii) $\psi$ is called a *semi-orthogonal wavelet* (or *s.o.* wavelet), if $\{\psi_{j,k}\}$ satisfies the condition:

$$\langle \psi_{j,k}, \psi_{\ell,m} \rangle = 0, \qquad j \neq \ell;\ j, k, \ell, m \in \mathbb{Z}. \qquad (3.6.2)$$

It is obvious that an o.n. wavelet is "*self-dual*" in the sense that

$$\psi^{j,k} = \psi_{j,k}, \qquad j, k \in \mathbb{Z}.$$

To determine the dual of an s.o. wavelet, let us first discuss the following equivalent statements of orthogonality.

**Theorem 3.23.** *For any function $\phi \in L^2(\mathbb{R})$, the following statements are equivalent:*

(i) $\{\phi(x - k): k \in \mathbb{Z}\}$ *is an orthonormal family in the sense that*

$$\langle \phi(\cdot - k),\ \phi(\cdot - \ell) \rangle = \delta_{k,\ell}, \qquad k, \ell \in \mathbb{Z}. \qquad (3.6.3)$$

(ii) *The Fourier transform $\hat{\phi}$ of $\phi$ satisfies*

$$\frac{1}{2\pi} \int_{-\infty}^{\infty} e^{-ijx} |\hat{\phi}(x)|^2 dx = \delta_{j,0}, \qquad j \in \mathbb{Z}. \qquad (3.6.4)$$

(iii) *The identity*

$$\sum_{k=-\infty}^{\infty} |\hat{\phi}(x + 2\pi k)|^2 = 1 \qquad (3.6.5)$$

*holds for almost all $x$.*

**Proof.** Since $|\hat{\phi}(x)|^2$ is in $L^1(\mathbb{R})$, it follows from Lemma 2.24 that the infinite series

$$G(x) := \sum_{k=-\infty}^{\infty} |\hat{\phi}(x + 2\pi k)|^2, \qquad (3.6.6)$$

that defines the function $G$, converges a.e. to $G$, and that $G \in L^1(0, 2\pi)$. Now, for each $j \in \mathbb{Z}$, the $j^{\text{th}}$ Fourier coefficient of $G$ is

$$
\begin{aligned}
c_j(G) &:= \frac{1}{2\pi} \int_0^{2\pi} e^{-ijx} G(x)\, dx \\
&= \frac{1}{2\pi} \sum_{k=-\infty}^{\infty} \int_0^{2\pi} e^{-ijx} |\hat{\phi}(x + 2\pi k)|^2 dx \\
&= \frac{1}{2\pi} \sum_{k=-\infty}^{\infty} \int_{2\pi k}^{2\pi(k+1)} e^{-ijy} |\hat{\phi}(y)|^2 dy \\
&= \frac{1}{2\pi} \int_{-\infty}^{\infty} e^{-ijx} |\hat{\phi}(x)|^2 dx.
\end{aligned}
$$

This establishes the equivalence of (ii) and (iii). The equivalence of (i) and (ii) follows by a direct application of the Parseval Identity with $j = k - \ell$, namely:

$$
\langle \phi(\cdot - k), \phi(\cdot - \ell) \rangle = \int_{-\infty}^{\infty} \phi(x - j)\overline{\phi(x)}\, dx
$$

$$
= \frac{1}{2\pi} \int_{-\infty}^{\infty} \hat{\phi}(x)e^{-ijx}\overline{\hat{\phi}(x)}\, dx
$$

$$
= \frac{1}{2\pi} \int_{-\infty}^{\infty} e^{-ijx}|\hat{\phi}(x)|^2\, dx. \qquad \blacksquare
$$

A somewhat weaker property than the property of orthonormality in the previous theorem is the "*Riesz* (or *unconditional*) *condition*", which we study in the following.

**Theorem 3.24.** *For any function $\phi \in L^2(\mathbb{R})$ and constants $0 < A \le B < \infty$, the following two statements are equivalent:*

(i) *$\{\phi(\cdot - k): k \in \mathbb{Z}\}$ satisfies the Riesz condition with Riesz bounds $A$ and $B$; that is, for any $\{c_k\} \in \ell^2$,*

$$
A\|\{c_k\}\|_{\ell^2}^2 \le \left\| \sum_{k=-\infty}^{\infty} c_k \phi(\cdot - k) \right\|_2^2 \le B\|\{c_k\}\|_{\ell^2}^2. \tag{3.6.7}
$$

(ii) *The Fourier transform $\hat{\phi}$ of $\phi$ satisfies*

$$
A \le \sum_{k=-\infty}^{\infty} |\hat{\phi}(x + 2\pi k)|^2 \le B, \text{ a.e.} \tag{3.6.8}
$$

**Proof.** For any $\{c_k\} \in \ell^2$, let $C(\omega)$ denotes its symbol; that is,

$$
C(\omega) := \sum_{k=-\infty}^{\infty} c_k e^{-ik\omega}. \tag{3.6.9}
$$

Then by the Parseval Identity, we may write

$$
\left\| \sum_{k=-\infty}^{\infty} c_k \phi(\cdot - k) \right\|_2^2 = \frac{1}{2\pi} \int_{-\infty}^{\infty} |C(\omega)\hat{\phi}(\omega)|^2\, d\omega \tag{3.6.10}
$$

$$
= \frac{1}{2\pi} \sum_{k=-\infty}^{\infty} \int_{2\pi k}^{2\pi(k+1)} |C(\omega)\hat{\phi}(\omega)|^2\, d\omega
$$

$$
= \frac{1}{2\pi} \sum_{k=-\infty}^{\infty} \int_{0}^{2\pi} |C(x)\hat{\phi}(x + 2\pi k)|^2\, dx.
$$

Let us make use of the notation

$$\Phi_{|\hat{\phi}|^2}(x) := \sum_{k=-\infty}^{\infty} |\hat{\phi}(x + 2\pi k)|^2 \qquad (3.6.11)$$

introduced in (2.5.4). Then by considering

$$g(\omega) = |C(\omega)|^2 / \|C\|_{L^2(0,2\pi)},$$

and appealing to the Parseval Identity

$$\|\{c_k\}\|_{\ell^2}^2 = \frac{1}{2\pi} \int_0^{2\pi} |C(\omega)|^2 d\omega = \|C\|_{L^2(0,2\pi)}^2$$

from (1.1.7), it follows from (3.6.10) that (3.6.7) can be formulated as:

$$A \le \frac{1}{2\pi} \int_0^{2\pi} g(x)\Phi_{|\hat{\phi}|^2}(x)dx \le B. \qquad (3.6.12)$$

It is clear that (3.6.8) implies (3.6.12). To see that (3.6.12) also implies (3.6.8), we again use the Gaussian function $g_\alpha(x - \omega)$ in place of $g(x)$ and allow $\alpha$ to tend to 0. ■

With the aid of the two theorems presented above, we can now formulate the dual of an s.o. wavelet.

**Theorem 3.25.** *Let $\psi \in L^2(\mathbb{R})$ be an s.o. wavelet and define $\tilde{\psi}$ via its Fourier transform:*

$$\widehat{\tilde{\psi}}(\omega) := \frac{\widehat{\psi}(\omega)}{\sum\limits_{k=-\infty}^{\infty} |\widehat{\psi}(\omega + 2\pi k)|^2}. \qquad (3.6.13)$$

*Then $\tilde{\psi}$ is the dual of $\psi$, in the sense that*

$$\langle \psi_{j,k}, \tilde{\psi}_{\ell,m} \rangle = \delta_{j,\ell}\delta_{k,m}, \qquad j,k,\ell,m \in \mathbb{Z}, \qquad (3.6.14)$$

*where*

$$\tilde{\psi}_{\ell,m}(x) := 2^{\ell/2}\tilde{\psi}(2^\ell x - m). \qquad (3.6.15)$$

*In other words, the dual basis $\{\psi^{j,k}\}$ relative to $\{\psi_{j,k}\}$ is given by $\psi^{j,k} = \tilde{\psi}_{j,k}$.*

**Proof.** Since $\psi$ is an s.o. wavelet, $\{\psi_{j,k}\}$ is a Riesz basis of $L^2(\mathbb{R})$ with Riesz bounds $A$ and $B$, say. Hence, by considering sequences $\{c_{j,k}\} \in \ell^2(\mathbb{Z}^2)$ of the form $c_{j,k} = c_k\delta_{j,0}$, $\{c_k\} \in \ell^2$, in (3.5.11), we see that (3.6.7) holds for $\psi$ in place of $\phi$. By Theorem 3.24, the denominator in (3.6.13) is bounded a.e. by $A$ and $B$ as in (3.6.8). This implies that $\tilde{\psi}$, as defined in (3.6.13), is in $L^2(\mathbb{R})$ and satisfies

$$\tilde{\psi}(x) = \sum_{k=-\infty}^{\infty} a_k\psi(x - k), \qquad \{a_k\} \in \ell^2, \qquad (3.6.16)$$

where

$$a_k = \frac{1}{2\pi} \int_0^{2\pi} e^{-ikx} \frac{1}{\sum\limits_{j=-\infty}^{\infty} |\hat{\psi}(x + 2\pi j)|^2} dx. \qquad (3.6.17)$$

So, since $\psi$ is an s.o. wavelet, the hypothesis in (3.6.2), with the notation in (3.6.15), immediately yields

$$\langle \psi_{j,k}, \widetilde{\psi}_{\ell,m} \rangle = 0, \qquad j \neq \ell; \quad j, k, \ell, m \in \mathbb{Z}.$$

For $j = \ell$, it follows from (3.6.13), by setting $p = k - m$, that

$$\langle \psi_{j,k}, \widetilde{\psi}_{j,m} \rangle = 2^j \int_{-\infty}^{\infty} \psi(2^j x - k) \overline{\widetilde{\psi}(2^j x - m)} \, dx$$

$$= \int_{-\infty}^{\infty} \psi(y - p) \overline{\widetilde{\psi}(y)} \, dy$$

$$= \frac{1}{2\pi} \int_{-\infty}^{\infty} e^{-ip\omega} \hat{\psi}(\omega) \overline{\hat{\widetilde{\psi}}(\omega)} \, d\omega$$

$$= \frac{1}{2\pi} \int_0^{2\pi} e^{-ip\omega} \left[ \sum_{k=-\infty}^{\infty} \hat{\psi}(\omega + 2\pi k) \overline{\hat{\widetilde{\psi}}(\omega + 2\pi k)} \right] d\omega$$

$$= \frac{1}{2\pi} \int_0^{2\pi} e^{-ip\omega} d\omega = \delta_{p,0} = \delta_{k,m}.$$

This completes the proof of the theorem. ∎

The foregoing result also indicates how an s.o. wavelet is changed to an o.n. wavelet. Indeed, by setting

$$\hat{\psi}^{\perp}(\omega) = \frac{\hat{\psi}(\omega)}{\left( \sum\limits_{k=-\infty}^{\infty} |\hat{\psi}(\omega + 2\pi k)|^2 \right)^{1/2}}, \qquad (3.6.18)$$

the dual $\widetilde{\psi}^{\perp}$ of $\psi^{\perp}$ is given by

$$\hat{\widetilde{\psi}}^{\perp}(\omega) = \frac{\hat{\psi}^{\perp}(\omega)}{\sum\limits_{k=-\infty}^{\infty} |\hat{\psi}^{\perp}(\omega + 2\pi k)|^2} = \hat{\psi}^{\perp}(\omega);$$

that is, $\widetilde{\psi}^{\perp} = \psi^{\perp}$, or $\psi^{\perp}$ is self-dual.

The formula in (3.6.18) is usually called an "*orthonormalization procedure*". However, if $\psi$ is any $\mathcal{R}$-function which is not an s.o. wavelet, this orthonormalization procedure is not effective in constructing o.n. wavelets. In fact, as already discussed in Section 1.4, there are $\mathcal{R}$-functions that do not have duals, in the sense that the dual basis $\{\psi^{j,k}\}$ relative to the Riesz basis $\{\psi_{j,k}\}$ is not given by $\{\widetilde{\psi}_{j,k}\}$ for some $\widetilde{\psi} \in L^2(\mathbb{R})$, where the notation in (3.6.15) is used. Since every Riesz basis is a frame, the dual of a frame may not be generated by a single $L^2(\mathbb{R})$ function either. This leads to the following definition of "*wavelets*".

**Definition 3.26.** *An $\mathcal{R}$-function $\psi \in L^2(\mathbb{R})$ is called an $\mathcal{R}$-wavelet (or wavelet), if it has a dual $\widetilde{\psi} \in L^2(\mathbb{R})$, in the sense that $\{\psi_{j,k}\}$ and $\{\widetilde{\psi}_{j,k}\}$, as defined by (3.5.12) and (3.6.15), satisfy the duality relationship (3.6.14).*

Since the duality relationship (3.6.14) is commutative, the dual $\widetilde{\psi}$ of a wavelet $\psi$ is itself a wavelet, with $\psi$ as its dual. That is, with the exception of o.n. wavelets which are self-duals, when we consider wavelets, we always consider pairs of wavelets.

If $\psi$ is a wavelet with dual $\widetilde{\psi}$, then by the definition of Riesz basis, every $f \in L^2(\mathbb{R})$ can be written as

$$f(x) = \sum_{j,k \in \mathbb{Z}} c_{j,k}\psi_{j,k}(x) = \sum_{j,k \in \mathbb{Z}} d_{j,k}\widetilde{\psi}_{j,k}(x). \tag{3.6.19}$$

These two (doubly) infinite series are called *"wavelet series"* and the convergence is in $L^2(\mathbb{R})$ (see Definition 3.19). By the duality relationship (3.6.14), it follows that

$$\begin{cases} c_{j,k} = \langle f, \widetilde{\psi}_{j,k} \rangle; \\ d_{j,k} = \langle f, \psi_{j,k} \rangle. \end{cases}$$

Hence, we have the following scheme for reconstructing finite-energy signals from discrete samples of their integral wavelet transforms.

**Theorem 3.27.** *Let $\psi$ be a wavelet with dual $\widetilde{\psi}$. For any $f \in L^2(\mathbb{R})$, consider its IWT, using both $\psi$ and $\widetilde{\psi}$ as basic wavelets, evaluated at $(b, a) = (\frac{k}{2^j}, \frac{1}{2^j})$, $j, k \in \mathbb{Z}$, namely:*

$$\begin{cases} d_{j,k} = \langle f, \psi_{j,k} \rangle = (W_\psi f)\left(\dfrac{k}{2^j}, \dfrac{1}{2^j}\right); \\ c_{j,k} = \langle f, \widetilde{\psi}_{j,k} \rangle = (W_{\widetilde{\psi}} f)\left(\dfrac{k}{2^j}, \dfrac{1}{2^j}\right). \end{cases} \tag{3.6.20}$$

*Then $f$ can be reconstructed from either $\{d_{j,k}\}$ or $\{c_{j,k}\}$, by using one of the two wavelet series in (3.6.19). Furthermore, the inner product of any two $L^2(\mathbb{R})$ functions can also be recovered from the analogous discrete samples of their IWT, by using the formula:*

$$\langle f, g \rangle = \sum_{j,k \in \mathbb{Z}} \langle f, \psi_{j,k} \rangle \langle \widetilde{\psi}_{j,k}, g \rangle. \tag{3.6.21}$$

We now return to the discussion of dyadic duals of dyadic wavelets introduced in Section 3.4. First, we must emphasize that dyadic wavelets are not necessarily wavelets in the sense of $\mathcal{R}$-wavelets, and dyadic duals are usually not dual wavelets.

Let $\psi$ be an s.o. wavelet with Riesz bounds $A$ and $B$. We have seen from the proof of Theorem 3.25 that $\widehat{\psi}$ satisfies the inequalities

$$A \le \sum_{k=-\infty}^{\infty} |\widehat{\psi}(x + 2\pi k)|^2 \le B \quad \text{a.e.} \tag{3.6.22}$$

for the same bounds $A$ and $B$. When $\{\psi_{j,k}\}$ is considered as a frame of $L^2(\mathbb{R})$ with sampling rate $b_0 = 1$, it follows from Theorems 3.20 and 3.21 that $\widehat{\psi}$ also satisfies the condition

$$A \le \sum_{j=-\infty}^{\infty} |\widehat{\psi}(2^{-j}\omega)|^2 \le B \quad \text{a.e.} \tag{3.6.23}$$

again for the same constants $A$ and $B$. The conclusions in (3.6.22) and (3.6.23) allow us to introduce two $L^2(\mathbb{R})$ functions $\psi^*$ and $\psi^\circ$, with Fourier transforms given by

$$\begin{cases} \widehat{\psi}^*(\omega) := \dfrac{\widehat{\psi}(\omega)}{\displaystyle\sum_{k=-\infty}^{\infty} |\widehat{\psi}(2^{-k}\omega)|^2}; \\[2em] \widehat{\psi}^\circ(\omega) := \dfrac{\widehat{\psi}(\omega)}{\displaystyle\sum_{k=-\infty}^{\infty} |\widehat{\psi}(\omega + 2\pi k)|^2}. \end{cases} \tag{3.6.24}$$

Let $\psi^\perp$ be the o.n. wavelet obtained by orthonormalization of $\psi$ using (3.6.18). Then $\psi^\perp$ is an $\mathcal{R}$-function with Riesz bounds $A = B = 1$. Hence, it follows from Theorems 3.20 and 3.21 that

$$\sum_{j=-\infty}^{\infty} |\widehat{\psi}^\perp(2^{-j}\omega)|^2 = 1 \quad \text{a.e.} \tag{3.6.25}$$

Since $\overline{\widehat{\psi}}\widehat{\psi}^\circ = |\widehat{\psi}^\perp|^2$, we also have

$$\sum_{j=-\infty}^{\infty} \overline{\widehat{\psi}(2^{-j}\omega)}\widehat{\psi}^\circ(2^{-j}\omega) = 1, \quad \text{a.e.} \tag{3.6.26}$$

So, in view of the fact that

$$\operatorname*{ess\,sup}_{-\infty < x < \infty} \sum_{j=-\infty}^{\infty} |\widehat{\psi}^\circ(2^{-j}x)|^2 < \infty, \tag{3.6.27}$$

which is a consequence of the mutual orthogonality among all of the $\psi^\circ(2^j x)$, $j \in \mathbb{Z}$, it follows from Theorem 3.16 that $\psi^\circ$ is a dyadic dual of $\psi$. Hence, by appealing to Theorem 3.15, we conclude that both $\psi^*$ and $\psi^\circ$ are dyadic duals of $\psi$. Since the two denominators in (3.6.24) are really quite different, one cannot expect these two dyadic duals to be the same, unless, of course, $\psi$ is an o.n. wavelet.

# 4 Cardinal Spline Analysis

In using basic wavelets such as dyadic wavelets, dyadic duals, frames, and $\mathcal{R}$-wavelets (which are simply called wavelets), for time-frequency analysis and other applications, several important points must be taken into consideration. Among them are: size of the time-frequency window, computational complexity and efficiency, simplicity in implementation, smoothness and symmetry of the basic wavelet, and order of approximation. One of the basic methods for constructing wavelets involves the use of "*cardinal B-spline functions*". These are probably the simplest functions with small supports that are most efficient for both software and hardware implementation. In addition, they possess a very nice property, called "*total positivity*", that controls zero-crossings and shapes of the "*spline curves*". This topic will be discussed in Chapter 6, where we will even see that minimally supported wavelets which are constructed by using a (finite) linear combination of translates of a cardinal $B$-spline produce time-frequency windows with "near-minimal" size as governed by the Uncertainty Principle. The present chapter is devoted to a study of cardinal spline functions with emphasis on their basic properties that are crucial to computation, graphical display, real-time (or on-line) processing of discrete data, and construction of wavelets.

## 4.1. Cardinal spline spaces

When we talk about "cardinal splines", we mean "polynomial spline functions with equally spaced simple knots". For convenience, let us first consider the set $\mathbb{Z}$ of all integers as the "knot sequence". As in (1.5.8), $\pi_n$ denotes the collection of all algebraic polynomials of degree at most $n$ and $C^n = C^n(\mathbb{R})$, the collection of all functions $f$ such that $f, f', \ldots, f^{(n)}$ are continuous everywhere, with the understanding that $C = C^0$, and $C^{-1}$ is the space of piecewise continuous functions as defined in the beginning of Section 2.1.

**Definition 4.1.** *For each positive integer $m$, the space $S_m$ of cardinal splines of order $m$ and with knot sequence $\mathbb{Z}$ is the collection of all functions $f \in C^{m-2}$ such that the restrictions of $f$ to any interval $[k, k+1)$, $k \in \mathbb{Z}$, are in $\pi_{m-1}$; that is,*

$$f|_{[k,k+1)} \in \pi_{m-1}, \quad k \in \mathbb{Z}.$$

The space $S_1$ of piecewise constant functions is easy to understand. The most convenient basis to use is $\{N_1(x - k): k \in \mathbb{Z}\}$ where $N_1$ is the characteristic function of $[0,1)$ defined in (3.2.2). To give a basis of $S_m$, $m \geq 2$, let us first consider the space $S_{m;N}$ consisting of the restrictions of functions $f \in S_m$ to the interval $[-N, N]$, where $N$ is a positive integer. In other words, we may consider $S_{m;N}$ as the subspace of functions $f \in S_m$ such that the restrictions

$$f|_{(-\infty, -N+1)} \quad \text{and} \quad f|_{[N-1, \infty)}$$

of $f$ are polynomials in $\pi_{m-1}$. This subspace is easy to characterize. Indeed, for an arbitrary function $f$ in $S_{m,N}$, by setting $p_{m,j} := f|_{[j,j+1)} \in \pi_{m-1}$, $j = -N, \ldots, N-1$, we have, in view of the fact that $f \in C^{m-2}$,

$$\left( p_{m,j}^{(\ell)} - p_{m,j-1}^{(\ell)} \right)(j) = 0, \quad \ell = 0, \ldots, m-2; m \geq 2.$$

That is, by considering the "jumps" of $f^{(m-1)}$ at the knot sequence $\mathbb{Z}$, namely:

$$\begin{aligned}
c_j &= p_{m,j}^{(m-1)}(j+0) - p_{m,j-1}^{(m-1)}(j-0) \\
&:= \lim_{\varepsilon \to 0+} [f^{(m-1)}(j+\varepsilon) - f^{(m-1)}(j-\varepsilon)],
\end{aligned} \tag{4.1.1}$$

the adjacent polynomial pieces of $f$ are related by the identity

$$p_{m,j}(x) = p_{m,j-1}(x) + \frac{c_j}{(m-1)!}(x-j)^{m-1}. \tag{4.1.2}$$

Hence, by introducing the notation

$$\begin{cases} x_+ := \max(0, x); \\ x_+^{m-1} := (x_+)^{m-1}, \quad m \geq 2, \end{cases} \tag{4.1.3}$$

it follows from (4.1.2), that for all $x \in [-N, N]$,

$$f(x) = f|_{[-N, -N+1)}(x) + \sum_{j=-N+1}^{N-1} \frac{c_j}{(m-1)!}(x-j)_+^{m-1}. \tag{4.1.4}$$

This holds for every $f \in S_{m,N}$, with the constants $c_j$ given by (4.1.1). Consequently, the collection

$$\{1, \ldots, x^{m-1}, (x+N-1)_+^{m-1}, \ldots, (x-N+1)_+^{m-1}\} \tag{4.1.5}$$

of $m + 2N - 1$ functions is a basis of $S_{m,N}$. This collection consists of both monomials and "*truncated powers*". Since we restrict our attention to the interval $[-N, N]$, it is also possible to replace the monomials $1, \ldots, x^{m-1}$ in (4.1.5) by the truncated powers:

$$(x+N+m-1)_+^{m-1}, \ldots, (x+N)_+^{m-1}. \tag{4.1.6}$$

That is, the following set of truncated powers, which are generated by using integer translates of a single function, $x_+^{m-1}$, is also a basis of $S_{m,N}$:

$$\{(x-k)_+^{m-1}\colon k = -N - m + 1, \ldots, N - 1\}. \tag{4.1.7}$$

This basis is more attractive than the basis in (4.1.5) for the following reasons. Firstly, each function $(x-j)_+^{m-1}$ vanishes to the left of $j$; secondly all the basis functions in (4.1.7) are generated by a single function $x_+^{m-1}$ which is independent of $N$. Moreover, since

$$S_m = \bigcup_{N=1}^{\infty} S_{m,N},$$

it follows that the basis in (4.1.7) can also be extended to be a "basis" $\mathcal{T}$ of the infinite dimensional space $S_m$, simply by taking the union of the bases in (4.1.7); that is, we have

$$\mathcal{T} := \{(x-k)_+^{m-1}\colon k \in \mathbb{Z}\}. \tag{4.1.8}$$

However, we must be more careful when we deal with infinite dimensional spaces. In this book, since we are mainly concerned with the Hilbert space $L^2(\mathbb{R})$, we are especially interested in cardinal splines that are in $L^2(\mathbb{R})$. Unfortunately, there is not a single function in $\mathcal{T}$ that qualifies to be a function in $L^2(\mathbb{R})$, and in fact, each $(x-k)_+^{m-1}$ grows to infinity fairly rapdily as $x \to +\infty$. To create $L^2(\mathbb{R})$ functions from $\mathcal{T}_N$, we must tame the polynomial growth of $(x-k)_+^{m-1}$. The only operation that is allowed in working with vector spaces is taking (finite) linear combinations. For instance, taking derivatives is not allowed but taking "differences" is certainly permissible. Since the effect of differences is the same as that of derivatives in taming polynomial growth, we will consider differences. More precisely, we will use "*backward differences*", defined recursively by

$$\begin{cases} (\Delta f)(x) := f(x) - f(x-1); \\ (\Delta^n f)(x) := (\Delta^{n-1}(\Delta f))(x), \quad n = 2, 3, \ldots . \end{cases} \tag{4.1.9}$$

Observe that just like the $m^{\text{th}}$ order differential operator, the $m^{\text{th}}$ order difference of any polynomial of degree $m-1$ or less is identically zero, that is,

$$\Delta^m f = 0, \quad f \in \pi_{m-1}. \tag{4.1.10}$$

This motivates the following definition.

**Definition 4.2.** *Let $M_1 := N_1$ be the characteristic function of $[0,1)$ as defined in (3.2.2), and for $m \geq 2$, let*

$$M_m(x) := \frac{1}{(m-1)!} \Delta^m x_+^{m-1}. \tag{4.1.11}$$

It is clear from the definition that $M_m$ is a linear combination of the basis functions in (4.1.8). In fact, it is easy to verify that

$$M_m(x) = \frac{1}{(m-1)!} \sum_{k=0}^{m} (-1)^k \binom{m}{k} (x-k)_+^{m-1}. \qquad (4.1.12)$$

From (4.1.10), it follows that $M_m(x) = 0$ for all $x \geq m$. Since $M_m(x)$ clearly vanishes for $x < 0$, we have supp $M_m \subseteq [0, m]$. By working a little harder, we can even conclude that

$$\text{supp } M_m = [0, m]. \qquad (4.1.13)$$

So, $M_m$ is certainly in $L^2(\mathbb{R})$. But is the collection

$$\mathcal{B} := \{M_m(x-k) : k \in \mathbb{Z}\} \qquad (4.1.14)$$

of integer-translates of $M_m$ a "basis" of $S_m$? Let us again return to $S_{m,N}$ which, according to (4.1.5) or (4.1.7), has dimension $m + 2N - 1$. Now, by using the support property (4.1.13), each function in the collection

$$\{M_m(x-k) : k = -N - m + 1, \dots, N - 1\} \qquad (4.1.15)$$

is nontrivial on the interval $[-N, N]$ and $M_m(x-k)$ vanishes identically on $[-N, N]$ for $k < -N - m + 1$ or $k > N - 1$. Since it can be shown that (4.1.15) is a linearly independent set, we have obtained another basis of $S_{m,N}$. So, analogous to (4.1.8), if we take the union of the bases in (4.1.15), $N = 1, 2, \dots$, we arrive at $\mathcal{B}$ in (4.1.14). One advantage of $\mathcal{B}$ over $\mathcal{T}$ in (4.1.8) is that we can now talk about a spline series

$$f(x) = \sum_{k=-\infty}^{\infty} c_k M_m(x-k) \qquad (4.1.16)$$

without worrying too much about convergence. Indeed, for each fixed $x \in \mathbb{R}$, since $M_m$ has compact support, all except for a finite number of terms in the infinite series (4.1.16) are zero.

As mentioned earlier, we are mainly interested in those cardinal splines that belong to $L^2(\mathbb{R})$, namely: $S_m \cap L^2(\mathbb{R})$. Let $V_0^m$ denote its $L^2(\mathbb{R})$-closure. That is, $V_0^m$ is the smallest closed subspace of $L^2(\mathbb{R})$ that contains $S_m \cap L^2(\mathbb{R})$. Since $M_m$ has compact support, we see that $\mathcal{B} \subset V_0^m$. In the next section, we will even show that $\mathcal{B}$ is a Riesz (or unconditional) basis of $V_0^m$.

So far, we have only considered cardinal splines with knot sequence $\mathbb{Z}$. More generally, we will also consider the spaces $S_m^j$ of cardinal splines with knot sequences $2^{-j}\mathbb{Z}$, $j \in \mathbb{Z}$. Since a spline function with knot sequence $2^{-j_1}\mathbb{Z}$ is also a spline function with knot sequence $2^{-j_2}\mathbb{Z}$, whenever $j_1 < j_2$, we have a (doubly-infinite) nested sequence

$$\cdots \subset S_m^{-1} \subset S_m^0 \subset S_m^1 \subset \cdots$$

of cardinal spline spaces, with $S_m^0 := S_m$. Analogous to the definition of $V_0^m$, we will let $V_j^m$ denote the $L^2(\mathbb{R})$-closure of $S_m^j \cap L^2(\mathbb{R})$. Hence, we have a nested sequence

$$\cdots \subset V_{-1}^m \subset V_0^m \subset V_1^m \subset \cdots \tag{4.1.17}$$

of closed cardinal spline subspaces of $L^2(\mathbb{R})$. It will be clear that this nested sequence of subspaces satisfies:

$$\begin{cases} \text{clos}_{L^2(\mathbb{R})} \left( \bigcup_{j \in \mathbb{Z}} V_j^m \right) = L^2(\mathbb{R}); \\ \bigcap_{j \in \mathbb{Z}} V_j^m = \{0\}. \end{cases} \tag{4.1.18}$$

Furthermore, it is clear that once we have shown that $\mathcal{B}$ is a Riesz basis of $V_0^m$, then for any $j \in \mathbb{Z}$, the collection

$$\{2^{j/2} M_m(2^j x - k) : k \in \mathbb{Z}\} \tag{4.1.19}$$

is also a Riesz basis of $V_j^m$ with the same Riesz bounds as those of $\mathcal{B}$.

## 4.2. $B$-splines and their basic properties

Let us return to the definition

$$N_m(x) := (N_{m-1} * N_1)(x) = \int_0^1 N_{m-1}(x - t)dt, \quad m \geq 2, \tag{4.2.1}$$

of the $m^{\text{th}}$ order cardinal $B$-spline introduced in (1.5.7), where $N_1$ is the characteristic function of the interval $[0,1)$. In Definition 4.2, we set $M_1 = N_1$; and in the following, we will see that $M_m = N_m$ for all $m \geq 2$ also. Hence, $N_m$ is an $m^{\text{th}}$ order cardinal spline function in $V_0^m \subset S_m$. While the definition of $M_m$ in (4.1.11) is explicit, the advantage of the definition of $N_m$ in (4.2.1) is that many important properties of $N_m$ can be derived from it very easily. Among them are the first seven of the eight properties listed in the following.

**Theorem 4.3.** *The $m^{\text{th}}$ order cardinal $B$-spline $N_m$ satisfies the following properties:*

(i) *For every $f \in C$,*

$$\int_{-\infty}^{\infty} f(x) N_m(x)dx = \int_0^1 \cdots \int_0^1 f(x_1 + \cdots + x_m)dx_1 \ldots dx_m. \tag{4.2.2}$$

(ii) *For every $g \in C^m$,*

$$\int_{-\infty}^{\infty} g^{(m)}(x) N_m(x)dx = \sum_{k=0}^m (-1)^{m-k} \binom{m}{k} g(k). \tag{4.2.3}$$

(iii) $N_m(x) = M_m(x)$, all $x$.

(iv) supp $N_m = [0, m]$.

(v) $N_m(x) > 0$, for $0 < x < m$.

(vi) $\sum\limits_{k=-\infty}^{\infty} N_m(x - k) = 1$, all $x$.

(vii) $N_m'(x) = (\Delta N_{m-1})(x) = N_{m-1}(x) - N_{m-1}(x - 1)$.

(viii) The cardinal B-splines $N_m$ and $N_{m-1}$ are related by the identity:

$$N_m(x) = \frac{x}{m-1} N_{m-1}(x) + \frac{m-x}{m-1} N_{m-1}(x - 1). \qquad (4.2.4)$$

(ix) $N_m$ is symmetric with respect to the center of its support, namely:

$$N_m\left(\frac{m}{2} + x\right) = N_m\left(\frac{m}{2} - x\right), \quad x \in \mathbb{R}.$$

**Proof.** (i) Assertion (4.2.2) certainly holds for $m = 1$. Suppose it also holds for $m-1$, then by the definition of $N_m$ in (4.2.1) and this induction hypothesis, we have

$$\int_{-\infty}^{\infty} f(x) N_m(x) dx = \int_{-\infty}^{\infty} f(x) \left\{ \int_0^1 N_{m-1}(x - t) dt \right\} dx$$

$$= \int_0^1 \left\{ \int_{-\infty}^{\infty} f(x) N_{m-1}(x - t) dx \right\} dt$$

$$= \int_0^1 \left\{ \int_{-\infty}^{\infty} f(y + t) N_{m-1}(y) dy \right\} dt$$

$$= \int_0^1 \int_0^1 \cdots \int_0^1 f(x_1 + \cdots + x_{m-1} + t) dx_1 \ldots dx_{m-1} dt$$

$$= \int_0^1 \cdots \int_0^1 f(x_1 + \cdots + x_m) dx_1 \ldots dx_m.$$

(ii) Assertion (4.2.3) follows from (4.2.2) since

$$\int_0^1 \cdots \int_0^1 g^{(m)}(x_1 + \cdots + x_m) dx_1 \ldots dx_m = \sum_{k=0}^{m} (-1)^{m-k} \binom{m}{k} g(k)$$

by direct integration.

(iii) Fix $x \in \mathbb{R}$. By selecting

$$g(t) = \frac{(-1)^m}{(m-1)!} (x - t)_+^{m-1},$$

the right-hand side of (4.2.3) agrees with the formula of $M_m(x)$ in (4.1.12). Since

$$g^{(m)}(t) = \delta(x - t),$$

where $\delta$ is the delta distribution (see (2.2.12) and (2.2.14)), the left-hand side of (4.2.3) becomes $N_m(x)$. That is, $N_m(x) = M_m(x)$ for any (fixed) $x \in \mathbb{R}$. Of course, one could avoid using the delta distribution to derive (iii) by relying on an induction argument without even appealing to (4.2.3), for instance.

Assertions (iv), (v), (vi) and (ix) can be easily derived by induction, using the definition of $N_m$ in (4.2.1).

(vii) Using (4.2.1) again, we have

$$N_m'(x) = \int_0^1 N_{m-1}'(x-t)dt = -N_{m-1}(x-1) + N_{m-1}(x) = (\Delta N_{m-1})(x).$$

(viii) To verify the identity in (viii), we use the definition of $M_m$ in (4.1.11) instead. Of course, we have already shown in (iii) that $N_m = M_m$. The idea is to represent $x_+^{m-1}$ as the product of a monomial and a truncated power, namely:

$$x_+^{m-1} = x \cdot x_+^{m-2},$$

and then apply the following "Leibniz Rule" for differences:

$$(\Delta^n fg)(x) = \sum_{k=0}^{n} \binom{n}{k} (\Delta^k f)(x)(\Delta^{n-k}g)(x-k). \qquad (4.2.5)$$

This identity for differences can be easily established by induction. It is almost exactly the same as the Leibniz Rule for derivatives. Now, if we set $f(x) = x$ and $g(x) = x_+^{m-2}$ in (4.2.5) and recall that $\Delta^k f = 0$ for $k \geq 2$ from (4.1.10), we then have

$$\begin{aligned}
N_m(x) = M_m(x) &= \frac{1}{(m-1)!} \Delta^m x_+^{m-1} \\
&= \frac{1}{(m-1)!} \{x \Delta^m x_+^{m-2} + m \Delta^{m-1}(x-1)_+^{m-2}\} \\
&= \frac{1}{(m-1)!} \{x[\Delta^{m-1} x_+^{m-2} - \Delta^{m-1}(x-1)_+^{m-2}] + m\Delta^{m-1}(x-1)_+^{m-2}\} \\
&= \frac{x}{m-1} M_{m-1}(x) + \frac{m-x}{m-1} M_{m-1}(x-1) \\
&= \frac{x}{m-1} N_{m-1}(x) + \frac{m-x}{m-1} N_{m-1}(x-1).
\end{aligned}$$

This completes the proof of Theorem 4.3. ∎

We next show that the cardinal $B$-spline basis

$$\mathcal{B} = \{N_m(x-k) \colon k \in \mathbb{Z}\}, \qquad (4.2.6)$$

which is the same as the basis introduced in (4.1.14), is a Riesz (or unconditional) basis of $V_0^m$ in the sense of (3.6.7). By Theorem 3.24, this is equivalent to investigating the existence of lower and upper bounds $A, B$ in (3.6.8). From (4.2.1), we see that $\widehat{N}_m = (\widehat{N}_1)^m$, so that

$$|\widehat{N}_m(\omega)|^2 = \left| \frac{1 - e^{-i\omega}}{i\omega} \right|^{2m}$$

(see (3.2.16)). Hence, replacing $\omega$ by $2x$, we have

$$\sum_{k=-\infty}^{\infty} |\widehat{N}_m(2x + 2\pi k)|^2 = 2^{2m} \sum_{k=-\infty}^{\infty} \frac{\sin^{2m}(x + \pi k)}{(2x + 2\pi k)^{2m}} \qquad (4.2.7)$$

$$= (\sin^{2m} x) \sum_{k=-\infty}^{\infty} \frac{1}{(x + \pi k)^{2m}}.$$

Recall from complex analysis that

$$\cot x = \lim_{n \to \infty} \sum_{k=-n}^{n} \frac{1}{x + \pi k}, \qquad (4.2.8)$$

which immediately yields

$$\sum_{k=-\infty}^{\infty} \frac{1}{(x + \pi k)^{2m}} = -\frac{1}{(2m - 1)!} \frac{d^{2m-1}}{dx^{2m-1}} \cot x. \qquad (4.2.9)$$

Therefore, substituting (4.2.9) into (4.2.7), we obtain

$$\sum_{k=-\infty}^{\infty} |\widehat{N}_m(2x + 2\pi k)|^2 = \frac{-\sin^{2m} x}{(2m - 1)!} \frac{d^{2m-1}}{dx^{2m-1}} \cot x. \qquad (4.2.10)$$

**Example 4.4.** For the first and second order cardinal $B$-splines $N_1$ and $N_2$, it follows from (4.2.10) that

$$\sum_{k=-\infty}^{\infty} |\widehat{N}_1(\omega + 2\pi k)|^2 = 1 \qquad (4.2.11)$$

and

$$\sum_{k=-\infty}^{\infty} |\widehat{N}_2(\omega + 2\pi k)|^2 = \frac{1}{3} + \frac{2}{3} \cos^2 \left( \frac{\omega}{2} \right). \qquad (4.2.12)$$

Hence, $\{N_1(\cdot - k)\}$ is orthonormal (see Theorem 3.23), and

$$\frac{1}{3} \leq \sum_{k=-\infty}^{\infty} |\widehat{N}_2(\omega + 2\pi k)|^2 \leq 1, \qquad (4.2.13)$$

where both the upper and lower (Riesz) bounds in (4.2.13) are best possible.

Although the formula in (4.2.10) is explicit and provides a tool for finding optimal Riesz bounds, the algebra in manipulating the trigonometric (sine and cosine) polynomials is quite involved for larger values of the spline order $m$. Another approach is to apply Theorem 2.28 to $f(x) = N_m(x)$. This requires knowledge of the values of

$$\int_{-\infty}^{\infty} N_m(y+k)\overline{N_m(y)}\, dy = N_{2m}(m+k). \tag{4.2.14}$$

The identity in (4.2.14) is an easy consequence of the definition (4.2.1), while the values of $N_{2m}$ at the knot sequence $\mathbb{Z}$ can be easily determined recursively, by applying (4.2.4) in Theorem 4.3, namely:

$$\begin{cases} N_2(k) = \delta_{k,1}, \quad k \in \mathbb{Z}; \text{ and} \\ N_{n+1}(k) = \dfrac{k}{n} N_n(k) + \dfrac{n-k+1}{n} N_n(k-1), \quad k = 1,\ldots,n. \end{cases} \tag{4.2.15}$$

Note that $N_{n+1}(k) = 0$ for $k \le 0$ or $k \ge n+1$. Hence, by applying (2.5.19) in Theorem 2.28, we have

$$\sum_{k=-\infty}^{\infty} |\widehat{N}_m(\omega + 2\pi k)|^2 = \sum_{k=-m+1}^{m-1} N_{2m}(m+k)e^{-ik\omega}. \tag{4.2.16}$$

An application of (v) and (vi) in Theorem 4.3 now yields

$$\sum_{k=-\infty}^{\infty} |\widehat{N}_m(\omega + 2\pi k)|^2 \le 1, \tag{4.2.17}$$

and the Riesz bound $B = 1$ here is the smallest possible.

To determine the greatest lower bound of the expression in (4.2.16), we consider the so-called "*Euler-Frobenius polynomials*"

$$E_{2m-1}(z) := (2m-1)! z^{m-1} \sum_{k=-m+1}^{m-1} N_{2m}(m+k)z^k \tag{4.2.18}$$

of order $2m - 1$ (or degree $2m - 2$). In Chapter 6, we will show that all the $2m - 2$ roots, $\lambda_1, \ldots, \lambda_{2m-2}$, of $E_{2m-1}$ are simple, real and in fact, negative; and furthermore, when they are arranged in decreasing order, say,

$$0 > \lambda_1 > \cdots > \lambda_{2m-2}, \tag{4.2.19}$$

these simple roots are in reciprocal pairs; that is,

$$\lambda_1 \lambda_{2m-2} = \cdots = \lambda_{m-1}\lambda_m = 1. \tag{4.2.20}$$

Hence, we have

$$A_m := \frac{1}{(2m-1)!} \prod_{k=1}^{m-1} \frac{(1+\lambda_k)^2}{|\lambda_k|} > 0. \tag{4.2.21}$$

Now, from (4.2.16) and (4.2.18), we may write

$$\sum_{k=-\infty}^{\infty} |\widehat{N}_m(\omega + 2\pi k)|^2 = \frac{1}{(2m-1)!} \prod_{k=1}^{2m-2} |e^{i\omega} - \lambda_k|$$

$$= \frac{1}{(2m-1)!} \prod_{k=1}^{m-1} \frac{|1 - \lambda_k e^{i\omega}| \, |1 - \lambda_k e^{-i\omega}|}{|\lambda_k|}$$

$$= \frac{1}{(2m-1)!} \prod_{k=1}^{m-1} \frac{1 - 2\lambda_k \cos\omega + \lambda_k^2}{|\lambda_k|},$$

whence it is clear that

$$\sum_{k=-\infty}^{\infty} |\widehat{N}_m(\omega + 2\pi k)|^2 \geq \sum_{k=-\infty}^{\infty} |\widehat{N}_m(\pi + 2\pi k)|^2 = A_m.$$

We have thus proved the following result.

**Theorem 4.5.** *For any integer $m \geq 2$, let $A_m$ be the positive number defined in (4.2.21). Then the cardinal B-spline basis $\mathcal{B}$ in (4.2.6) is a Riesz basis of $V_0^m$ with Riesz bounds $A = A_m$ and $B = 1$. Furthermore, these bounds are best possible.*

## 4.3. The two-scale relation and an interpolatory graphical display algorithm

Let us first return to Section 4.1 and study the relationship between any two consecutive subspaces of the nested sequence $\{V_j^m : j \in \mathbb{Z}\}$ of closed subspaces of $L^2(\mathbb{R})$ as discussed in (4.1.17)–(4.1.19). In view of the fact that $M_m = N_m$, we may and will always use the notation $N_m$ instead of $M_m$. Observe that the following result, via a simple change of variables, is a trivial consequence of Theorem 4.5.

**Corollary 4.6.** *For any pair of integers $m$ and $j$, with $m \geq 2$, the family*

$$\mathcal{B}_j := \{2^{j/2} N_m(2^j x - k) : k \in \mathbb{Z}\}$$

*is a Riesz basis of $V_j^m$ with Riesz bounds $A = A_m$ and $B = 1$. Furthermore, these bounds are optimal.*

Note that the basis $\mathcal{B}_j$ defined above reduces to $\mathcal{B}$ for $j = 0$, and in the construction of computational algorithms, it is more convenient to drop the normalization constant $2^{j/2}$ in $\mathcal{B}_j$. This only changes the Riesz bounds by a factor of $2^{-j}$. Hence, for each $j$, since $N_m(2^j x) \in V_j^m$ and $V_j^m \subset V_{j+1}^m$, we have, from Corollary 4.6, that

$$N_m(2^j x) = \sum_{k=-\infty}^{\infty} p_{m,k} N_m(2^{j+1} x - k), \qquad (4.3.1)$$

where $\{p_{m,k} : k \in \mathbb{Z}\}$ is some sequence in $\ell^2$. Now, replacing $2^j x$ by $y$ and taking the Fourier transform on both sides of (4.3.1), we obtain the following equivalent formulation of (4.3.1):

$$\widehat{N}_m(\omega) = \frac{1}{2} \left( \sum_{k=-\infty}^{\infty} p_{m,k} e^{-ik\omega/2} \right) \widehat{N}_m \left( \frac{\omega}{2} \right). \qquad (4.3.2)$$

This formula can be applied to determine the sequence $\{p_{m,k}\}$ in (4.3.1). Indeed, since

$$\widehat{N}_m(\omega) = \left( \frac{1 - e^{-i\omega}}{i\omega} \right)^m$$

(see (3.2.16)), we have

$$\frac{1}{2} \sum_{k=-\infty}^{\infty} p_{m,k} e^{-ik\omega/2} = \left( \frac{1 - e^{-i\omega}}{i\omega} \right)^m \left( \frac{i\omega/2}{1 - e^{-i\omega/2}} \right)^m$$

$$= \left( \frac{1 + e^{-i\omega/2}}{2} \right)^m = 2^{-m} \sum_{k=0}^{m} \binom{m}{k} e^{-ik\omega/2},$$

and this yields

$$p_{m,k} = \begin{cases} 2^{-m+1} \binom{m}{k} & \text{for } 0 \le k \le m; \\ 0 & \text{otherwise.} \end{cases} \qquad (4.3.3)$$

Consequently, the precise formulation of (4.3.1) is given by

$$N_m(x) = \sum_{k=0}^{m} 2^{-m+1} \binom{m}{k} N_m(2x - k), \qquad (4.3.4)$$

which is called the *"two-scale relation"* for cardinal $B$-splines of order $m$.

As already discussed in Section 1.6, this two-scale relation for cardinal $B$-splines (and more generally, in (1.6.2), for any scaling function $\phi$) is one of a pair of two-scale relations that gives rise to the so-called (wavelet) reconstruction algorithm described by (1.6.10) and Figure 1.6.2. (The other formula describes the relation between the wavelet $\psi(x)$ and $\phi(2x - k)$, $k \in \mathbb{Z}$, as in (1.6.3).) Observe that in the wavelet decomposition (1.6.1), if all the wavelet components $g_{N-M}, \ldots, g_{N-1}$ of $f_N$ are identically zero, then we do not need the formula (1.6.3) to write any $f_{N-M} \in V_{N-M}$ as an $f_N \in V_N$. In other words, "half" of the reconstruction algorithm in (1.6.10) can be used to express any function $f_{N-M}$ at the $(N - M)^{\text{th}}$ *"resolution level"* (with $2^{N-M}$

pixels per unit length) as a function $f_N$ at the (higher) $N^{\text{th}}$ resolution level (with $2^N$ pixels per unit length). Of course, $f_{N-M} = f_N$ identically; but we obtain a "better picture" of the same function at a higher resolution.

We now restrict our attention to cardinal splines and incorporate this procedure with the algorithm in (4.2.15) for calculating $B$-spline values at the knots to give a very efficient algorithm for displaying the graph of any cardinal spline function at any desirable resolution level, *exactly*. Let us first formulate the objective of this "interpolatory graphical display algorithm" precisely, as follows:

Consider a cardinal spline function

$$f_{j_0}(x) = \sum_{\ell} a_{\ell}^{(j_0)} N_m(2^{j_0} x - \ell) \tag{4.3.5}$$

of order $m$ and with knot sequence $2^{-j_0}\mathbb{Z}$, where $j_0$ is any (fixed) integer. Suppose that $\{a_{\ell}^{(j_0)}\}$ is a "*causal*" sequence of (known) real numbers, where causality means that $a_{\ell}^{(j_0)} = 0$ for all $\ell < \ell_0$, say. The objective is to compute *all* the values of the sequence

$$f_{j_0}\left(\frac{k}{2^{j_1}}\right), \quad k \in \mathbb{Z}, \tag{4.3.6}$$

*exactly*, for any pre-assigned integer $j_1 \geq j_0$ in "*real-time*", meaning that the sequence in (4.3.6) is computed for increasing values of $k$ as soon as the coefficient "data" sequence $\{a_{\ell}^{(j_0)}\}$ has been registered in increasing values of $\ell$. Observe that to display the graph of $f(x)$, it is adequate to display the sequence $f(k/2^{j_1})$, $k \in \mathbb{Z}$, provided that the (fixed) integer $j_1$ is sufficiently large. Of course the size of $j_1$ is limited by the performance of the available equipment.

For each $j \geq j_0$, let us use the notation

$$\begin{cases} f_j(x) = \sum_{\ell} a_{\ell}^{(j)} N_m(2^j x - \ell); \\ \mathbf{a}^j := \{a_{\ell}^{(j)}\}, \quad \ell \in \mathbb{Z}. \end{cases} \tag{4.3.7}$$

By applying the two-scale relation (4.3.1), we see that the identity $f_{j+1}(x) = f_j(x)$ is equivalent to the identity

$$\sum_{\ell} a_{\ell}^{(j+1)} N_m(2^{j+1} x - \ell) = \sum_{\ell} a_{\ell}^{(j)} N_m(2^j x - \ell)$$

$$= \sum_{\ell} a_{\ell}^{(j)} \sum_{k} p_{m,k} N_m(2^{j+1} x - 2\ell - k)$$

$$= \sum_{\ell} \left\{ \sum_{k} p_{m,\ell-2k} a_k^{(j)} \right\} N_m(2^{j+1} x - \ell).$$

Hence, since the collection $N_m(2^{j+1}x - \ell)$, $\ell \in \mathbb{Z}$, is a Riesz basis of $V_{j+1}^m$, the identity $f_{j+1}(x) = f_j(x)$ is precisely described by the formula

$$a_\ell^{(j+1)} = \sum_k p_{m,\ell-2k} a_k^{(j)}, \quad \ell \in \mathbb{Z}, \tag{4.3.8}$$

where $\mathbf{a}^j = \{a_k^{(j)}\}$ and $\mathbf{a}^{j+1} = \{a_\ell^{(j+1)}\}$ are the coefficient sequences of $f_j(x)$ and $f_{j+1}(x)$ in (4.3.7), respectively. Finally, from the sequence $\mathbf{a}^{j_1} = \{a_k^{(j_1)}\}$, we still have to compute the values of $f(k/2^{j_1})$, $k \in \mathbb{Z}$. This is possible by convolving the sequence $\mathbf{a}^{j_1}$ with the sequence $\{N_m(k)\}$, $k \in \mathbb{Z}$. Recall that the values of $N_m(k)$ can be computed by applying the algorithm in (4.2.15). Indeed, for any $k \in \mathbb{Z}$, we have

$$f_{j_0}\left(\frac{k}{2^{j_1}}\right) = \sum_\ell a_\ell^{(j_1)} N_m\left(2^{j_1}\frac{k}{2^{j_1}} - \ell\right) \tag{4.3.9}$$

$$= \sum_\ell a_\ell^{(j_1)} N_m(k - \ell) = \sum_\ell w_{m,k-\ell} a_\ell^{(j_1)},$$

where

$$w_{m,k} := N_m(k), \quad k \in \mathbb{Z}. \tag{4.3.10}$$

Note that both (4.3.8) and (4.3.9) are only *"moving average"* (MA) formulas, except that the sequence $\mathbf{a}^j$ in (4.3.8) needs *"upsampling"*. This means that a zero term must be inserted in between any two consecutive terms of the sequence $\mathbf{a}^j$. To be precise, let us set

$$\begin{cases} \tilde{\mathbf{a}}^j = \{\tilde{a}_\ell^{(j)}\}, \quad \ell \in \mathbb{Z}; \text{ with} \\ \tilde{a}_{2k}^{(j)} := a_k^{(j)}, \quad \text{and} \\ \tilde{a}_{2k+1}^{(j)} := 0, \quad k \in \mathbb{Z}. \end{cases} \tag{4.3.11}$$

Then the formula in (4.3.8) becomes an MA formula:

$$a_\ell^{(j+1)} = \sum_k p_{m,\ell-k} \tilde{a}_k^{(j)}, \quad \ell \in \mathbb{Z}. \tag{4.3.12}$$

We summarize the procedure derived above as follows.

**Algorithm 4.7** (Interpolatory graphical display algorithm).
    Let $f_{j_0}$ be a cardinal spline function with causal coefficient sequence

$$\mathbf{a}^{j_0} = \{a_\ell^{(j_0)}: \ell = \ell_0, \ell_0 + 1, \ldots\}$$

as in (4.3.5). Select any $j_1 \geq j_0$. Then for $j = j_0, \ldots, j_1 - 1$, compute
    ($1°$) $\tilde{\mathbf{a}}^j$ using (4.3.11), and
    ($2°$) $\mathbf{a}^{j+1}$ using ($1°$) and (4.3.12).

Finally, compute

(3°) $\{f_{j_0}(\frac{k}{2^{j_1}}): k \in \mathbb{Z}\}$ using (4.3.9) and (2°) for $j = j_1 - 1$.
(Skip (1°) and (2°) if $j_1 = j_0$).

This algorithm can be described by the following schematic diagram, where $\uparrow$ means upsampling by applying (4.3.11), $\searrow$ means MA with weight sequence $\{p_{m,k}\}$, and $\rightarrow$ means MA with weight sequence $\{w_{m,k}\}$. Since the weight sequences $\{p_{m,k}\}$ and $\{w_{m,k}\}$ are very simple symmetric finite sequences whose terms are integer multiples of $2^{-m+1}$ and $1/(m-1)!$, respectively, the implementation of this algorithm is indeed very simple.

$$
\begin{array}{ccccccccc}
\tilde{\mathbf{a}}^{j_0} & & \tilde{\mathbf{a}}^{j_0+1} & & & \tilde{\mathbf{a}}^{j_1-1} & & & \\
\uparrow & \searrow & \uparrow & \searrow \cdots & & \uparrow & \searrow & & \\
\mathbf{a}^{j_0} & & \mathbf{a}^{j_0+1} & & & \mathbf{a}^{j_1-1} & & \mathbf{a}^{j_1} & \rightarrow & \{f_{j_0}(k/2^{j_1})\}
\end{array}
$$

**Figure 4.3.1.** Interpolatory graphical display.

**Example 4.8.** To display graphically a cubic spline curve

$$f_0(x) = \sum_{\ell=0}^{\infty} a_\ell N_4(x - \ell) \tag{4.3.13}$$

at the resolution level of 1024 pixels per unit length without any error, we use $m = 4$, $j_0 = 0$, $a_\ell^{(0)} = a_\ell$, and $j_1 = 10$ (since $2^{10} = 1024$) in Algorithm 4.7. Furthermore, the weight sequences $\{p_{4,k}\}$ and $\{w_{4,k}\}$ can be easily computed by applying (4.3.3) and (4.2.15). The nonzero values are given by

$$\{p_{4,0}, \ldots, p_{4,4}\} = \left\{ \frac{1}{8}, \frac{4}{8}, \frac{6}{8}, \frac{4}{8}, \frac{1}{8} \right\} \tag{4.3.14}$$

and

$$\{w_{4,1}, w_{4,2}, w_{4,3}\} = \left\{ \frac{1}{6}, \frac{4}{6}, \frac{1}{6} \right\}. \tag{4.3.15}$$

(Observe that the sum of the sequence $\{p_{4,k}\}$ is 2 instead of 1, since the data sequence $\{a_\ell^0\}$ has to be upsampled. Of course, the sum of the other weight sequence $\{w_{4,k}\}$ is 1.) In applying (4.3.12) and (4.3.9) in Steps (2°) and (3°) in Algorithm 4.7, the common denominators 8 and 6 in (4.3.14) and (4.3.15), respectively, could be dropped in order to achieve integer operations. Of course the final output must then be divided by

$$8^{j_1-j_0} \times 6 = 8^{10} \times 6 = 3 \times 2^{31}.$$

Symmetry of the sequences in (4.3.14) and (4.3.15) should also be applied in implementation to save processing time.

### 4.4. B-net representations and computation of cardinal splines

The interpolatory graphical display algorithm described in the previous section can also be applied to determine all the polynomial pieces of any cardinal spline function exactly. It requires an additional operation such as matrix inversion to calculate the polynomials. In this section we introduce a more direct scheme to compute these analytical expressions. By using the Bernstein representation of a polynomial, the values of the Bernstein coefficients (or better known as "B-nets") do not change with the position and length of the interval to which the polynomial is confined. This is a very important feature in implementation with cardinal spline functions, since it is often necessary to shift a cardinal B-spline series and to scale the series to different resolution levels.

Let $n$ be any nonnegative integer. We first observe that the collection of polynomials

$$\phi_k^n(x) := \binom{n}{k}(1-x)^{n-k}x^k, \qquad 0 \le k \le n, \tag{4.4.1}$$

is a basis of the polynomial space $\pi_n$. This basis is used to define the $n^{\text{th}}$ degree Bernstein polynomial operator:

$$(B_n f)(x) := \sum_{k=0}^{n} f\left(\frac{k}{n}\right) \phi_k^n(x). \tag{4.4.2}$$

If $f$ is a continuous function on the interval [0,1], then it is clear that $B_n f$ interpolates $f$ at the end-points of this interval, namely:

$$\begin{cases} (B_n f)(0) = f(0); \\ (B_n f)(1) = f(1). \end{cases} \tag{4.4.3}$$

However, $B_n f$ does not interpolate $f$ at the interior points $\frac{1}{n}, \ldots, \frac{n-1}{n}$ of the interval in general. Instead, the graph of the polynomial curve $y = (B_n f)(x)$ lies in the convex hull of the set

$$\left\{ \left(\frac{k}{n}, f\left(\frac{k}{n}\right)\right) : k = 0, \ldots, n \right\}. \tag{4.4.4}$$

More precisely, this set "controls" the graph of $y = (B_n f)(x)$. We do not intend to go into any details in this direction, but only wish to point out that the "shape" of the graph of $y = (B_n f)(x)$ is governed by the "control net" (4.4.4), which is in turn governed by the graph of $y = f(x)$. In particular, we have:

(i) if $f \ge 0$ on [0,1], then $B_n f \ge 0$ on [0,1];
(ii) if $f \uparrow$ on [0,1], then $B_n f \uparrow$ on [0,1]; and
(iii) if $f$ is concave upward on [0,1], then $B_n f$ is concave upward on [0,1].

The two main reasons for (i)-(iii) to hold are:

(1°) $B_n f$ is a positive linear operator that preserves all linear polynomials in the sense that

$$B_n(f) = f, \quad f \in \pi_1; \text{ and}$$

(2°) the "Descartes Rule of Signs" applies to the monomial basis $\{1, x, \ldots, x^n\}$ on the interval $(0, \infty)$.

In general, the Bernstein polynomial operator in (4.4.2) can be replaced by any Bernstein polynomial

$$P_n(x) = \sum_{k=0}^{n} a_k^n \phi_k^n(x) \tag{4.4.5}$$

with coefficient sequence

$$\mathbf{a}^n := \{a_k^n : k = 0, \ldots, n\} \tag{4.4.6}$$

without losing any of the nice geometric properties of $B_n f$, simply by considering $f$ to be the piecewise linear (or second order cardinal spline) interpolant

$$f_{\mathbf{a}^n}(x) = \sum_{k=0}^{n} a_k^n N_2(nx - k + 1) \tag{4.4.7}$$

of the data in (4.4.6) at $\{k/n\}$. The graph of $y = f_{\mathbf{a}^n}(x)$ (or for simplicity, the coefficient sequence $\mathbf{a}^n = \{a_k^n\}$ itself) is called the *B-net representation* of the Bernstein polynomial $P_n$ in (4.4.5).

In the following, we will make use of the operations:

$$\begin{cases} \partial a_k^n := a_{k+1}^n - a_k^n, \quad \text{and} \\ \sigma a_k^n := \dfrac{1}{n+1} \displaystyle\sum_{j=0}^{k-1} a_j^n \end{cases} \tag{4.4.8}$$

to differentiate and integrate Bernstein polynomials. Here and throughout, an empty sum is always assumed to be zero.

**Theorem 4.9.** *For each $n \in \mathbb{Z}$, $n \geq 0$, let $P_n$ be an $n^{\text{th}}$ degree Bernstein polynomial with B-net $\mathbf{a}^n$ as defined in (4.4.5) and (4.4.6). Then the derivative of $P_n$ is given by*

$$P_n'(x) = \sum_{k=0}^{n-1} (n \partial a_k^n) \phi_k^{n-1}(x); \tag{4.4.9}$$

*and if $P_{n+1}'(x) = P_n(x)$, then the integral of $P_n$ is given by*

$$\int_0^x P_n(t) dt = \sum_{k=0}^{n+1} (a_0^{n+1} + \sigma a_k^n) \phi_k^{n+1}(x) - a_0^{n+1}. \tag{4.4.10}$$

**Proof.** By using the notation

$$\phi_{-1}^{n-1} := 0 \quad \text{and} \quad \phi_n^{n-1} := 0,$$

we have, from (4.4.1),

$$\frac{d}{dx}\phi_k^n(x) = \binom{n}{k}\{-(n-k)x + k(1-x)\}(1-x)^{n-k-1}x^{k-1}$$

$$= n\left\{-\binom{n-1}{k}x + \binom{n-1}{k-1}(1-x)\right\}(1-x)^{n-k-1}x^{k-1}$$

$$= n\{\phi_{k-1}^{n-1}(x) - \phi_k^{n-1}(x)\},$$

for $k = 0, \ldots, n$. Hence,

$$P_n'(x) = \sum_{k=0}^n na_k^n\{\phi_{k-1}^{n-1}(x) - \phi_k^{n-1}(x)\}$$

$$= \sum_{k=0}^n n(a_{k+1}^n - a_k^n)\phi_k^{n-1}(x) = \sum_{k=0}^{n-1}(n\partial a_k^n)\phi_k^{n-1}(x).$$

This verifies (4.4.9). By applying this formula to $P_{n+1}$ and using the hypothesis $P_{n+1}' = P_n$, we have

$$a_0^{n+1} + \frac{1}{n+1}\sum_{j=0}^{k-1}a_j^n = a_k^{n+1}.$$

Hence, the assertion in (4.4.10) follows from the integral formula

$$\int_0^x P_n(t)dt = P_{n+1}(x) - P_{n+1}(0) = P_{n+1}(x) - a_0^{n+1}, \tag{4.4.11}$$

completing the proof of the theorem. ∎

We now turn to the study of $B$-net representation of the $m^{\text{th}}$ order cardinal $B$-splines $N_m$. Recall that $N_m$ consists of $m$ nontrivial polynomial pieces of degree $m-1$, which we denote by

$$P_{m-1,k} := N_m|_{[k-1,k)}, \quad k = 1, \ldots, m. \tag{4.4.12}$$

Also, observe that the restriction of $N_m(x-1)$ to the same interval $[k-1,k)$ is the polynomial $P_{m-1,k-1}(x)$. So, if we let $\mathbf{a}^{m-1}(k) = \{a_\ell^{m-1}(k)\}$, $0 \le \ell \le m-1$, denote the $B$-net of $P_{m-1,k}$, then the restriction of $N_m(x) - N_m(x-1)$ to the interval $[k-1,k)$ is given by the Bernstein polynomial

$$P_{m-1,k}(x) - P_{m-1,k-1}(x) = \sum_{\ell=0}^{m-1}\{a_\ell^{m-1}(k) - a_\ell^{m-1}(k-1)\}\phi_\ell^{m-1}(x-k+1).$$

$$\tag{4.4.13}$$

Therefore, by applying the identity in (vii) of Theorem 4.3, or equivalently,

$$P'_{m,k}(x) = P_{m-1,k}(x) - P_{m-1,k-1}(x), \quad x \in [k-1, k), \tag{4.4.14}$$

we have

$$\sum_{\ell=0}^{m-1} \{a_\ell^{m-1}(k) - a_\ell^{m-1}(k-1)\} \phi_\ell^{m-1}(x-k+1) = \frac{d}{dx} \left\{ \sum_{\ell=0}^{m} a_\ell^m(k) \phi_\ell^m(x-k+1) \right\}.$$

Now, if we integrate both sides over the interval $[0, x]$ and apply (4.4.11) and (4.4.10), we obtain the following relation between the $B$-nets of $N_m$ and $N_{m+1}$, namely:

$$a_\ell^m(k) = a_0^m(k) + \sigma\{a_\ell^{m-1}(k) - a_\ell^{m-1}(k-1)\} \tag{4.4.15}$$

$$= a_0^m(k) + \frac{1}{m} \sum_{j=0}^{\ell-1} \{a_j^{m-1}(k) - a_j^{m-1}(k-1)\},$$

$\ell = 0, \ldots, m$. For $m \geq 2$, since $N_m$ is continuous, we even have $a_0^m(k) = P_{m,k}(k-1) = P_{m,k-1}(k-1) = a_m^m(k-1)$. Thus, we have derived the following scheme for computing the $B$-nets of all the polynomial pieces of $N_m$ for any integer $m \geq 2$.

**Algorithm 4.10.** (Cardinal $B$-spline $B$-net algorithm)
   Let $m \geq 2$ be any integer and set

$$P_{m-1,k}(x) = N_m|_{[k-1,k)}(x) = \sum_{\ell=0}^{m-1} a_\ell^{m-1}(k) \phi_\ell^{m-1}(x-k+1), \tag{4.4.16}$$

$k = 1, \ldots, m$. Also set

$$a_j^{m-1}(0) = 0 \quad \text{and} \quad a_j^{m-1}(m+1) = 0, \quad j = 0, \ldots, m-1, \tag{4.4.17}$$

and consider the initial conditions

$$a_0^1(1) = 0, \quad a_1^1(1) = 1 = a_0^1(2), \quad a_1^1(2) = 0. \tag{4.4.18}$$

Compute (1°) and (2°) below by using (4.4.17) and (4.4.18) for $m = 2$, and then repeat the same process by using (4.4.17) and the previous result for $m = 3, 4, \ldots$; where

(1°)                          $b_j^{m-1}(k) := a_j^{m-1}(k) - a_j^{m-1}(k-1),$

   for $j = 0, \ldots, m-1$ and $k = 1, \ldots, m+1$; and

$$(2°) \qquad a_\ell^m(k) = a_m^m(k-1) + \frac{1}{m} \sum_{j=0}^{\ell-1} b_j^{m-1}(k),$$

for $\ell = 0, \ldots, m$, and $k = 1, \ldots, m+1$.

The schematic diagram for Algorithm 4.10 is shown in Figure 4.4.1, where the condition (4.4.17) is used in each step to compute the $b_j^\ell(k)$ values.

$$
\begin{array}{ccccccc}
\mathbf{b}^1(1),\ldots,\mathbf{b}^1(3) & & \mathbf{b}^2(1),\ldots,\mathbf{b}^2(4) & & \mathbf{b}^{m-2}(1),\ldots,\mathbf{b}^{m-2}(m) & & \\
\uparrow & \searrow & \uparrow & \searrow \cdots & \uparrow & \searrow & \\
\mathbf{a}^1(1),\mathbf{a}^1(2) & & \mathbf{a}^2(1),\ldots,\mathbf{a}^2(3) & & \mathbf{a}^{m-2}(1),\ldots,\mathbf{a}^{m-2}(m-1) & & \mathbf{a}^{m-1}(1),\ldots,\mathbf{a}^{m-1}(m)
\end{array}
$$

**Figure 4.4.1.** Computation of $B$-nets for $N_m$.

**Example 4.11.** The $B$-nets for the quadratic ($m = 3$), cubic ($m = 4$), and quartic ($m = 5$) cardinal $B$-splines are given below.

(i) For $m = 3$, $\mathbf{a}^2(1), \ldots, \mathbf{a}^2(3)$

$$= \left\{0,0,\frac{1}{2}\right\}, \left\{\frac{1}{2},\frac{2}{2},\frac{1}{2}\right\}, \left\{\frac{1}{2},0,0\right\}$$

(ii) For $m = 4$, $\mathbf{a}^3(1), \ldots, \mathbf{a}^3(4)$

$$= \left\{0,0,0,\frac{1}{6}\right\}, \left\{\frac{1}{6},\frac{2}{6},\frac{4}{6},\frac{4}{6}\right\}, \left\{\frac{4}{6},\frac{4}{6},\frac{2}{6},\frac{1}{6}\right\}, \left\{\frac{1}{6},0,0,0\right\}$$

(iii) For $m = 5$, $\mathbf{a}^4(1), \ldots, \mathbf{a}^4(5)$

$$= \left\{0,0,0,0,\frac{1}{24}\right\}, \left\{\frac{1}{24},\frac{2}{24},\frac{4}{24},\frac{8}{24},\frac{11}{24}\right\}, \left\{\frac{11}{24},\frac{14}{24},\frac{16}{24},\frac{14}{24},\frac{11}{24}\right\},$$

$$\left\{\frac{11}{24},\frac{8}{24},\frac{4}{24},\frac{2}{24},\frac{1}{24}\right\}, \left\{\frac{1}{24},0,0,0,0\right\}.$$

Once the $B$-nets for $N_m$ are known, it is easy to determine the $B$-nets for any cardinal spline function $f_j$ as shown in (4.3.7). By a change of variable, we may restrict our attention to cardinal spline functions

$$f_0(x) = \sum_j c_j N_m(x-j) \qquad (4.4.19)$$

with knot sequence $\mathbb{Z}$. Let $k \in \mathbb{Z}$ and consider the restriction

$$f_0|_{[k-1,k)}(x) =: \sum_{\ell=0}^{m-1} d_\ell(k)\phi_\ell^{m-1}(x - k + 1) \qquad (4.4.20)$$

of $f_0$ on the interval $[k-1, k)$. By applying (4.4.16) and (4.4.19), we also have

$$f_0|_{[k-1,k)}(x) = \sum_j c_j N_m(\cdot - j)|_{[k-1,k)}(x) \qquad (4.4.21)$$

$$= \sum_j c_j P_{m-1,k-j}|_{[k-1,k)}(x)$$

$$= \sum_j c_j \sum_\ell a_\ell^{m-1}(k - j)\phi_\ell^{m-1}(x - k + 1).$$

Consequently, equating (4.4.20) and (4.4.21) yields

$$d_\ell(k) = \sum_{j=k-m}^{k} a_\ell^{m-1}(k - j)c_j, \quad \ell = 0, \ldots, m - 1 \quad \text{and} \quad k \in \mathbb{Z}, \qquad (4.4.22)$$

since $a_\ell^{m-1}(k - j) = 0$ for $j > k$ or $j < k - m$. Let us summarize the preceding derivation by stating that, for each fixed $k \in \mathbb{Z}$ and $\ell = 0, \ldots, m - 1$, the moving average (MA) formula in (4.4.22), with weight sequence

$$\{a_\ell^{m-1}(j)\colon\ j = 0, \ldots, m\}$$

(whose terms can be computed by applying Algorithm 4.10, and are given in Example 4.11 for $m = 3, 4, 5$), can be used to compute the $B$-net

$$\mathbf{d}(k) = \{d_0(k), \ldots, d_{m-1}(k)\}$$

of the restriction, on the interval $[k-1, k)$, of the cardinal spline series $f_0$ with coefficient sequence $\{c_j\}$.

## 4.5. Construction of spline approximation formulas

We begin by writing down a useful formula on Fourier transforms which can be easily verified by taking the $j^{\text{th}}$ order derivatives of both sides of (2.1.6), namely:

$$(\mathcal{F}[t^j f(t)])(\omega) = i^j D^j \hat{f}(\omega), \quad j = 0, 1, \ldots, \qquad (4.5.1)$$

where the notation

$$D^j g(x) := g^{(j)}(x) \qquad (4.5.2)$$

is used. Applying this formula to a monomial multiple of the $m^{\text{th}}$ order cardinal $B$-spline $N_m$ and a shift of its reflection $N_m^-$ by some $x \in \mathbb{R}$ yields

$$\int_{-\infty}^{\infty} e^{-i\omega t}[(x - t)^j N_m(t)]dt = (x - iD)^j \widehat{N_m^-}(\omega); \qquad (4.5.3)$$

and

$$\int_{-\infty}^{\infty} e^{-i\omega t}[t^j N_m(x-t)]dt \tag{4.5.4}$$

$$= e^{-ix\omega} \int_{-\infty}^{\infty} e^{i\omega t}[(x-t)^j N_m(t)]dt$$

$$= e^{-ix\omega}((x-iD)^j \widehat{N}_m)(-\omega).$$

On the other hand, from the formula

$$\widehat{N}_m(\omega) = \left(\frac{1-e^{-i\omega}}{i\omega}\right)^m,$$

it is clear that $\widehat{N}_m$ satisfies the property

$$\begin{cases} \widehat{N}_m(0) = 1; \\ (D^j \widehat{N}_m)(2\pi k) = 0, \quad j = 0, \ldots, m-1; \quad 0 \neq k \in \mathbb{Z}. \end{cases} \tag{4.5.5}$$

Hence, since (4.5.3) and (4.5.4) certainly agree at $\omega = 0$, an application of (4.5.5) to these two formulas gives:

$$\int_{-\infty}^{\infty} e^{-i2\pi kt}[(x-t)^j N_m(t)]dt \tag{4.5.6}$$

$$= \int_{-\infty}^{\infty} e^{-i2\pi kt}[t^j N_m(x-t)]dt, \quad 0 \leq j \leq m-1,\ k \in \mathbb{Z},\ \text{and } x \in \mathbb{R}.$$

As a consequence of (4.5.6), the following result can be easily derived by applying a version of the Poisson Summation Formula in (2.5.11).

**Theorem 4.12.** *Let $m \geq 1$ be any integer. Then*

$$\sum_{k=-\infty}^{\infty} p(k)N_m(x-k) = \sum_{k=0}^{m-1} N_m(k)p(x-k), \quad p \in \pi_{m-1}. \tag{4.5.7}$$

The identity in (4.5.7) says that if the coefficient sequence $\{c_k\}$ of a cardinal $B$-spline series of order $m$ is a "*polynomial sequence*" of degree $m-1$, in the sense that $c_k = p(k)$ for some $p \in \pi_{m-1}$, then the cardinal spline function reduces to a polynomial in $\pi_{m-1}$. We remark that the lower limit on the right-hand side of (4.5.7) can be changed to 1 for $m \geq 2$.

Let us now digress from the above discussion for the time being and consider the following problem of "cardinal spline interpolation" using the "centered" cardinal $B$-spline $N_m\left(x + \frac{m}{2}\right)$, namely: For any given "admissible" data sequence $\{f_j\}$ determine the solution $\{c_k\}$ in

$$\sum_{k=-\infty}^{\infty} c_k N_m\left(x + \frac{m}{2} - k\right)\bigg|_{x=j} = f_j, \quad j \in \mathbb{Z}. \tag{4.5.8}$$

Here, $\{f_j\}$ is said to be admissible if it has at most polynomial growth. Using the "*symbol*" notation:

$$\begin{cases} \widetilde{N}_m(z) := \sum_k N_m\left(k + \dfrac{m}{2}\right) z^k; \\[2mm] \widetilde{C}(z) := \sum_k c_k z^k; \\[2mm] \widetilde{F}(z) := \sum_k f_k z^k, \end{cases} \tag{4.5.9}$$

we may write (4.5.8), at least formally, as

$$\widetilde{C}\widetilde{N}_m = \widetilde{F}. \tag{4.5.10}$$

Observe that $\widetilde{N}_m$ is a symmetric Laurent polynomial. Also, note that as a consequence of Theorem 4.3, (vi), the cosine polynomial

$$D(\omega) := \widetilde{D}(z) := 1 - \widetilde{N}_m(z), \quad z = e^{-i\omega}, \tag{4.5.11}$$

is nonnegative for all $\omega$. The introduction of $\widetilde{D}$ allows us to rewrite (4.5.10) as

$$\widetilde{C} = \frac{1}{1 - \widetilde{D}}\widetilde{F}. \tag{4.5.12}$$

So, at least formally, we have

$$\widetilde{C} = (1 + \widetilde{D} + \widetilde{D}^2 + \cdots)\widetilde{F}. \tag{4.5.13}$$

(We remark here without proof that since we do have $0 \leq D(\omega) < 1$, the "Neumann series" in (4.5.13) actually converges. More details will be discussed in Section 4.6.) In any event, the formal expression in (4.5.13) motivates the consideration of the finite sequences $\Lambda_k = \{\lambda_j^{(k)}\}$, defined by:

$$\widetilde{\Lambda}_k(z) := \sum_j \lambda_j^{(k)} z^j := 1 + \widetilde{D} + \cdots + \widetilde{D}^k. \tag{4.5.14}$$

Each of these sequences, in turn, defines a convolution operator on the data sequence $\{f_j\}$, namely:

$$(\Lambda_k * \{f_j\})(\ell) := \sum_j \lambda_{\ell-j}^{(k)} f_j, \tag{4.5.15}$$

whose symbol is given by $\widetilde{\Lambda}_k \widetilde{F}$.

Now suppose that the data sequence $\{f_j\}$ is obtained from the measurement of some continuous function $f$; that is, $f_j = f(j)$. Then to simplify the notation in (4.5.15), it is more convenient to write

$$(\Lambda_k f)(\ell) := (\Lambda_k * \{f(j)\})(\ell) = \sum_j \lambda_{\ell-j}^{(k)} f(j). \qquad (4.5.16)$$

This sequence defines a linear spline operator

$$(Q_k f)(x) := \sum_{\ell=-\infty}^{\infty} (\Lambda_k f)(\ell) N_m \left( x + \frac{m}{2} - \ell \right), \quad f \in C, \qquad (4.5.17)$$

that maps $C = C(\mathbb{R})$ to the cardinal spline space $S_m$. We must emphasize that since $\Lambda_k = \{\lambda_j^{(k)}\}$ is a finite sequence, each $(\Lambda_k f)(\ell)$ depends only on the values of $f(j)$ in a neighborhood of $j = \ell$ and this neighborhood is independent of $\ell$. In other words, $Q_k$ is a "*bounded linear local spline operator*" defined on $C$. The importance of $Q_k$ is that it preserves all polynomials in $\pi_{m-1}$ for any sufficiently large $k$. Precisely, we have the following result.

**Theorem 4.13.** *Let $m \geq 1$ be any integer. Then for each $k > \frac{m-3}{2}$, the linear operator $Q_k$ defined in (4.5.17) satisfies*

$$(Q_k p)(x) = p(x), \quad p \in \pi_{m-1}. \qquad (4.5.18)$$

**Proof.** Let $p \in \pi_{m-1}$. Then since $\{\lambda_j^k\}$ is a finite sequence, $\Lambda_k p$ is a polynomial sequence of degree $m - 1$. More precisely, $(\Lambda_k p)(\ell) = q(\ell)$ where $q$ is the polynomial

$$q(x) = \sum_j \lambda_j^{(k)} p(x - j).$$

Furthermore, for each $j \in \mathbb{Z}$, we have

$$(Q_k p)(j) = \sum_{\ell} (\Lambda_k p)(\ell) N_m \left( j - \ell + \frac{m}{2} \right); \qquad (4.5.19)$$

and using the symbol notation, it follows from (4.5.14) and (4.5.11) that

$$(Q_k p)^\sim = (\Lambda_k p)^\sim \widetilde{N}_m \qquad (4.5.20)$$
$$= (1 + \widetilde{D} + \cdots + \widetilde{D}^k) \widetilde{P} \widetilde{N}_m$$
$$= (1 + \widetilde{D} + \cdots + \widetilde{D}^k) \widetilde{N}_m \widetilde{P}$$
$$= (1 + \widetilde{D} + \cdots + \widetilde{D}^k)(1 - \widetilde{D}) \widetilde{P}$$
$$= (1 - \widetilde{D}^{k+1}) \widetilde{P},$$

where $\widetilde{P}$ is the symbol of $\{p(j)\}$. Now, recall from (4.5.11) and (4.5.9) that $\widetilde{D}$ is the symbol of the sequence $\{d_j\}$, where

$$d_j = \delta_{j,0} - N_m \left( \frac{m}{2} - j \right), \quad j \in \mathbb{Z}, \qquad (4.5.21)$$

so that $\widetilde{D}^{k+1}$ is the symbol of the $(k+1)$-fold convolution of $\{d_j\}$. Since $N_m\left(\frac{m}{2}+\ell\right) = N_m\left(\frac{m}{2}-\ell\right)$ for all $\ell \in \mathbb{Z}$, we have

$$(\{d_j\} * \{p(j)\})(\ell) = \sum_j d_{\ell-j}p(j) \tag{4.5.22}$$

$$= \left(1 - N_m\left(\frac{m}{2}\right)\right)p(\ell) - N_m\left(\frac{m}{2}-1\right)(p(\ell+1)+p(\ell-1))$$

$$- \cdots - N_m\left(\frac{m}{2}-\left[\frac{m}{2}\right]\right)\left(p\left(\ell+\left[\frac{m}{2}\right]\right)+p\left(\ell-\left[\frac{m}{2}\right]\right)\right)$$

$$= -\sum_{j=1}^{[m/2]} N_m\left(\frac{m}{2}-j\right)[p(\ell+j)-2p(\ell)+p(\ell-j)],$$

where $[x]$ denotes the greatest integer not exceeding $x$, and the property $\sum N_m\left(\frac{m}{2}-j\right) = 1$ has been used. The importance of the formulation in (4.5.22) is that the convolution of $\{d_j\}$ with $\{p(j)\}$ is written as a (finite) linear combination of second central differences of $p(j)$. Hence, $\widetilde{D}^{k+1}\widetilde{P}$ is the symbol of a (finite) linear combination of $2(k+1)^{\text{st}}$ order differences of $\{p(j)\}$, so that for any $p \in \pi_{m-1}$, we have

$$\widetilde{D}^{k+1}\widetilde{P} = 0, \quad k > \frac{m-3}{2}. \tag{4.5.23}$$

Putting (4.5.23) into (4.5.20), we obtain

$$(Q_k p)(\ell) = p(\ell), \quad \ell \in \mathbb{Z}, \quad k > \frac{m-3}{2}, \quad p \in \pi_{m-1}. \tag{4.5.24}$$

Consequently, since $\Lambda_k p$ in (4.5.19) is a polynomial sequence of degree $m-1$, it follows from Theorem 4.12 that $(Q_k p)(x)$ is a polynomial in $\pi_{m-1}$. Hence, (4.5.18) immediately follows by applying (4.5.24).  ∎

**Example 4.14.** For $m = 4$, we may choose $k = 1$. Then the linear local cubic spline operator $Q_1$ is given by

$$(Q_1 f)(x) = \sum_{\ell=-\infty}^{\infty} \frac{1}{6}(-f(\ell+1)+8f(\ell)-f(\ell-1))N_4(x+2-\ell). \tag{4.5.25}$$

By Theorem 4.13, we have

$$(Q_1 p)(x) = p(x), \quad p(x) = 1, x, x^2, x^3.$$

**Proof.** From (4.5.14) and (4.5.11), it follows that $\tilde{\Lambda}_1 = 2 - \widetilde{N}_4$, so that

$$\lambda_j^{(1)} = 2\delta_{j,0} - N_4(2+j).$$

Since $N_4(2) = \frac{2}{3}$, $N_4(1) = N_4(3) = \frac{1}{6}$, and $N_4(\ell) = 0$ for all other integers $\ell$, we have, from (4.5.16),

$$(\Lambda_1 f)(\ell) = \sum_j \lambda^{(1)}_{\ell-j} f(j) = -\frac{1}{6} f(\ell+1) + \left(2 - \frac{2}{3}\right) f(\ell) - \frac{1}{6} f(\ell-1).$$

This gives (4.5.25). ∎

The operators $Q_k$ are called "quasi-interpolation" operators. More generally, we will adopt the following definition. Throughout the following discussion, we will consider the space

$$C_b(\mathbb{R}) = \{f \in C(\mathbb{R}) : \sup_{-\infty < x < \infty} |f(x)| < \infty\}. \tag{4.5.26}$$

**Definition 4.15.** *A bounded linear operator $Q$ that maps $C_b(\mathbb{R})$ into the cardinal spline space $S_m$ is called a quasi-interpolation operator, if on one hand, it preserves all of $\pi_{m-1}$, in the sense that $(Qp)(x) = p(x)$ for all $p \in \pi_{m-1}$, and on the other hand, it is local, in the sense that a compact set $J$ exists such that for any $f \in C$ and $x \in \mathbb{R}$, $(Qf)(x)$ depends only on $f(y)$ for $y$ in*

$$J + x := \{y + x : y \in J\}. \tag{4.5.27}$$

**Remark.** In the previous definition, when $Q$ is defined to be a bounded linear operator on $C_b(\mathbb{R})$, it is implicit that only function-value data are considered. To handle data consisting of certain derivative values, we must consider bounded linear operators $Q$ on a corresponding sub-class of smooth functions, such as some "Sobolev space".

In Theorem 4.13, we have formulated a sequence of quasi-interpolation operators $Q_k$ whose local supports $J$ increase in size as $k$ increases. We point out, without going into any details here, that $\{Q_k\}$ actually converges to the cardinal spline interpolation operator $Q_\infty$, which is uniquely determined by the interpolation property:

$$(Q_\infty f - f)(\ell) = 0, \quad \ell \in \mathbb{Z}, \tag{4.5.28}$$

for any $f \in C$ (see the next section for a discussion of cardinal spline interpolation). The reason for studying quasi-interpolation operators is that they provide simple and computationally efficient schemes for constructing cardinal spline approximants that provide the highest order of approximation, namely: the order $m$ is achieved, when $m^{\text{th}}$ order cardinal splines are used. In addition, the local structure of a quasi-interpolation operator makes it possible for being adopted for real-time (or on-line) applications.

To see how a quasi-interpolation operator is used to give optimal-order approximation, we simply scale the $B$-splines from the "zeroth resolution level" (that is, $x = k, k \in \mathbb{Z}$) to the "$j^{\text{th}}$ resolution level" (that is, $x = k/2^j$,

$k \in \mathbb{Z}$). In the $L^2(\mathbb{R})$ setting, this means that we process the discrete data in a high-resolution spline space $V_j^m$ where the sampling period can be as small as $h = 2^{-j}$. In general, we may consider any small scaling parameter $h > 0$, and measure the order of approximation in terms of the power $m$ of $h$, viz., $O(h^m)$, which means that the approximation error is bounded by a constant multiple of $h^m$ as $h \to 0^+$, when the data samples represent a sufficiently smooth function. Let us use the notation

$$(s_h f)(x) := f\left(\frac{1}{h}x\right) \tag{4.5.29}$$

to describe the scaling process. Then any quasi-interpolation operator $Q$, as described in Definition 4.15, gives rise to an "*approximation formula*", using the composition of scalings and quasi-interpolation, namely:

$$Q^h := s_h \circ Q \circ s_{h^{-1}}. \tag{4.5.30}$$

**Theorem 4.16.** *Suppose that $Q$ is a quasi-interpolation operator from $C_b(\mathbb{R})$ to $S_m$, $K$ any compact set in $\mathbb{R}$, and $\Omega$ any open set containing $K$. Then for each $f \in C_b(\mathbb{R}) \cap C^m(\Omega)$, there exists some positive constant $C$, depending only on $f$ and $K$, such that*

$$\max_{x \in K} |(Q^h f - f)(x)| \leq C h^m, \tag{4.5.31}$$

*for all sufficiently small $h > 0$.*

**Proof.** From the definition of $Q$, it is easy to see that $Q^h p = p$ for all $p \in \pi_{m-1}$ and $\|Q^h\| = \|Q\|$, $h > 0$. Let

$$\rho := m + \max_{x \in J} |x|,$$

and suppose that

$$\max_{x \in K} |(Q^h f - f)(x)| = |(Q^h f - f)(x_0)|, \quad x_0 \in K,$$

so that

$$\max_{x \in K} |(Q^h f - f)(x)| = \max_{|x - x_0| \leq h\rho} |(Q^h f - f)(x)|.$$

Then by the local nature of $Q$ and $N_m$, we have

$$\max_{x \in K} |(Q^h f - f)(x)| = \max_{|x - x_0| \leq h\rho} |(Q^h(f - p) + (p - f))(x)|$$

$$\leq \max_{|x - x_0| \leq h\rho} |f(x) - p(x)|(1 + \|Q\|),$$

where $p$ is any polynomial in $\pi_{m-1}$. Noting that $f \in C^m(\Omega)$ and $\Omega$ is an open set containing $x_0 \in K$, and selecting the $(m-1)^{\text{st}}$ degree Taylor polynomial of $f$ at $x_0$ to be the polynomial $p$ in the above estimate, we obtain (4.5.31) with

$$C = \frac{\rho^m}{m!} \sup_{x \in \Omega^0} |f^{(m)}(x)|,$$

where $\Omega^0$ is any open set containing $K$ such that the closure of $\Omega^0$ lies in $\Omega$. ∎

In order to give a more general scheme for constructing quasi-interpolation formulas, let us return to Theorem 4.13, where $Q_k$, as defined in (4.5.17), is determined by the convolution of the finite sequence $\Lambda_k = \{\lambda_j^k\}$ with the data sequence $\{f_j\}$. Let $k$ be fixed and consider the bounded linear functional $\lambda^*$ defined on $C$ by

$$\lambda^* f := (\Lambda_k f)(0) = \sum_j \lambda_j^{(k)} f(-j).$$

For any fixed $\ell \in \mathbb{Z}$, if $f(\cdot + \ell)$ is considered as a function of the variable represented by the dot, it is clear that

$$\lambda^* f(\cdot + \ell) = (\Lambda_k f)(\ell) = \sum_j \lambda_j^{(k)} f(\ell - j).$$

That is, the quasi-interpolation formula (4.5.17) is determined by a single linear functional $\lambda^*$ on $C$, namely:

$$(Q_k f)(x) = \sum_\ell \lambda^* f(\cdot + \ell) N_m \left( x + \frac{m}{2} - \ell \right). \tag{4.5.32}$$

To generalize this formulation, let us examine if we can use a family of bounded linear functionals $\lambda^\ell, \ell \in \mathbb{Z}$, on $C$ to replace $\lambda^*$; for instance, we try to replace $\lambda^* f(\cdot + \ell)$ by $\lambda^\ell f(\cdot + \ell)$ in (4.5.32). That is, we are interested in studying the conditions on the family $\{\lambda^\ell\}$, such that the linear spline operator $Q^0$, defined by

$$(Q^0 f)(x) = \sum_\ell \lambda^\ell f(\cdot + \ell) N_m \left( x + \frac{m}{2} - \ell \right) \tag{4.5.33}$$

is a quasi-interpolation operator. For $Q^0$ to be a bounded operator, it is sufficient to assume that the norms $\|\lambda^\ell\|$ of these linear functionals satisfy

$$\sup_j \sum_{\ell=j}^{m+j} \|\lambda^\ell\| < \infty. \tag{4.5.34}$$

For instance, if $\{\lambda^\ell\}$ is a finite family, then (4.5.34) is certainly satisfied. On the other hand, it is more difficult to demonstrate the local nature for a general $Q^0$. Since only translation-invariant bounded linear functionals are of interest in many applications, we will consider those $\lambda^\ell$ of the form

$$\lambda^\ell f(\cdot + \ell) = \sum_j c_j^{(\ell)} f(\ell - j), \qquad (4.5.35)$$

where $\{c_j^{(\ell)}\}$, $\ell \in \mathbb{Z}$, are finite sequences. Hence, if the union of the supports of these sequences is a finite set, then $Q^0$ also has the local property. Again, this property is certainly possessed by any finite family $\{\lambda^\ell\}$. Finally, how is polynomial reproduction achieved by $Q^0$? A very simple assumption on the $\lambda^\ell$'s to yield this property is to require each $\lambda^\ell$ to satisfy

$$\lambda^\ell p = \lambda^* p, \quad p \in \pi_{m-1}. \qquad (4.5.36)$$

We remark that (4.5.35) may be replaced by a more general formula that involves derivative data information, if we are willing to consider bounded linear operators $Q^0$ on the corresponding subspace of differentiable functions.

We end this section with an example to demonstrate the effectiveness of the freedom achieved by allowing more than one $\lambda^\ell$.

**Example 4.17.** Observe that any of the quasi-interpolation operators $Q_k$ in (4.5.17) requires the data information of $f(\ell)$, for all $\ell \in \mathbb{Z}$. We have already formulated the cubic spline operator $Q_1$ in Example 4.14, where

$$\lambda^* f = -\frac{1}{6} f(1) + \frac{8}{6} f(0) - \frac{1}{6} f(-1).$$

Let us derive a cubic spline quasi-interpolation operator $Q^0$ that only requires data information $f(2\ell)$, $\ell \in \mathbb{Z}$.

**Solution.** We consider two bounded linear functionals $\lambda_e$ and $\lambda_o$ defined by

$$\begin{cases} \lambda_e f := \sum_j c_j^{(1)} f(2j); \\[2mm] \lambda_o f := \sum_j c_j^{(2)} f(2j + 1). \end{cases} \qquad (4.5.37)$$

Then by setting:

$$\lambda^{2\ell} := \lambda_e \quad \text{and} \quad \lambda^{2\ell-1} := \lambda_o, \quad \ell \in \mathbb{Z},$$

we have

$$\begin{cases} \lambda^{2\ell} f(\cdot + 2\ell) = \sum_j c_j^{(1)} f(2j + 2\ell); \\[2mm] \lambda^{2\ell-1} f(\cdot + 2\ell - 1) = \sum_j c_j^{(2)} f(2j + 2\ell). \end{cases} \qquad (4.5.38)$$

Hence, the only required data on $f$ in the quasi-interpolation formula

$$(Q^0 f)(x) = \sum_k \lambda^k f(\cdot + k) N_4(x + 2 - k) \tag{4.5.39}$$

$$= \sum_\ell \left[ \sum_j c_j^{(1)} f(2j + 2\ell) \right] N_4(x - 2\ell + 2)$$

$$+ \sum_\ell \left[ \sum_j c_j^{(2)} f(2j + 2\ell) \right] N_4(x - 2\ell + 3)$$

are $f(2\ell)$, $\ell \in \mathbb{Z}$. To determine $\{c_j^{(1)}\}$ and $\{c_j^{(2)}\}$, we apply (4.5.36). First, we must calculate $\lambda^* p_n$ where $p_n(x) = x^n$, $n = 0, \ldots, 3$. These values are

$$(\lambda^* p_0, \ldots, \lambda^* p_3) = \left( 1, 0, -\frac{1}{3}, 0 \right).$$

Hence, to satisfy (4.5.36), we have to solve two sets of linear equations:

$$\begin{cases} \sum_j c_j^{(1)} = 1, & \sum_j 2j c_j^{(1)} = 0, \\ \sum_j (2j)^2 c_j^{(1)} = -\frac{1}{3}, & \sum_j (2j)^3 c_j^{(1)} = 0; \end{cases} \tag{4.5.40}$$

and

$$\begin{cases} \sum_j c_j^{(2)} = 1, & \sum_j (2j + 1) c_j^{(2)} = 0 \\ \sum_j (2j + 1)^2 c_j^{(2)} = -\frac{1}{3}, & \sum_j (2j + 1)^3 c_j^{(2)} = 0. \end{cases} \tag{4.5.41}$$

Of course there are no unique solutions, but solutions of (4.5.40) and (4.5.41) with smallest supports are given by:

$$c_0^{(1)} = \frac{26}{24}; \quad c_{-1}^{(1)} = c_1^{(1)} = -\frac{1}{24}; \quad c_\ell^{(1)} = 0, \text{ for } \ell \neq -1, 0, 1, \tag{4.5.42}$$

and

$$c_0^{(2)} = c_{-1}^{(2)} = \frac{7}{12}; \quad c_1^{(2)} = c_{-2}^{(2)} = -\frac{1}{12}; \quad c_\ell^{(2)} = 0, \text{ for } \ell \neq -2, -1, 0, 1. \tag{4.5.43}$$

These are the coefficients in (4.5.39). ∎

## 4.6. Construction of spline interpolation formulas

The general scheme for constructing approximation formulas introduced in the previous section does not produce spline functions that interpolate the given discrete data in general. To construct a spline interpolation operator, it is also very important to require the operator to reproduce polynomials at least up to some desirable degree. This requirement not only helps in achieving a tolerable order of approximation, but is also critical in preserving certain shapes of the given data. After all, to interpolate a set of constant data, one expects to use a (horizontal) straight line.

We will first give a brief discussion of the cardinal spline interpolation problem introduced in (4.5.8) and point out that although the highest order of approximation is achieved in this case, the corresponding spline interpolation operator cannot be local. This limits its applications to many engineering problems, such as real-time (or on-line) data interpolation. The main objective of this section is to introduce a constructive scheme that produces quasi-interpolation operators which have the additional interpolation property.

Central to our discussion is the goal of constructing so-called *"fundamental splines"* that interpolate the data $\{\delta_{j,0}\}$. With a fundamental spline on hand, an interpolation operator may be readily obtained by using any given data sequence as the coefficient sequence of the spline series formed by integer translates of the fundamental spline.

Let us first investigate the cardinal spline interpolation problem stated in (4.5.8) with data sequence $\{\delta_{j,0}\}$. By solving the bi-infinite system

$$\sum_{k=-\infty}^{\infty} c_k^{(m)} N_m \left(\frac{m}{2} + j - k\right) = \delta_{j,0}, \quad j \in \mathbb{Z}, \tag{4.6.1}$$

for $\{c_k^{(m)}\}$, we have an $m^{\text{th}}$ order "fundamental cardinal spline function"

$$L_m(x) = \sum_{k=-\infty}^{\infty} c_k^{(m)} N_m \left(x + \frac{m}{2} - k\right), \tag{4.6.2}$$

that has the interpolation property

$$L_m(j) = \delta_{j,0} \tag{4.6.3}$$

as given by (4.6.1). In contrast to the cardinal $B$-spline $N_m$ which has compact support, we will see that the coefficient sequence $\{c_k^{(m)}\}$ is not finite for each $m \geq 3$, so that the fundamental cardinal spline $L_m$ does not vanish identically outside any compact set. Hence, when it is applied to interpolate a given data sequence $\{f_j\}$, where $f_j = f(j)$ for some $f \in C$, say, one has to be careful about the convergence of the infinite spline series

$$(J_m f)(x) := \sum_{k=-\infty}^{\infty} f(k) L_m(x - k). \tag{4.6.4}$$

Fortunately, as we will see in a moment, $\{c_k^m\}$ decays to zero exponentially fast as $k \to \pm\infty$. This implies that the fundamental cardinal spline function $L_m(x)$ also decays to zero at the same rate as $x \to \pm\infty$. Thus, if $\{f(k)\}$ is of at most polynomial growth, then the spline series in (4.6.4) certainly converges at every $x \in \mathbb{R}$; and in view of the interpolation property (4.6.3), we have

$$(J_m f - f)(j) = 0, \quad j \in \mathbb{Z}. \tag{4.6.5}$$

That is, the interpolation spline operator $J_m$ gives a spline function $J_m f$ that interpolates the given data function $f$ at every $x = j$, $j \in \mathbb{Z}$.

To study the fundamental cardinal spline functions $L_m(x)$, we must return to the system (4.6.1) of linear equations whose coefficients are given by the $B$-spline values $N_m\left(\frac{m}{2} + k\right)$. As in (4.5.9), we consider the symbol

$$\widetilde{N}_m(z) = \sum_k N_m \left( \frac{m}{2} + k \right) z^k,$$

and note that this symmetric Laurent polynomial can be easily transformed into an algebraic polynomial with integer coefficients by considering

$$E_{m-1}(z) := (m-1)! z^{[(m-1)/2]} \widetilde{N}_m(z), \tag{4.6.6}$$

where, as before, $[x]$ denotes the largest integer not exceeding $x$. This notion generalizes the definition of Euler-Frobenius polynomials from even-order cardinal $B$-splines to those of arbitrary orders. (See (4.2.18) and the next two chapters for more details and generality.) The most important property of the Euler-Frobenius polynomial $E_{m-1}$ in (4.6.6) for our purpose here is that it does not vanish on the unit circle $|z| = 1$ (see Theorem 5.10 in the next chapter for a more general result). Hence, it follows that $\widetilde{N}_m(z) \neq 0$ for all $z = e^{-i\omega}$, $\omega \in \mathbb{R}$. Now, as in (4.5.10), the system of linear equations (4.6.1) can be written as

$$\widetilde{C}_m(z) = \frac{1}{\widetilde{N}_m(z)}, \tag{4.6.7}$$

where $\widetilde{C}_m(z)$ is the symbol of $\{c_k^{(m)}\}$. By using partial fractions, it is easy to see that the sequence $\{c_k^{(m)}\}$ has exponential decay as $k \to \pm\infty$, and the decay rate is given by the magnitude of the root of $E_{m-1}$ in $|z| < 1$ which is closest to the unit circle $|z| = 1$.

By applying the Poisson Summation Formula (2.5.8), we may also write

$$\widetilde{N}_m(z) = \sum_{k=-\infty}^{\infty} \left( \widehat{N}_m \left( \cdot + \frac{m}{2} \right) \right)(\omega + 2\pi k), \quad z = e^{-i\omega}, \tag{4.6.8}$$

where $\widehat{N}_m\left( \cdot + \frac{m}{2} \right)$ denotes the Fourier transform of $N_m\left(x + \frac{m}{2}\right)$, and this Fourier transform is evaluated at $\omega + 2\pi k$ in (4.6.8). Hence, by taking the Fourier transform of both sides of (4.6.2) and applying (4.6.7) and (4.6.8), we have two formulas of (the Fourier transform of) the fundamental cardinal spline function, namely:

$$\hat{L}_m(\omega) = \frac{\left(\widehat{N_m\left(\cdot + \frac{m}{2}\right)}\right)(\omega)}{\widetilde{N}_m(e^{-i\omega})} \tag{4.6.9}$$

$$= \frac{\left(\widehat{N_m\left(\cdot + \frac{m}{2}\right)}\right)(\omega)}{\sum\limits_{k=-\infty}^{\infty} \left(\widehat{N_m\left(\cdot + \frac{m}{2}\right)}\right)(\omega + 2\pi k)}.$$

Each of these two formulas can be used for computing $L_m(x)$.

**Example 4.18.** Determine the cubic fundamental cardinal spline $L_4(x)$.

**Solution.** By applying the recursive algorithm described in (4.2.15), the nonzero values of $N_4(k)$, $k \in \mathbb{Z}$, are found to be

$$\{N_4(1), N_4(2), N_4(3)\} = \left\{\frac{1}{6}, \frac{4}{6}, \frac{1}{6}\right\},$$

as already observed in (4.3.15). Hence, the corresponding Euler-Frobenius polynomial is given by

$$E_3(z) = 1 + 4z + z^2 = (z + 2 - \sqrt{3})(z + 2 + \sqrt{3}).$$

Consequently, we have

$$C_4(z) = \frac{(4-1)! z^{\lfloor (4-1)/2 \rfloor}}{E_3(z)} = \frac{6z}{(z + 2 - \sqrt{3})(z + 2 + \sqrt{3})} \tag{4.6.10}$$

$$= \frac{6}{(-2 + \sqrt{3}) - (-2 - \sqrt{3})} \left( \frac{-2 + \sqrt{3}}{z + 2 - \sqrt{3}} - \frac{-2 - \sqrt{3}}{z + 2 + \sqrt{3}} \right)$$

$$= \sqrt{3} \left( \sum_{n=0}^{\infty} (-2 + \sqrt{3})^{n+1} z^{-n-1} + \sum_{n=0}^{\infty} (-2 - \sqrt{3})^{-n} z^n \right)$$

$$= \sqrt{3} \sum_{n=-\infty}^{\infty} (-2 + \sqrt{3})^{|n|} z^n,$$

so that the sequence $\{c_k^{(4)}\}$ is given by

$$c_k^{(4)} = (-1)^k \sqrt{3}(2 - \sqrt{3})^{|k|}, \quad k \in \mathbb{Z}. \tag{4.6.11}$$

This yields the cubic fundamental cardinal spline

$$L_4(x) = \sum_{k=-\infty}^{\infty} (-1)^k \sqrt{3}(2 - \sqrt{3})^{|k|} N_4(x + 2 - k). \tag{4.6.12}$$

Observe that the rate of decay of $L_4(x)$ is

$$O((2 - \sqrt{3})^{|x|}), \quad \text{as} \quad x \to \pm\infty, \tag{4.6.13}$$

in view of the fact that supp $N_4(\cdot + 2 - k) = [k - 2, k + 2]$. ∎

As for computing the spline interpolant $J_m f$ in (4.6.4), rather than doing so directly, it is more efficient to approximate $J_m f$ by $Q_k f$ for large values of $k$, where $Q_k$ is the quasi-interpolation operator introduced in (4.5.17). To analyze the resulting error, let us first return to (4.5.11) and (4.5.21), and consider

$$|(\underbrace{\{d_j\} * \cdots * \{d_j\}}_{k})(\ell)| = \frac{1}{2\pi}\left|\int_0^{2\pi} [\widetilde{D}(e^{-i\omega})]^k e^{i\ell\omega} d\omega\right| \tag{4.6.14}$$

$$= \frac{1}{2\pi}\left|\int_0^{2\pi} (1 - \widetilde{N}_m(e^{-i\omega}))^k e^{i\ell\omega} d\omega\right|$$

$$\leq r_m^k,$$

where

$$r_m = 1 - \min_\omega \widetilde{N}_m(e^{-i\omega}). \tag{4.6.15}$$

Hence, by repeating the telescoping argument in (4.5.20), we have, for each $\ell \in \mathbb{Z}$ and $p \in \pi_{m-1}$,

$$|(Q_k f - J_m f)(\ell)| = |(Q_k f - f)(\ell)|$$
$$= |(Q_k(f - p))(\ell) - (f - p)(\ell)|$$
$$= |(\underbrace{\{d_j\} * \cdots * \{d_j\}}_{k+1} * \{f(j) - p(j)\})(\ell)|$$
$$\leq r_m^{k+1} \max_{j\in\mathbb{Z}} |f(j) - p(j)|.$$

This gives

$$|(Q_k f - J_m f)(\ell)| \leq r_m^{k+1} \text{dist}_{\ell^\infty}(f, \pi_{m-1}), \quad \ell \in \mathbb{Z}, \tag{4.6.16}$$

where

$$\text{dist}_{\ell^\infty}(f, \pi_{m-1}) := \min_{p\in\pi_{m-1}} \max_{k\in\mathbb{Z}} |f(k) - p(k)|.$$

**Example 4.19.** Estimate the error between $(Q_k f)(\ell)$ and $(J_4 f)(\ell), \ell \in \mathbb{Z}$, cubic spline interpolation (i.e., $m = 4$).

**Solution.** It is easy to see that

$$\widetilde{N}_4(e^{-i\omega}) = \frac{e^{i\omega} + 4 + e^{-i\omega}}{6} = \frac{2 + \cos\omega}{3},$$

and this yields

$$r_4 = 1 - \min_{\omega} \left( \frac{2 + \cos \omega}{3} \right) = \frac{2}{3}.$$

Hence, by (4.6.16), we have

$$|(Q_k f - J_4 f)(\ell)| \leq \left( \frac{2}{3} \right)^{k+1} \mathrm{dist}_{\ell^\infty}(f, \pi_3), \quad \ell \in \mathbb{Z}. \tag{4.6.17}$$

For instance, if $\{f(k)\}$ is a bounded sequence, then by observing $\mathrm{dist}_{\ell^\infty}(f, \pi_3) \leq \mathrm{dist}_{\ell^\infty}(f, \pi_0)$ it follows from (4.6.17) that

$$\sup_{\ell \in \mathbb{Z}} |(Q_k f - J_4 f)(\ell)| \leq (\sup_{\ell \in \mathbb{Z}} f(\ell) + \inf_{\ell \in \mathbb{Z}} f(\ell)) \frac{1}{2} \left( \frac{2}{3} \right)^{k+1} . \qquad \blacksquare$$

In replacing $J_m f$ by the quasi-interpolant $Q_k f$, we do not have exact interpolation in general, although a good estimate is given in (4.6.16). One of the compelling reasons for choosing $Q_k f$ over $J_m f$, however, is that the local nature of the quasi-interpolation operator makes it possible to employ it for real-time applications. If we must insist on exact interpolation, then one way to obtain a local interpolation formula is to use a spline space on a finer grid, such as $S_m^j$, for some $j > 0$, instead of $S_m$ (where $S_m^j$ is the space of $m^{\mathrm{th}}$ order cardinal splines with knot sequence $2^{-j}\mathbb{Z}$, as discussed in Section 4.1). For instance, to interpolate a given data sequence $\{f(j): j \in \mathbb{Z}\}$, it is clear that with

$$\delta_m := \text{the smallest integer bounded below by } \log_2 m - 1, \tag{4.6.18}$$

then the cardinal spline function

$$\xi_m(x) := \frac{1}{N_m(m/2)} N_m \left( 2^{\delta_m} x + \frac{m}{2} \right) \tag{4.6.19}$$

satisfies the requirement

$$\xi_m(j) = \delta_{j,0}, \quad j \in \mathbb{Z}. \tag{4.6.20}$$

Hence, the spline operator

$$(R_m f)(x) := \sum_{k=-\infty}^{\infty} f(k) \xi_m(x - k) \tag{4.6.21}$$

is both local and interpolatory. Unfortunately, $R_m f$ is a *very bad* representation of $f$, since even the constant data function is not reproduced. For instance, when cubic splines are used, we have $\delta_4 = 1$ and

$$\xi_4(x) = \frac{3}{2} N_4(2x + 2) \tag{4.6.22}$$

so that for the data function $f(x) = 1$, $x \in \mathbb{R}$, we have $(R_4 f)\left(\frac{1}{2}\right) = \frac{3}{2} N_4(3) + \frac{3}{2} N_4(1) = \frac{1}{2}$, but not 1.

We now have two local methods: the method of constructing quasi-interpolation operators $Q$ as studied in the previous section, and the local interpolation formula (4.6.21). The first method reproduces all polynomials in the corresponding spline space, and hence provides the optimal order of approximation (see Theorem 4.16), while the second method provides an interpolatory spline. To construct a bounded linear *local* operator $P$ that possesses both the polynomial-reproduction property of $Q$ and the interpolatory property of $R_m$, we consider the following "blending" operation:

$$P := R_m + Q - R_m Q. \tag{4.6.23}$$

Since both $Q$ and $R_m$ are bounded linear local operators on $C$, so is $P$. Now, for any $x \in \mathbb{R}$ and $p \in \pi_{m-1}$, in view of the fact that $(Qp)(x) = p(x)$, we have

$$(Pp)(x) = (R_m p)(x) + (Qp)(x) - (R_m Qp)(x)$$
$$= (R_m p)(x) + p(x) - (R_m p)(x) = p(x),$$

so that $P$ also preserves all of $\pi_{m-1}$. On the other hand, for any $f \in C$ and $j \in \mathbb{Z}$, since $(R_m f)(j) = f(j)$, we also have

$$(Pf)(j) = (R_m f)(j) + (Qf)(j) - (R_m(Qf))(j)$$
$$= f(j) + (Qf)(j) - (Qf)(j) = f(j).$$

Hence, we have proved that, indeed, $P$ is a quasi-interpolation operator which is interpolatory on $\mathbb{Z}$. Two important points must still be observed. Firstly, since we are interested in interpolation of the data $f(j)$, $j \in \mathbb{Z}$, the local operator $P$ should only depend on this data set, but nothing else. This is certainly true for $R_m$ in (4.6.21), but the quasi-interpolation operator $Q$ must also have this property. Fortunately, the general formulation of $Q^0$ in (4.5.33) by imposing the condition (4.5.36) on the defining linear functionals, as introduced in the above section, can be applied to change the data set. Secondly, since the range of $R_m$ is the spline space $S_m^{\delta_m}$, we should also restrict the range of $Q = Q^0$ to $S_m^{\delta_m}$. We have thus established the following result.

**Theorem 4.20.** *Let $Q^0$ be a quasi-interpolation operator from $C_b(\mathbb{R})$ into $S_m^{\delta_m}$ in terms of the data sequences $f(j)$, $j \in \mathbb{Z}$, $f \in C_b(\mathbb{R})$. Then the operator $P$ defined in (4.6.23), with $R_m$ given by (4.6.21), is also a quasi-interpolation operator on $C_b(\mathbb{R})$ with range $S_m^{\delta_m}$ in terms of the data sequences $f(j)$, $j \in \mathbb{Z}$, and satisfies the additional property*

$$(Pf - f)(\ell) = 0, \quad \ell \in \mathbb{Z}.$$

Of course the interpolatory quasi-interpolation operator $P$ can be "scaled" to yield the operators

$$P^h := s_h \circ P \circ s_{h^{-1}}, \quad h > 0, \tag{4.6.24}$$

by choosing $h = 2^{-j}a$ (for any fixed positive constant $a$ and large positive integer $j$) to take care of other data sequences

$$f(2^{-j}a\ell), \quad \ell \in \mathbb{Z}, \tag{4.6.25}$$

and to achieve the optimal order $O(h^m)$ of approximation (see Theorem 4.16). We end this chapter by providing an example of cubic spline interpolation and quasi-interpolation operators.

**Example 4.21.** Construct a local cubic spline interpolation operator $P_4^0$ that depends only on the data set $f(j)$, $j \in \mathbb{Z}$, and preserves all cubic polynomials.

**Solution.** By scaling the quasi-interpolation operator $Q^0$ in (4.5.39) in Example 4.17 by $1/2$, we obtain the quasi-interpolation formula

$$(Q_4^0 f)(x) = \sum_{\ell} \left\{ \sum_j c_j^{(1)} f(j + \ell) \right\} N_4(2x - 2\ell + 2) \tag{4.6.26}$$

$$+ \sum_{\ell} \left\{ \sum_j c_j^{(2)} f(j + \ell) \right\} N_4(2x - 2\ell + 3)$$

that reproduces all of $\pi_3$, where $\{c_j^{(1)}\}$ and $\{c_j^{(2)}\}$ are given by (4.5.42) and (4.5.43). Hence, by applying (4.6.23) with $Q = Q_4^0$ and $R_m = R_4$ as given by (4.6.21) and (4.6.22), we have

$$(P_4^0 f)(x) = \sum_k \frac{3}{2} f(k) N_4(2x + 2 - 2k) + (Q_4^0 f)(x) \tag{4.6.27}$$

$$- \sum_k \frac{3}{2} \left\{ \sum_{j,\ell} c_j^{(1)} f(j + \ell) N_4(2k - 2\ell + 2) \right.$$

$$\left. + \sum_{j,\ell} c_j^{(2)} f(j + \ell) N_4(2k - 2\ell + 3) \right\} N_4(2x + 2 - 2k)$$

$$= \sum_{\ell} \left\{ \sum_j c_j^{(2)} f(j + \ell) \right\} N_4(2x + 2 - (2\ell - 1))$$

$$+ \sum_k \left\{ \frac{3}{2} f(k) - \sum_j \frac{1}{4} c_j^{(2)} (f(j + k) + f(j + k + 1)) \right\}$$

$$\times N_4(2x + 2 - 2k) = \sum_{\ell} \left\{ \sum_n v_{\ell - 2n}^{(4)} f(n) \right\} N_4(2x + 2 - \ell),$$

where

$$v_n^{(4)} := \begin{cases} \frac{29}{24} & \text{for} & n = 0 \\[2mm] \frac{7}{12} & \text{for} & n = \pm 1 \\[2mm] -\frac{1}{8} & \text{for} & n = \pm 2 \\[2mm] -\frac{1}{12} & \text{for} & n = \pm 3 \\[2mm] \frac{1}{48} & \text{for} & n = \pm 4 \\[2mm] 0 & \text{otherwise.} \end{cases} \qquad (4.6.28)$$

Observe that to determine the cardinal $B$-spline series

$$(P_4^0 f)(x) = \sum_\ell \gamma_\ell(f) N_4(2x + 2 - \ell), \qquad (4.6.29)$$

we can simply apply the MA formula

$$\gamma_\ell(f) = \sum_{n=\ell-4}^{\ell+4} v_{\ell-n}^{(4)} \tilde{f}(n), \qquad (4.6.30)$$

where $\{\tilde{f}(n)\}$ is obtained from the data sequence $\{f(n)\}$ by upsampling, namely: $\tilde{f}(2n) = f(n)$ and $\tilde{f}(2n+1) = 0$, $n \in \mathbb{Z}$. (See (4.3.11).)  ∎

# 5  Scaling Functions and Wavelets

Any $\mathcal{R}$-wavelet (or simply, wavelet) gives rise to some decomposition of the Hilbert space $L^2(\mathbb{R})$ into a direct sum of closed subspaces $W_j$, $j \in \mathbb{Z}$; in the sense that each subspace $W_j$ is the closure in $L^2(\mathbb{R})$ of the linear span of the collection of functions

$$\psi_{j,k}(x) = 2^{j/2}\psi(2^j x - k), \quad k \in \mathbb{Z}.$$

Hence, the corresponding subspaces

$$V_j := \cdots \dot{+} W_{j-2} \dot{+} W_{j-1}, \quad j \in \mathbb{Z},$$

form a nested sequence of subspaces of $L^2(\mathbb{R})$, whose union is dense in $L^2(\mathbb{R})$ and whose intersection is the null space $\{0\}$.

This observation motivates the following introduction of a very useful technique for constructing the wavelet $\psi$ and its corresponding dual $\tilde{\psi}$, namely: the investigation of the existence, and a study of the structure, of some "*scaling function*" $\phi$ that generates the spaces $V_j$, $j \in \mathbb{Z}$, in the same manner as $\psi$ generates the spaces $W_j$, $j \in \mathbb{Z}$. In particular, the collection of functions

$$\phi(x - k), \quad k \in \mathbb{Z},$$

is to form a Riesz (or unconditional) basis of $V_0$; and hence, $\phi$ generates a multiresolution analysis (MRA) $\{V_j\}$ of $L^2(\mathbb{R})$. Since $\phi \in V_0 \subset V_1$, there exists a unique sequence $\{p_n\} \in \ell^2$ that relates $\phi(x)$ with the functions $\phi(2x - k)$, $k \in \mathbb{Z}$; and the structure of $\phi$ is governed by that of this "two-scale sequence" $\{p_n\}$. For instance, a finite two-scale sequence characterizes a scaling function $\phi$ with compact support. In this respect, $\phi$ has minimum support if the length of this finite sequence is the shortest.

We will see that there is quite a lot of freedom in choosing the corresponding wavelet $\psi$ and its dual $\tilde{\psi}$, and another objective of this chapter is to investigate the structure of the complementary spaces $W_j$ (in the sense that $V_{j+1} = V_j \dot{+} W_j$, $j \in \mathbb{Z}$), and the corresponding "two-scale sequences" relating $W_j$ with $V_{j+1}$, that describe this freedom. With full knowledge of what the freedom is, it is then possible to construct the wavelet $\psi$ and its dual $\tilde{\psi}$ to meet certain specifications. Among those specifications of special interest, particularly to the engineer, are: decomposition of the space $L^2(\mathbb{R})$ as an orthogonal sum of the subspaces $W_j$; an orthonormal basis of $L^2(\mathbb{R})$ generated by $\psi$; finite reconstruction and decomposition sequences as a result of compactly supported $\psi$ and $\widetilde{\psi}$; and symmetry or anti-symmetry of $\psi$ and $\widetilde{\psi}$. In addition to a study of these features, we will also discuss the relation between symmetric wavelets and linear-phase filtering.

## 5.1. Multiresolution analysis

If some wavelet $\psi \in L^2(\mathbb{R})$ has to be constructed, then it is advisable to study the structure of the $L^2(\mathbb{R})$ decomposition it generates. As usual, let $\psi_{j,k}(x) := 2^{j/2}\psi(2^j x - k)$ and

$$W_j := \operatorname{clos}_{L^2(\mathbb{R})} \langle \psi_{j,k} \colon k \in \mathbb{Z} \rangle. \tag{5.1.1}$$

Then this family of subspaces of $L^2(\mathbb{R})$ gives a direct-sum decomposition of $L^2(\mathbb{R})$ in the sense that every $f \in L^2(\mathbb{R})$ has a *unique* decomposition

$$f(x) = \cdots + g_{-1}(x) + g_0(x) + g_1(x) + \cdots, \tag{5.1.2}$$

where $g_j \in W_j$ for all $j \in \mathbb{Z}$, and we shall describe this by writing

$$L^2(\mathbb{R}) = \sum_{j \in \mathbb{Z}}^{\bullet} W_j := \cdots \dot{+} W_{-1} \dot{+} W_0 \dot{+} W_1 + \cdots \tag{5.1.3}$$

(see (1.4.3)-(1.4.5)). Being in $W_j$, the component $g_j$ of $f$ has a unique wavelet series representation, where the coefficient sequence gives localized spectral information of $f$ in the $j^{\text{th}}$ octave (or frequency band) in terms of the integral wavelet transform of $f$ with the dual $\widetilde{\psi}$ of $\psi$ as the basic wavelet (see Theorem 3.27). We will return to this topic in Section 5.4. Using the decomposition of $L^2(\mathbb{R})$ in (5.1.3), we also have a nested sequence of closed subspaces $V_j$, $j \in \mathbb{Z}$, of $L^2(\mathbb{R})$ defined by

$$V_j := \cdots \dot{+} W_{j-2} \dot{+} W_{j-1}. \tag{5.1.4}$$

Let us summarize the properties of $\{V_j\}$, which are simple consequences of (5.1.1), (5.1.3), and (5.1.4), in the following (see Section 1.5).

**Lemma 5.1.** *The subspaces $V_j$ defined by (5.1.4) satisfy:*
(1°) $\cdots \subset V_{-1} \subset V_0 \subset V_1 \subset \cdots$;
(2°) $\operatorname{clos}_{L^2}\left( \bigcup_{j \in \mathbb{Z}} V_j \right) = L^2(\mathbb{R})$;
(3°) $\bigcap_{j \in \mathbb{Z}} V_j = \{0\}$;
(4°) $V_{j+1} = V_j \dot{+} W_j$, $j \in \mathbb{Z}$; *and*
(5°) $f(x) \in V_j \Leftrightarrow f(2x) \in V_{j+1}$, $j \in \mathbb{Z}$.

Now, suppose that a function $\phi \in V_0$ exists such that

$$\{\phi(\cdot - k): \ k \in \mathbb{Z}\} \tag{5.1.5}$$

is a Riesz basis of $V_0$ with Riesz bounds $A$ and $B$ (see (3.6.7)). Then by setting

$$\phi_{j,k}(x) := 2^{j/2}\phi(2^j x - k), \tag{5.1.6}$$

it follows from (5.1.4), (5.1.1), and (5°) above that for each $j \in \mathbb{Z}$, the family

$$\{\phi_{j,k}: \ k \in \mathbb{Z}\}$$

is also a Riesz basis of $V_j$ with the same Riesz bounds $A$ and $B$. As a consequence, the spaces $V_j$ also possess the following property:

(6°) $f(x) \in V_j \Leftrightarrow f\left(x + \frac{1}{2^j}\right) \in V_j, \quad j \in \mathbb{Z}.$

We have seen that in order to construct a wavelet $\psi$, we always end up with a nested sequence $\{V_j\}$ of subspaces of $L^2(\mathbb{R})$ that satisfies the properties (1°)–(5°). So, these properties may be considered as *necessary conditions* for the existence of a wavelet $\psi$. The technique we are going to study is first to construct a so-called "*scaling function*" $\phi \in L^2(\mathbb{R})$ that "generates" a sequence of closed subspaces of $L^2(\mathbb{R})$ (which we will also call $V_j$, with the understanding that these subspaces are *no longer* defined by (5.1.4) through some $\psi$ whose existence is still under investigation). More precisely, we have the following.

**Definition 5.2.** *A function $\phi \in L^2(\mathbb{R})$ is called a scaling function, if the subspaces $V_j$ of $L^2(\mathbb{R})$, defined by*

$$V_j := clos_{L^2(\mathbb{R})}\langle \phi_{j,k}: \ k \in \mathbb{Z}\rangle, \quad j \in \mathbb{Z}, \tag{5.1.7}$$

*(where the notation in (5.1.6) is used) satisfy the properties (1°), (2°), (5°), and (6°) stated above in this section, and if $\{\phi(\cdot - k): \ k \in \mathbb{Z}\}$ is a Riesz basis of $V_0$. We also say that the scaling function $\phi$ generates a multiresolution analysis $\{V_j\}$ of $L^2(\mathbb{R})$.*

**Remark.** If $\phi \in L^2(\mathbb{R})$ is a scaling function that generates an MRA $\{V_j\}$ of $L^2(\mathbb{R})$, then the nested sequence $\{V_j\}$ of MRA subspaces necessarily satisfies the property in (3°). The proof of this fact requires a little work and will not be discussed here. (See (7.2.29) in the proof of Lemma 7.13 in Chapter 7.) In addition, it is always possible to introduce complementary subspaces $W_j$ as in (4°). However, we will always assume that these subspaces are chosen "consistently" for all $j \in \mathbb{Z}$. For instance, if $W_0 \perp V_0$, then we require $W_j \perp V_j$, $j \in \mathbb{Z}$. In general, if $W_0$ is generated by some $\psi$ in the sense of (5.1.1) for $j = 0$, then all the other subspaces $W_j$ are assumed to be generated analogously by the same $\psi$. In summary, all the properties (1°)-(6°) will be assumed in any MRA $\{V_j\}$ of $L^2(\mathbb{R})$.

If $\phi$ generates an MRA, then since $\phi \in V_0$ is also in $V_1$ and since $\{\phi_{1,k}: k \in \mathbb{Z}\}$ is a Riesz basis of $V_1$, there exists a unique $\ell^2$-sequence $\{p_k\}$ that describes the "*two-scale relation*"

$$\phi(x) = \sum_{k=-\infty}^{\infty} p_k \phi(2x - k) \tag{5.1.8}$$

of the scaling function $\phi$. (See (4.3.1) and (4.3.3)-(4.3.4) for the two-scale relation of the $m^{\text{th}}$ order cardinal $B$-spline.) This sequence $\{p_k\}$ is called the "*two-scale sequence*" of $\phi$. Corresponding to this $\ell^2$-sequence, let us introduce the notation

$$P(z) = P_\phi(z) := \frac{1}{2} \sum_{k=-\infty}^{\infty} p_k z^k, \tag{5.1.9}$$

which differs from the symbol notation in (4.5.9) in that a normalization constant of $\frac{1}{2}$ is used to define $P$. This normalization simplifies the following Fourier transform formulation:

$$\hat{\phi}(\omega) = P(z)\hat{\phi}\left(\frac{\omega}{2}\right), \quad z = e^{-i\omega/2}, \tag{5.1.10}$$

of the identity (5.1.8). We will call $P = P_\phi$ the "*two-scale symbol*" of the scaling function $\phi$.

In order to be able to derive certain desirable properties of the $L^2(\mathbb{R})$ scaling function $\phi$, and later, its corresponding wavelet $\psi$ and dual wavelet $\tilde{\psi}$, we will make the following assumptions on $\phi$ and its two-scale sequence:

(A1) $\qquad\qquad\qquad \phi \in L^1(\mathbb{R});$

(A2) $\qquad\qquad\qquad \sum_{k=-\infty}^{\infty} \phi(x - k) = 1, \quad \text{a.e.};$

(A3) $\qquad\qquad\qquad \{p_k\} \in \ell^1.$

The assumption in (A2) is called the property of "partition of unity" of $\phi$. It is a standard (although not necessary) hypothesis for deriving the density property of the spaces $V_j$ in (2°). (See the proof of Theorem 4.16.) Observe that every cardinal $B$-spline satisfies (A2). Assumption (A1) implies that $\hat{\phi}$ is a continuous function on $\mathbb{R}$, as guaranteed by Theorem 2.2, (ii). Hence, it follows from Corollary 2.27 and the Poisson Summation Formula in (2.5.11) that (A2) is a consequence of the following conditions on $\hat{\phi}$:

$$\begin{cases} \hat{\phi}(0) = 1; \\ \hat{\phi}(2\pi k) = 0, \quad 0 \neq k \in \mathbb{Z}. \end{cases} \tag{5.1.11}$$

(See (4.5.5) for a more qualitative description of the $m^{\text{th}}$ order cardinal $B$-splines.) In general, since $\phi$ generates a Riesz basis of $V_0$, it follows from

(5.1.8) that by the normalization $\hat{\phi}(0) = 1$, (A2) already follows. Finally, the assumption in (A3) guarantees that $P = P_\phi$ is a continuous function on the unit circle $|z| = 1$. It is a very weak hypothesis; and in applications, we are interested in finite sequences so that the corresponding $P = P_\phi$ are Laurent polynomials. This further assumption will be made in the next section.

From the continuity of $P = P_\phi$ on $|z| = 1$ and the first condition in (5.1.11), and by applying (5.1.10), we have

$$P(1) = \frac{1}{2}\sum_k p_k = 1. \tag{5.1.12}$$

On the other hand, it follows from the assumption that $\{\phi(\cdot - k): k \in \mathbb{Z}\}$ is a Riesz basis of $V_0$ and the second condition of $\hat{\phi}$ in (5.1.11) that $P(z)$ also satisfies

$$P(-1) = \frac{1}{2}\sum_k (-1)^k p_k = 0. \tag{5.1.13}$$

Indeed, by Theorem 3.24 and the continuity of $\hat{\phi}$, we have

$$\sum_k |\hat{\phi}(x + 2\pi k)|^2 \geq A > 0, \quad x \in \mathbb{R},$$

so that $\hat{\phi}((2k_0 + 1)\pi) \neq 0$ for some $k_0 \in \mathbb{Z}$; and hence, evaluation of both sides of (5.1.10) at $\omega = 2(2k_0 + 1)\pi$ yields (5.1.13). Of course, an equivalent statement of (5.1.12) and (5.1.13) is

$$\sum_k p_{2k} = \sum_k p_{2k+1} = 1. \tag{5.1.14}$$

As another consequence of the continuity of $\hat{\phi}$ and the condition $\hat{\phi}(0) = 1$, we observe, by repeated application of (5.1.10), that as $n \to \infty$,

$$\hat{\phi}(\omega) = \left(\prod_{k=1}^{n} P(e^{-i\omega/2^k})\right) \hat{\phi}\left(\frac{\omega}{2^n}\right) \tag{5.1.15}$$

$$\to \prod_{k=1}^{\infty} P(e^{-i\omega/2^k}), \quad \omega \in \mathbb{R},$$

pointwise, provided that the infinite product converges. We will return to the convergence argument after considering the following example.

**Example 5.3.** For the $m^{\text{th}}$ order cardinal $B$-spline $N_m$, we have

$$P(z) = P_{N_m}(z) = \left(\frac{1+z}{2}\right)^m \tag{5.1.16}$$

(see (4.3.3)), so that

$$
\begin{aligned}
\prod_{k=1}^{n} P(e^{-i\omega/2^k}) &= \prod_{k=1}^{n} \left( \frac{1 + e^{-i\omega/2^k}}{2} \right)^m \\
&= \prod_{k=1}^{n} \left( \frac{1 + e^{-i\omega/2^k}}{2} \cdot \frac{1 - e^{-i\omega/2^k}}{1 - e^{-i\omega/2^k}} \right)^m \\
&= \prod_{k=1}^{n} \left( \frac{1}{2} \cdot \frac{1 - e^{-i\omega/2^{k-1}}}{1 - e^{-i\omega/2^k}} \right)^m \\
&= \frac{1}{2^{mn}} \left( \frac{1 - e^{-i\omega}}{1 - e^{-i\omega/2}} \frac{1 - e^{-i\omega/2}}{1 - e^{-i\omega/2^2}} \cdots \frac{1 - e^{-i\omega/2^{n-1}}}{1 - e^{-i\omega/2^n}} \right)^m \\
&= \frac{1}{2^{mn}} \left( \frac{1 - e^{-i\omega}}{1 - e^{-i\omega/2^n}} \right)^m \to \left( \frac{1 - e^{-i\omega}}{i\omega} \right)^m,
\end{aligned}
$$

as $n \to \infty$, and this limit agrees with the formulation of $\widehat{N}_m(\omega)$ in (3.2.16).  ∎

In view of the preceding spline example, we will restrict our attention to two-scale equations with governing sequences $\{p_k\}$ given by

$$
P(z) = \frac{1}{2} \sum_k p_k z^k = \left( \frac{1+z}{2} \right)^N S(z), \tag{5.1.17}
$$

where $N$ is some positive integer, $S(1) = 1$, and $S(z)$ is sufficiently smooth on the unit circle $|z| = 1$. More precisely, we consider the following.

**Definition 5.4.** *A Laurent series $P(z)$ of the form (5.1.17) is called an "admissible two-scale symbol" if $S$ is a continuous function on the unit circle satisfying*
(i) $S(1) = 1$, and
(ii) *as a function of $\omega$, the $L^\infty(0, 2\pi)$ modulus of continuity of $S(e^{-i\omega})$ as in (2.4.23) is of order $O(\eta^\alpha)$, for some $\alpha$, with $0 < \alpha \le 1$, as $\eta \to 0^+$.*

For any admissible two-scale symbol $P$ with factor $S$ as in (5.1.17), let us consider the bounds $B_j = B_j(S)$ and $b_j = b_j(S)$ defined by

$$
\begin{cases}
B_j = B_j(S) := \sup_{\omega \in \mathbb{R}} \left| \prod_{k=1}^{j} S(e^{-i\omega/2^k}) \right|; \\
b_j = b_j(S) := \frac{1}{j} \log_2 B_j = \frac{1}{j \ln 2} \ln B_j.
\end{cases} \tag{5.1.18}
$$

We have the following convergence result.

**Theorem 5.5.** *Let* $P$ *be an admissible two-scale symbol of the form* (5.1.17).
*Then the infinite product*

$$g(\omega) := \prod_{k=1}^{\infty} P(e^{-i\omega/2^k}) \qquad (5.1.19)$$

*converges pointwise everywhere to some function* $g$. *Furthermore, for every
positive integer* $n_0$, *there exists some positive constant* $C_{n_0}$, *such that the
limit function* $g$ *satisfies*

$$|g(\omega)| \le C_{n_0}(1+|\omega|)^{-N+b_{n_0}}, \quad \omega \in \mathbb{R}, \qquad (5.1.20)$$

*where* $b_{n_0}$ *is defined as in* (5.1.18). *In particular, if there is some* $n_0$ *such that*
$b_{n_0} < N - \frac{1}{2}$, *then there exists a function* $\phi \in L^2(\mathbb{R})$ *such that* $\hat{\phi} = g$, $\hat{\phi}(0) = 1$,
*and* $\hat{\phi}$ *satisfies the two-scale relation* (5.1.10).

**Proof.** For any fixed $\omega$, since $S(1) = 1$ and the $L^{\infty}(0, 2\pi)$ modulus of continuity
of $S(e^{-i\omega})$ is of order $O(\eta^{\alpha})$, $0 < \alpha \le 1$, we have

$$|1 - S(e^{-i\omega/2^k})| = O\left(\frac{|\omega|^{\alpha}}{2^{k\alpha}}\right), \quad k \to \infty.$$

Hence, since $\sum(|\omega|^{\alpha}/2^{k\alpha}) < \infty$ and

$$\prod_{k=1}^{K} |S(e^{-i\omega/2^k})| = \exp\left\{\sum_{k=1}^{K} \ln|1 - (1 - S(e^{-i\omega/2^k}))|\right\} \qquad (5.1.21)$$

$$= \exp\left\{O\left(\sum_{k=1}^{K} \frac{|\omega|^{\alpha}}{2^{k\alpha}}\right)\right\},$$

it follows that the infinite product

$$\prod_{k=1}^{\infty} S(e^{-i\omega/2^k})$$

converges. Therefore, in view of Example 5.3, the infinite product in (5.1.19)
converges for every $\omega$.

To establish the estimate in (5.1.20), we first observe, again from Example 5.3, that

$$\left|\prod_{k=1}^{\infty} \left(\frac{1 + e^{-i\omega/2^k}}{2}\right)^N\right| = \left(\frac{\sin\frac{\omega}{2}}{\omega/2}\right)^N \le C'(1+|\omega|)^{-N}. \qquad (5.1.22)$$

Next, for any fixed $\omega$, there corresponds a unique $n \in \mathbb{Z}$ such that $2^{n-1} < 1 + |\omega| \le 2^n$. So, on one hand, since $|\omega/2^k| \le 1$ for all $k \ge n$, the above estimates give

$$\left| \prod_{k=n+1}^{\infty} S(e^{-i\omega/2^k}) \right| \le C'',$$

where $C''$ is independent of $\omega$. Hence, with $C''' := C'C''$, we conclude, using (5.1.22), that

$$|g(\omega)| \le C'''(1 + |\omega|)^{-N} \prod_{k=1}^{n} |S(e^{-i\omega/2^k})|. \tag{5.1.23}$$

On the other hand, using (5.1.18), we have, for any fixed positive integer $n_0$ and for all large $n > 0$,

$$\prod_{k=1}^{n} |S(e^{-i\omega/2^k})| = \prod_{k=1}^{n_0} |S(e^{-i\omega/2^k})| \prod_{k=n_0+1}^{2n_0} |S(e^{-i\omega/2^k})| \cdots \tag{5.1.24}$$

$$\prod_{k=\left[\frac{n}{n_0}\right]n_0+1}^{n} |S(e^{-i\omega/2^k})|$$

$$\le C'_{n_0} B_{n_0}^{\left[\frac{n}{n_0}\right]} \le C''_{n_0} B_{n_0}^{n/n_0}.$$

Since

$$n - 1 < \log_2(1 + |\omega|) \le n,$$

it follows that

$$B_{n_0}^{n/n_0} \le C'''_{n_0} B_{n_0}^{\log_2(1+|\omega|)^{1/n_0}} = C'''_{n_0}(1 + |\omega|)^{b_{n_0}}.$$

Hence, with $C_{n_0} := C'''C''_{n_0}C'''_{n_0}$, the assertion in (5.1.20) is a consequence of (5.1.22) and (5.1.24).

If $b_{n_0} < N - \frac{1}{2}$, then from (5.1.20) we see that $g \in L^2(\mathbb{R})$; and by the $L^2(\mathbb{R})$ isometry of the Fourier transform established in Theorem 2.17, we have $g = \hat{\phi}$ for some $\phi \in L^2(\mathbb{R})$. In addition, it is clear from the estimates (5.1.21) and (5.1.22) that $\hat{\phi} \in C$, so that $\hat{\phi}(0) = g(0) = P(1) = 1$, and

$$P(e^{-i\omega/2})\hat{\phi}\left(\frac{\omega}{2}\right) = P(e^{-i\omega/2}) \prod_{k=1}^{\infty} P(e^{-i\omega/2^{k+1}})$$

$$= \prod_{k=1}^{\infty} P(e^{-i\omega/2^k})$$

$$= \hat{\phi}(\omega).$$

This completes the proof of the theorem. ∎

Observe that the foregoing theorem does not give any information on the smoothness of $\phi$ and on whether or not it generates a Riesz basis of $V_0$. In fact, without imposing any additional conditions on the two-scale symbol $P$, it is very difficult to draw any conclusion concerning whether or not the scaling function $\phi$ generates a Riesz basis of $V_0$. A discussion of this issue is delayed to Chapter 7, where $\{\phi(\cdot - k): \ k \in \mathbb{Z}\}$ is required to be an orthonormal family. In what follows, we will only be concerned with the smoothness of the scaling function $\phi$.

**Theorem 5.6.** *Under the hypotheses in Theorem 5.5, if*

$$b := \inf\{b_j: \ j \geq 1\} \tag{5.1.25}$$

*satisfies $b < N - 1$, then the limit function $g$ in (5.1.19) is in $L^2(\mathbb{R}) \cap L^1(\mathbb{R})$, and the function $\phi \in L^2(\mathbb{R})$, satisfying $\hat{\phi} = g$ as stated in Theorem 5.5, is in $C^\beta(\mathbb{R})$, where $\beta$ is the largest integer which is strictly smaller than $N - b - 1$. Furthermore, for any $\alpha > 0$, with $0 < \beta + \alpha < N - b - 1$, $\phi^{(\beta)}$ satisfies:*

$$\sup_{0 < h \leq \eta} \sup_{x \in \mathbb{R}} |\phi^{(\beta)}(x + h) - \phi^{(\beta)}(x)| = O(\eta^\alpha), \quad \eta \to 0^+.$$

**Definition 5.7.** *The class of all functions $f \in C = C(\mathbb{R})$ satisfying*

$$\sup_{0 < h \leq \eta} \sup_{x \in \mathbb{R}} |f(x + h) - f(x)| = O(\eta^\alpha), \quad \eta \to 0^+, \tag{5.1.26}$$

*where $0 < \alpha \leq 1$, is denoted by Lip $\alpha$; and the class of functions $f \in C^m = C^m(\mathbb{R})$, where $m$ is a positive integer such that $f^{(m)} \in \text{Lip } \alpha$, $0 < \alpha \leq 1$, will be denoted by $\text{Lip}^m \alpha$.*

**Proof of Theorem 5.6.** Select a positive integer $n_0$ such that

$$0 < \beta + \alpha < N - b_{n_0} - 1.$$

Note that by the definition of $\beta$, we have $0 < \alpha < 1$; and by (5.1.20) in Theorem 5.5, we see that

$$(1 + |\omega|)^\beta |\hat{\phi}(\omega)| \leq C_{n_0} (1 + |\omega|)^{-1-\alpha}. \tag{5.1.27}$$

Hence, the Lebesgue Dominated Convergence Theorem permits us to differentiate inside the integral of the formula

$$\phi(x) = \frac{1}{2\pi} \int_{-\infty}^{\infty} e^{ix\omega} \hat{\phi}(\omega) d\omega,$$

$\beta$ times, yielding $\phi \in C^\beta$ and

$$\phi^{(\beta)}(x) = \frac{1}{2\pi} \int_{-\infty}^{\infty} (i\omega)^\beta e^{ix\omega} \hat{\phi}(\omega) d\omega. \tag{5.1.28}$$

Now, from the estimate

$$|e^{i(x+h)\omega} - e^{ix\omega}| \leq \min(2, |h\omega|)$$
$$\leq 2^{1-\alpha}|h\omega|^{\alpha} \leq 2|h|^{\alpha}(1 + |\omega|)^{\alpha},$$

along with (5.1.27) and (5.1.28), it follows that

$$|\phi^{(\beta)}(x+h) - \phi^{(\beta)}(x)| \leq \frac{1}{2\pi}\int_{-\infty}^{\infty} |\omega|^{\beta}|e^{i(x+h)\omega} - e^{ix\omega}||\hat{\phi}(\omega)|d\omega$$
$$\leq C_{n_0}\frac{|h|^{\alpha}}{\pi}\int_{-\infty}^{\infty}(1 + |\omega|)^{\beta+\alpha-N+b_0}d\omega,$$

where the integral is finite, since $\beta + \alpha - N + b_0 < -1$. Hence, we have proved that $\phi \in \mathrm{Lip}^{\beta}\alpha$. ∎

## 5.2. Scaling functions with finite two-scale relations

In this section, we restrict our attention to two-scale relations (5.1.8) described by finite sums. A very important consequence of this restriction is that the corresponding scaling functions necessarily have compact supports. Hence, as we will see in this section, the graphical display algorithm for cardinal $B$-splines in Section 4.3 also applies to plotting the graph of any causal series

$$f(x) = \sum_{\ell} a_{\ell}\phi(x - \ell), \tag{5.2.1}$$

where $\phi$ is any such scaling function, in real-time. We will also study the class of all scaling functions with finite two-scale relations that generate the same multiresolution analysis, and investigate the ones with minimum supports. This is important in revealing the basic structure of the MRA under investigation, and minimally supported scaling functions will be instrumental in constructing wavelets with smaller supports. It will be clear that the smaller the supports of a scaling function $\phi$ and its corresponding wavelet $\psi$ are, the shorter the reconstruction sequences (see (1.6.2)) used in the wavelet reconstruction algorithm become. (For more details, see Section 5.4 later in this chapter.)

Let $\phi$ be a scaling function described by the two-scale relation

$$\phi(x) = \sum_{k=0}^{N_{\phi}} p_k^{\phi}\phi(2x - k), \quad p_0^{\phi}, p_{N_{\phi}}^{\phi} \neq 0. \tag{5.2.2}$$

When no possible confusion arises, we will drop the subscript or superscript $\phi$, by writing

$$\begin{cases} p_k := p_k^{\phi}; \\ N := N_{\phi}. \end{cases} \tag{5.2.3}$$

We remark that by a change of index in $p_k$, any finite two-scale relation can be written as in (5.2.2). Of course, the scaling function $\phi$ must also be shifted accordingly.

Let us first take care of the cases $N_\phi = 0, 1$ in (5.2.2).

(i) For $N_\phi = 0$, we have, by (5.1.12),

$$\phi(x) = 2\phi(2x),$$

so that the two-scale symbol is $P(z) = 1$ and the infinite product in (5.1.19) is $g(\omega) = 1$ for all $\omega$. So if $g = \hat{\phi}$, then $\phi$ must be the delta distribution.

(ii) For $N_\phi = 1$, we have, by (5.1.14),

$$\phi(x) = \phi(2x) + \phi(2x - 1),$$

which is the same as the two-scale relation of the first order cardinal $B$-spline $N_1$ in (4.3.4). So $\phi = N_1$.

Hence, we will always assume that $N_\phi \geq 2$. In the following, an iterative procedure will be introduced to construct the scaling functions $\phi$. For this purpose, we will only be concerned with scaling functions which are continuous everywhere. Under this additional assumption, it is also possible to show that for $N_\phi = 2$, the two-scale relation must be given by

$$\phi(x) = \frac{1}{2}\phi(2x) + \phi(2x - 1) + \frac{1}{2}\phi(2x - 2), \tag{5.2.4}$$

which is identical with the two-scale relation of the second order cardinal $B$-spline $N_2$ as in (4.3.4) for $m = 2$, and hence, $\phi = N_2$. For $N_\phi \geq 3$, however, we are going to see some very interesting varieties in Chapter 7. For instance, when $N_\phi = 3$, of course we always have the quadratic cardinal $B$-spline $N_3$ whose two-scale equation is

$$N_3(x) = \frac{1}{4}N_3(2x) + \frac{3}{4}N_3(2x - 1) + \frac{3}{4}N_3(2x - 2) + \frac{1}{4}N_3(2x - 3). \tag{5.2.5}$$

However, there is another alternative, namely: Daubechies' scaling function $\phi_3^D$ governed by

$$\phi_3^D(x) = \frac{1 + \sqrt{3}}{4}\phi_3^D(2x) + \frac{3 + \sqrt{3}}{4}\phi_3^D(2x - 1) \tag{5.2.6}$$

$$+ \frac{3 - \sqrt{3}}{4}\phi_3^D(2x - 2) + \frac{1 - \sqrt{3}}{4}\phi_3^D(2x - 3).$$

More details on $\phi_3^D$ will be given in Chapter 7. Here, we only point out two essential features of the two-scale sequence in (5.2.6). Firstly, as required by (5.1.14), we have

$$\begin{cases} p_0 + p_2 = \dfrac{1 + \sqrt{3}}{4} + \dfrac{3 - \sqrt{3}}{4} = 1; \\[2mm] p_1 + p_3 = \dfrac{3 + \sqrt{3}}{4} + \dfrac{1 - \sqrt{3}}{4} = 1, \end{cases} \tag{5.2.7}$$

and secondly,

$$P(z) = \frac{1}{2} \sum_{k=0}^{3} p_k z^k \tag{5.2.8}$$

$$= \frac{1}{2} \left\{ \frac{1+\sqrt{3}}{4} + \frac{3+\sqrt{3}}{4}z + \frac{3-\sqrt{3}}{4}z^2 + \frac{1-\sqrt{3}}{4}z^3 \right\}$$

$$= \left( \frac{1+z}{2} \right)^2 \left( \frac{(1+\sqrt{3}) + (1-\sqrt{3})z}{2} \right),$$

which satisfies the admissibility condition in Definition 5.4, with $N = 2$ and $S(z)$ being a trigonometric polynomial in $\omega$ such that $S(1) = 1$.

To understand a scaling function $\phi$ better, we consider the recursive scheme

$$\phi_n(x) = \sum_k p_k \phi_{n-1}(2x - k), \quad n = 1, 2, \ldots, \tag{5.2.9}$$

for some suitable initial function $\phi_0$. By considering the Fourier transform formulation of (5.2.9) (see (5.1.10)) and following the same process as in (5.1.15), we have

$$\hat{\phi}_n(\omega) = P(e^{-i\omega/2})\hat{\phi}_{n-1}\left(\frac{\omega}{2}\right) = \cdots \tag{5.2.10}$$

$$= \left\{ \prod_{k=1}^{n} P(e^{-i\omega/2^k}) \right\} \hat{\phi}_0\left(\frac{\omega}{2^n}\right).$$

Hence, if $P$ is an admissible two-scale symbol and the Fourier transform $\hat{\phi}_0$ of the initial choice $\phi_0$ is continuous at $\omega = 0$ and satisfies $\hat{\phi}_0(0) = 1$, then by Theorem 5.5 both sides of (5.2.10) converge for any $\omega \in \mathbb{R}$, and

$$\lim_{n \to \infty} \hat{\phi}_n(\omega) = g(\omega),$$

where $g$ is the infinite product given in (5.1.19). In addition, if the two-scale symbol $P$ also satisfies $b < N - 1$ where $b$ and $N$ are given in (5.1.25) and (5.1.17), respectively, then by Theorems 5.5 and 5.6, we have $g = \hat{\phi}$, where $\phi \in L^1(\mathbb{R}) \cap L^2(\mathbb{R})$ is in $\mathrm{Lip}^\beta \alpha$, with $0 < \alpha < 1$ and $\beta$ is the largest integer satisfying $0 < \beta + \alpha < N - b - 1$ (see Definition 5.7). Therefore, under these conditions, a scaling function $\phi$ can be obtained by taking the limit of $\phi_n$ in the recursive scheme (5.2.9). We will sketch a proof of this in a moment. In view of the foregoing discussions on the cases in (i) and (ii), we see that for a two-scale sequence with at least three non-zero terms the second order cardinal $B$-spline $N_2$, being a continuous spline function with lowest order, provides a good choice as the initial function in (5.2.9) for producing the scaling function

$\phi$. That is, we recommend the following recursive scheme:

$$\begin{cases} \phi(x) = \lim_{n \to \infty} \phi_n(x), \quad \text{where} \\ \phi_n(x) = \sum_{k=0}^{N_\phi} p_k^\phi \phi_{n-1}(2x - k), \quad n = 1, 2, \dots, \quad \text{and} \\ \phi_0(x) = N_2(x). \end{cases} \quad (5.2.11)$$

In fact, under the above assumption on the two-scale symbol $P$, this recursive scheme is uniformly convergent.

**Sketch of Proof.** Let $\varepsilon > 0$. Since $\hat{\phi} \in L^1(\mathbb{R})$, we have

$$\int_{|\omega| \geq M} |\hat{\phi}(\omega)| d\omega < \varepsilon,$$

for all sufficiently large values of $M > 0$, and

$$\lim_{n \to \infty} \|\chi_{[-M,M]}(\hat{\phi}_n - g)\|_{L^1(\mathbb{R})} = 0$$

for any fixed value of $M$. On the other hand, since

$$\hat{\phi}_n(\omega) = \left\{ \prod_{k=1}^{n} P(e^{-i\omega/2^k}) \right\} \left( \frac{\sin \omega/2^{n+1}}{\omega/2^{n+1}} \right)^2 e^{-i\omega/2^n},$$

we have, by applying the same estimate as in (5.1.24),

$$\int_{M \leq |\omega| \leq 2^n \pi} |\hat{\phi}_n(\omega)| d\omega \geq C \int_{|\omega| \geq M} \frac{d\omega}{(1 + |\omega|)^{1+\eta}} \leq \frac{C'}{M^\eta} < \varepsilon,$$

where $0 < \eta < N - b - 1$ and $M$ is sufficiently large. Finally, the periodicity of

$$\prod_{k=1}^{n} P(e^{-i\omega/2^k})$$

may be used to yield

$$\int_{|\omega| > 2^n \pi} |\hat{\phi}(\omega)| d\omega = \sum_{k \neq 0} \int_{|\omega| \leq 2^n \pi} \left| \prod_{k=1}^{n} P(e^{-i\omega/2^k}) \right| \left( \frac{\sin(\omega/2^{n+1})}{2^{-n-1}\omega + k\pi} \right)^2 d\omega \to 0;$$

and uniform convergence follows from the observation that $|\phi_n(x) - \phi(x)| \leq \frac{1}{2\pi} \|\hat{\phi}_n - \hat{\phi}\|_{L^1(\mathbb{R})}$. ∎

As a consequence of the process in (5.2.11), we see that $\phi$ has compact support, and in fact, we can find its support exactly, provided that $\phi$ is continuous. It is interesting to observe that, indeed, supp $\phi_n$ increases monotonically with $n$. More precisely, by simple computations, we have

$$
\begin{cases}
\text{supp } \phi_0 = [0, 2], \\[2mm]
\text{supp } \phi_1 = \left[0, \dfrac{1}{2}(2 + N_\phi)\right] = \left[0, \dfrac{2 + N_\phi}{2}\right], \\[3mm]
\text{supp } \phi_2 = \left[0, \dfrac{1}{2}\left(\dfrac{2 + N_\phi}{2} + N_\phi\right)\right] = \left[0, \dfrac{2 + (2^2 - 1)N_\phi}{2^2}\right], \\[3mm]
\quad \cdots\cdots\cdots\cdots\cdots\cdots\cdots\cdots\cdots\cdots \\[2mm]
\text{supp } \phi_n = \left[0, \dfrac{2 + (2^n - 1)N_\phi}{2^n}\right];
\end{cases}
\tag{5.2.12}
$$

and hence, since $N_\phi \geq 2$, we have

$$
\text{supp } \phi_n \subseteq [0, N_\phi], \quad n = 1, 2, \ldots,
$$

and it follows from (5.2.11) and (5.2.12) that

$$
\text{supp } \phi = [0, N_\phi].
\tag{5.2.13}
$$

Knowing that the support of $\phi$ is $[0, N_\phi]$, as in (5.2.13), is a tremendous help in computing $\phi(x)$, at least at all the dyadic points $x = k/2^j$, where $j, k \in \mathbb{Z}$. This is evident by referring to the two-scale relation (5.2.2). In fact, if the values of $\phi(1), \ldots, \phi(N_\phi - 1)$ are known, then since $\phi(k) = 0$ for all $k \leq 0$ or $k \geq N_\phi$, the relations

$$
\begin{cases}
\phi\left(\dfrac{k}{2}\right) = \displaystyle\sum_\ell p_\ell^\phi \phi(k - \ell), \\[4mm]
\phi\left(\dfrac{k}{2^2}\right) = \displaystyle\sum_\ell p_\ell^\phi \phi\left(\dfrac{k}{2} - \ell\right), \\[2mm]
\quad \cdots\cdots\cdots\cdots\cdots\cdots\cdots
\end{cases}
$$

uniquely determine all the values of $\phi(x)$ at $x = k/2^j$, $j, k \in \mathbb{Z}$.

To determine the values of $\phi(k)$, $k \in \mathbb{Z}$, we again use the two-scale relation (5.2.2) with $x$ being an integer. That is, in matrix notation, we have

$$
\mathbf{m} = M\mathbf{m},
\tag{5.2.14}
$$

where $\mathbf{m}$ is the column vector

$$
\mathbf{m} := [\phi(1) \ldots \phi(N_\phi - 1)]^T,
\tag{5.2.15}
$$

and $M$ the $(N_\phi - 1) \times (N_\phi - 1)$ matrix

$$M := [p^\phi_{2j-k}]_{1 \le j, k \le N_\phi - 1}, \tag{5.2.16}$$

with $j$ being the row index and $k$ the column index. Recalling that $\phi$ generates a partition of unity (see (A.2) in Section 5.1), we can determine the values of $\phi(k)$, $k \in \mathbb{Z}$, simply by finding the eigenvector $\mathbf{m}$ in (5.2.14) corresponding to the eigenvalue 1 and imposing the normalization condition

$$\phi(1) + \cdots + \phi(N_\phi - 1) = 1. \tag{5.2.17}$$

**Example 5.8.** Determine the values of $\phi^D_3(k)$, $k \in \mathbb{Z}$, where the two-scale relation of $\phi^D_3$ is given by (5.2.6).

**Solution.** By (5.2.6), we have $N_\phi = 3$ and the matrix $M$ in (5.2.16) becomes

$$M = \begin{bmatrix} p^\phi_1 & p^\phi_0 \\ p^\phi_3 & p^\phi_2 \end{bmatrix} = \frac{1}{4} \begin{bmatrix} 3 + \sqrt{3} & 1 + \sqrt{3} \\ 1 - \sqrt{3} & 3 - \sqrt{3} \end{bmatrix}. \tag{5.2.18}$$

It is easy to see that the solution space of (5.2.14) is

$$\mathbf{m} = a[1 + \sqrt{3} \quad 1 - \sqrt{3}]^T, \qquad a \in \mathbb{R}.$$

So, by the normalization condition (5.2.17), we have $a = \frac{1}{2}$ and

$$\begin{cases} \phi^D_3(1) = \dfrac{1 + \sqrt{3}}{2}; \\[2mm] \phi^D_3(2) = \dfrac{1 - \sqrt{3}}{2}. \end{cases} \blacksquare \tag{5.2.19}$$

Having computed the values of $\phi(k)$, $k \in \mathbb{Z}$, it is now very easy to compute

$$\phi\left(\frac{k}{2^j}\right), \qquad j, k \in \mathbb{Z}. \tag{5.2.20}$$

In fact, the Interpolatory Graphical Display Algorithm (see Algorithm 4.7) can be applied, without any change, even to compute any causal series

$$f_{j_0}(x) = \sum_{\ell=0}^{\infty} a^{(j_0)}_\ell \phi(2^{j_0} x - \ell), \tag{5.2.21}$$

in real-time, for any fixed $j_0 \in \mathbb{Z}$, at $x = k/2^{j_1}$ for any $k \in \mathbb{Z}$ and any $j_1 \ge j_0$. Hence, to compute $\phi(k/2^j)$ in (5.2.20), we simply apply this algorithm to $j_0 = 0$ and $a^{(0)}_\ell = \delta_{\ell,0}$. Of course, one must set

$$\begin{cases} p_{m,k} = p_k \\ w_{m,k} = \phi(k) \end{cases} \tag{5.2.22}$$

in Algorithm 4.7 for computing (or displaying the graph of) $f_{j_0}$ in (5.2.21).

We now turn to a study of the class $\Phi$ of all the scaling functions $\phi$ with finite two-scale relations that generate the same MRA $\{V_j\}$ of $L^2(\mathbb{R})$. Again, without loss of generality, we may assume that the two-scale relation of any $\phi \in \Phi$ takes on the form (5.2.2), and hence, by (5.2.13), the support of $\phi$ is precisely the interval $[0, N_\phi]$. So, $\phi^* \in \Phi$ has minimum support if and only if

$$N_{\phi^*} \leq N_\phi, \quad \phi \in \Phi. \tag{5.2.23}$$

Corresponding to any $\phi \in \Phi$, let us consider the autocorrelation function

$$F_\phi(x) := \int_{-\infty}^{\infty} \phi(x+y)\overline{\phi(y)}\,dy \tag{5.2.24}$$

introduced in Definition 2.9, and the symbol of the sequence $\{F_\phi(k)\}$, namely:

$$E_\phi(z) := \sum_{k \in \mathbb{Z}} F_\phi(k)z^k. \tag{5.2.25}$$

Since $F_\phi$ clearly satisfies:

$$\begin{cases} F_\phi(-x) = \overline{F_\phi(x)}, & x \in \mathbb{R}; \\ \operatorname{supp} F_\phi \subseteq [-N_\phi, N_\phi], \end{cases} \tag{5.2.26}$$

it follows that $E_\phi$ is a Laurent polynomial. Let $k_\phi$ denote the "*one-sided degree*" of $E_\phi$; that is, $k_\phi$ is the largest integer for which $F_\phi(k_\phi) \neq 0$. Then

$$\Pi_\phi(z) := z^{k_\phi} E_\phi(z) \tag{5.2.27}$$

is an (algebraic) polynomial (in $z$) of degree $2k_\phi$, and the "*reciprocal polynomial*" of $\Pi_\phi$ is given by

$$\Pi_\phi^r(z) := z^{2k_\phi}\overline{\Pi_\phi\left(\frac{1}{\overline{z}}\right)}. \tag{5.2.28}$$

In view of the first property in (5.2.26), it is clear that

$$\Pi_\phi^r(z) = \Pi_\phi(z), \text{ all } z. \tag{5.2.29}$$

We call $\Pi_\phi$ the "*generalized Euler-Frobenius polynomial*" and $E_\phi$ the "*generalized Euler-Frobenius Laurent polynomial*" relative to $\phi$. (Recall that a multiplicative normalization constant is used to give integer coefficients for the ordinary Euler-Frobenius polynomials relative to the cardinal $B$-splines in (4.2.18), and more generally in (4.6.6).)

We need the following terminology.

**Definition 5.9.** *Let $z_0$ be a zero (or root) of an algebraic polynomial $p(z)$. Then we call $z_0$ a symmetric zero (or symmetric root) of $p(z)$ if (i) $z_0 \neq 0$ and (ii) $p(-z_0) = p(z_0) = 0$.*

In the following theorem concerning scaling functions $\phi$ with finite two-scale sequences, recall that $P_\phi(z)$ denotes the two-scale symbol of $\phi$.

**Theorem 5.10.** *Let $\phi \in \Phi$ be any scaling function governed by (5.2.2). Then*
(i) *both $E_\phi(z)$ and $\Pi_\phi(z)$ never vanish on $|z| = 1$;*
(ii) *for all $\omega \in \mathbb{R}$,*

$$E_\phi(e^{-i\omega}) = \sum_{k=-\infty}^{\infty} |\hat{\phi}(\omega + 2\pi k)|^2; \qquad (5.2.30)$$

(iii) *for all $\omega \in \mathbb{R}$,*

$$|P_\phi(e^{-i\omega/2})|^2 E_\phi(e^{-i\omega/2}) + |P_\phi(-e^{-i\omega/2})|^2 E_\phi(-e^{-i\omega/2}) = E_\phi(e^{-i\omega}); \qquad (5.2.31)$$

(iv) *for all complex numbers $z$,*

$$P_\phi(z)P_\phi^r(z)\Pi_\phi(z) + (-1)^{N_\phi - k_\phi} P_\phi(-z)P_\phi^r(-z)\Pi_\phi(-z) \qquad (5.2.32)$$
$$= z^{N_\phi - k_\phi}\Pi_\phi(z^2); \quad and$$

(v) *$P_\phi$ has no symmetric zeros that lie on $|z| = 1$.*

**Proof.** The identity in (5.2.30) follows from the Poisson Summation Formula (2.5.19). Hence, by applying Theorem 3.24, since $\{\phi(\cdot - k): \ k \in \mathbb{Z}\}$ is a Riesz basis of $V_0$, we have $E_\phi(e^{-i\omega}) \neq 0$ for all $\omega \in \mathbb{R}$. In particular, assertion (i) follows from (5.2.27). To derive (5.2.31), we start from (5.2.30) and apply the Fourier transform formulation of the two-scale relation (5.2.2), obtaining:

$$E_\phi(e^{-i\omega}) = \sum_{k=-\infty}^{\infty} \left| P_\phi(e^{-i\frac{\omega + 2\pi k}{2}}) \hat{\phi}\left(\frac{\omega + 2\pi k}{2}\right) \right|^2$$

$$= |P_\phi(e^{-i\omega/2})|^2 \sum_{k=-\infty}^{\infty} \left| \hat{\phi}\left(\frac{\omega + 4\pi k}{2}\right) \right|^2$$

$$+ |P_\phi(-e^{-i\omega/2})|^2 \sum_{k=-\infty}^{\infty} \left| \hat{\phi}\left(\frac{\omega + 2\pi(2k + 1)}{2}\right) \right|^2$$

$$= |P_\phi(e^{-i\omega/2})|^2 E_\phi(e^{-i\omega/2}) + |P_\phi(-e^{-i\omega/2})|^2 E_\phi(-e^{-i\omega/2}).$$

This establishes (iii). To prove (iv), we appeal to the formulas

$$\overline{P_\phi(z)} = z^{-N_\phi} P_\phi^r(z) \quad \text{and} \quad E_\phi(z) = z^{-k_\phi}\Pi_\phi(z)$$

which hold for $|z| = 1$, and verify that (5.2.31) is equivalent to (5.2.32) for $z = e^{-i\omega/2}$. Now, since both sides of (5.2.32) are entire functions (being algebraic polynomials in $z$), they must be identical for all $z$.

Finally, if $z_0 \neq 0$ is a symmetric zero of $P_\phi$, then by (5.2.32), we have $\Pi_\phi(z_0^2) = 0$, so that $|z_0| \neq 1$ by applying (i). ∎

We are now ready to give a characterization of those $\phi \in \Phi$ that have minimum support.

**Theorem 5.11.** *A scaling function* $\phi^* \in \Phi$ *has minimum support if and only if its two-scale symbol* $P_{\phi^*}$ *has no symmetric zeros.*

**Proof.** Let $\phi^* \in \Phi$ be given arbitrarily and consider the factorization of its two-scale symbol $P_{\phi^*}$ into the form

$$P_{\phi^*}(z) = m_{\phi^*}(z)n_{\phi^*}(z^2), \qquad (5.2.33)$$

where $m_{\phi^*}$ and $n_{\phi^*}$ are polynomials satisfying

$$\begin{cases} m_{\phi^*}(1) = n_{\phi^*}(1) = 1 \\ n_{\phi^*}(0) \neq 0 \\ m_{\phi^*} \text{ has no symmetric zeros.} \end{cases} \qquad (5.2.34)$$

Recall that supp $\phi^* = [0, N_{\phi^*}]$ and $N_{\phi^*} = \deg P_{\phi^*}$. Also, observe from (5.2.33) and (5.2.34) that $P_{\phi^*}$ has no symmetric zeros if and only if

$$\deg n_{\phi^*} = 0.$$

Now, by Theorem 5.10, (v), since $P_{\phi^*}$ has no symmetric zeros that lie on $|z| = 1$, the polynomial $n_{\phi^*}$ is zero-free on $|z| = 1$, so that $n_{\phi^*}^{-1}$ is analytic on $|z| = 1$, and has a Laurent expansion

$$\frac{1}{n_{\phi^*}(z)} = \sum_{n=-\infty}^{\infty} r_n z^n, \qquad \{r_n\} \in \ell^1.$$

Define a function $\phi^{**} \in V_0$ by

$$\phi^{**}(x) = \sum_{n=-\infty}^{\infty} r_n \phi^*(x - n). \qquad (5.2.35)$$

Then using the notation $z = e^{-i\omega/2}$, we may reformulate (5.2.35) as

$$\hat{\phi}^{**}(\omega) = \frac{1}{n_{\phi^*}(z^2)}\hat{\phi}^*(\omega), \qquad (5.2.36)$$

or

$$\hat{\phi}^*\left(\frac{\omega}{2}\right) = n_{\phi^*}(z)\hat{\phi}^{**}\left(\frac{\omega}{2}\right). \qquad (5.2.37)$$

Now, by applying both (5.2.36) and (5.2.37) and using the two-scale relation of $\phi^*$, we have

$$\begin{aligned} \hat{\phi}^{**}(\omega) &= \frac{1}{n_{\phi^*}(z^2)}P_{\phi^*}(z)\hat{\phi}^*\left(\frac{\omega}{2}\right) \\ &= \frac{1}{n_{\phi^*}(z^2)}m_{\phi^*}(z)n_{\phi^*}(z^2)\hat{\phi}^*\left(\frac{\omega}{2}\right) \\ &= m_{\phi^*}(z)\hat{\phi}^*\left(\frac{\omega}{2}\right) = m_{\phi^*}(z)n_{\phi^*}(z)\hat{\phi}^{**}\left(\frac{\omega}{2}\right). \end{aligned}$$

This shows that $\phi^{**} \in \Phi$ and its two-scale symbol is given by

$$P_{\phi^{**}}(z) = m_{\phi^*}(z)n_{\phi^*}(z).$$

So, in view of (5.2.33), we have

$$\begin{aligned}
\deg P_{\phi^{**}} &= \deg m_{\phi^*} + \deg n_{\phi^*} \\
&\leq \deg m_{\phi^*} + 2\deg n_{\phi^*} \\
&= \deg P_{\phi^*}.
\end{aligned}$$

Hence, for $\phi^*$ to have minimum support, we must have $\deg P_{\phi^{**}} = \det P_{\phi^*}$, or $\deg n_{\phi^*} = 0$, or equivalently, $P_{\phi^*}$ has no symmetric zeros.

To prove the converse, suppose that $\phi^* \in \Phi$ and that $P_{\phi^*}$ has no symmetric zeros. Then in view of (5.2.13), in order to prove that $\phi^*$ has minimum support, it suffices to prove that

$$\deg P_\phi \geq \deg P_{\phi^*}, \tag{5.2.38}$$

for any $\phi \in \Phi$. Since $\phi^* \in \Phi$, we may write

$$\phi(x) = \sum_{n=-\infty}^{\infty} s_n \phi^*(x-n) \tag{5.2.39}$$

for some sequence $\{s_n\} \in \ell^2$. The Fourier transform formulation of (5.2.39) is

$$\begin{cases} \hat{\phi}(\omega) = C(z^2)\hat{\phi}^*(\omega); \\ C(z) := \displaystyle\sum_{n=-\infty}^{\infty} s_n z^n, \quad z = e^{-i\omega/2}. \end{cases} \tag{5.2.40}$$

By taking the discrete Fourier transform of both sides of (5.2.39), we see that $C(z)$ is a rational function, being the quotient of the (polynomial) symbol of $\{\phi(k)\}$ and that of $\{\phi^*(k)\}$. In addition, since it follows from (5.2.40) that

$$|C(z^2)|^2 = \frac{E_\phi(z^2)}{E_{\phi^*}(z^2)},$$

we observe, by applying Theorem 5.10, (i), that the rational function $C(z)$ is both zero-free and pole-free on $|z| = 1$. Now, let us write

$$\begin{cases} \displaystyle\sum_n \phi(n)z^n = \sum_{n=0}^{N_\phi} \phi(n)z^n = q_\phi(z)d(z); \\ \displaystyle\sum_n \phi^*(n)z^n = \sum_{n=0}^{N_{\phi^*}} \phi^*(n)z^n = q_{\phi^*}(z)d(z), \end{cases} \tag{5.2.41}$$

where $d, q_\phi, q_{\phi^*}$ are polynomials with $q_\phi(0) \neq 0$, $q_{\phi^*}(0) \neq 0$, and $q_\phi, q_{\phi^*}$ have no common zeros. Then it follows from the two-scale relations of $\phi$ and $\phi^*$, and also from (5.2.40) that for $z = e^{-i\omega/2}$,

$$\hat{\phi}(\omega) = P_\phi(z)\hat{\phi}\left(\frac{\omega}{2}\right) = C(z)P_\phi(z)\hat{\phi}^*\left(\frac{\omega}{2}\right)$$

$$= C(z)P_\phi(z)\frac{1}{P_{\phi^*}(z)}\hat{\phi}^*(\omega)$$

$$= \frac{C(z)}{C(z^2)}\frac{P_\phi(z)}{P_{\phi^*}(z)}\hat{\phi}(\omega),$$

whence

$$P_\phi(z) = \frac{C(z^2)}{C(z)}P_{\phi^*}(z). \tag{5.2.42}$$

On the other hand, from (5.2.41) and our earlier observation on $C(z)$, we have

$$\frac{C(z^2)}{C(z)} = \frac{\left(\sum_n \phi(n)z^{2n}\right) \Big/ \left(\sum_n \phi^*(n)z^{2n}\right)}{\left(\sum_n \phi(n)z^n\right) \Big/ \left(\sum_n \phi^*(n)z^n\right)} \tag{5.2.43}$$

$$= \frac{q_\phi(z^2)q_{\phi^*}(z)}{q_\phi(z)q_{\phi^*}(z^2)}.$$

So, since $q_\phi(z^2)$ and $q_{\phi^*}(z^2)$ are relatively prime, and since $P_\phi(z)$ in (5.2.42) is a polynomial, we may conclude that the polynomial $q_{\phi^*}(z)P_{\phi^*}(z)$ is divisible by $q_{\phi^*}(z^2)$; that is,

$$q_{\phi^*}(z)P_{\phi^*}(z) = r(z)q_{\phi^*}(z^2), \tag{5.2.44}$$

where $r(z)$ is some polynomial.

Let us assume that $\deg q_{\phi^*} \geq 1$ and let $\{z_1, \ldots, z_p\}$ be the zeros of $q_{\phi^*}$. Since $q_{\phi^*}(0) \neq 0$, we have $z_1, \ldots, z_p \neq 0$. In addition, since

$$C(z) = \frac{\sum_n \phi(n)z^n}{\sum_n \phi^*(n)z^n} = \frac{q_\phi(z)}{q_{\phi^*}(z)}, \tag{5.2.45}$$

where $q_\phi$ and $q_{\phi^*}$ are relatively prime, and $C(z)$ is zero-free and pole-free on $|z| = 1$, we conclude that none of the $z_j$, $j = 1, \ldots, p$, lie on $|z| = 1$. Hence, there exists some $j_0$, $1 \leq j_0 \leq p$, such that neither branch $\pm z'_{j_0}$ of the square-root of $z_{j_0}$ belongs to the set $\{z_1, \ldots, z_p\}$. That is, while each of the two polynomials, $(z - z'_{j_0})$ and $(z + z'_{j_0})$, does not divide $q_{\phi^*}(z)$, their product $(z - z'_{j_0})(z + z'_{j_0}) = (z^2 - z_{j_0})$ is a factor of $q_{\phi^*}(z^2)$. Therefore, it follows from (5.2.44) that $(z^2 - z_{j_0})$ is a factor of $P_{\phi^*}(z)$. Since $z_{j_0} \neq 0$, $P_{\phi^*}$ now has a symmetric root, and this is a contradiction to the hypothesis. Hence, $q_{\phi^*}$ must be a constant. Consequently, we have, from (5.2.42) and (5.2.43), that

$$P_\phi(z) = \frac{q_\phi(z^2)}{q_\phi(z)} P_{\phi^*}(z),$$

and this implies that $\deg P_\phi \geq \deg P_{\phi^*}$. ∎

In our proof of Theorem 5.11, we have actually derived several nice properties of any $\phi \in \Phi$, two of which are stated in the following.

**Theorem 5.12.** *For any $\phi_1, \phi_2 \in \Phi$, the symbol $C(z)$ of the sequence $\{s_n\}$ relating $\phi_1$ and $\phi_2$, in the sense that*

$$\phi_2(x) = \sum_{n=-\infty}^{\infty} s_n \phi_1(x - n),$$

*is a rational function which is both zero-free and pole-free on the unit circle $|z| = 1$. In addition, if $\phi_1 \in \Phi$ has minimum support, then $C(z)$ is a polynomial; that is, every $\phi_2 \in \Phi$ is a finite linear combination of integer translates of the minimally supported $\phi_1 \in \Phi$. In particular, the minimally supported $\phi_1 \in \Phi$ is unique.*

**Proof.** The first statement has been proved previously. If $\phi_1$ has minimum support, then by the preceding theorem, the two-scale symbol $P_{\phi_1}$ of $\phi_1$ has no symmetric roots. Hence, by the foregoing derivation, we note that $q_{\phi_1}$ is a constant, say $q_{\phi_1}(z) = q_{\phi_1}(0) \neq 0$; so in view of (5.2.45), with $\phi^* = \phi_1$, $C(z)$ is a polynomial and this establishes the second statement in the theorem. Finally, suppose that both $\phi_1$ and $\phi_2$ have minimum supports. Then by the definition of $\Phi$ and (5.2.13), we have

$$\text{supp } \phi_1 = \text{supp } \phi_2 = [0, N_{\phi_1}]$$

and

$$\phi_2(x) = \sum_{j=0}^{p} c_j \phi_1(x - j), \quad c_p \neq 0. \tag{5.2.46}$$

Suppose $p \geq 1$. Then for $x \in [N_{\phi_1} + p - 1, N_{\phi_1} + p]$, we have

$$0 = \phi_2(x) = \sum_{j=0}^{p} c_j \phi_1(x - j) = c_p \phi_1(x - p).$$

Since $\phi_1$ is non-trivial on $[N_{\phi_1} - 1, N_{\phi_1}]$, so is $\phi_1(\cdot - p)$ on $[N_{\phi_1} + p - 1, N_{\phi_1} + p]$. Hence, $c_p = 0$, which is a contradiction to (5.2.46); and this implies

$$\phi_2(x) = c_0 \phi_1(x).$$

That $c_0 = 1$ is a consequence of $\hat{\phi}_2(0) = \hat{\phi}_1(0) = 1$.  ∎

We end this section with the following example.

**Example 5.13.** For any positive integer $m$, the $m^{\text{th}}$ order cardinal $B$-spline $N_m$ is a scaling function that generates the MRA $\{V_j^m \colon j \in \mathbb{Z}\}$ of $L^2(\mathbb{R})$ as defined in Section 4.1. The two-scale relation of $N_m$ is given by (4.3.4), and the Riesz bounds of $N_m$ are $A = A_m$ and $B = 1$, where $A_m$ is defined in (4.2.21). Let $\Phi_m$ denote the class of all compactly supported $\phi \in V_0^m$ that generate the same MRA $\{V_j^m\}$. Then since the two-scale symbol $P_{N_m}$ of $N_m$ is $(1 + z)^m / 2^m$ which has no symmetric zeros, $N_m$ is the unique function in $\Phi_m$ with minimum support.

Without going into any details, we remark that $N_m$ is the only function in $\Phi_m$, although in general the cardinality of $\Phi$ may be infinite.

## 5.3. Direct-sum decompositions of $L^2(\mathbb{R})$

In the last section, we only considered scaling functions with finite two-scale sequences. To develop a more general theory, we will allow the two-scale sequences to be in $\ell^1$ (see the assumption (A3) in Section 5.1), so that the corresponding two-scale symbols belong to the so-called "*Wiener Class*".

**Definition 5.14.** *A Laurent series is said to belong to the Wiener Class $\mathcal{W}$ if its coefficient sequence is in $\ell^1$.*

Since the discrete convolution of two $\ell^1$-sequences is again a sequence in $\ell^1$, it is clear that $\mathcal{W}$ is an "*algebra*". The truth is that $\mathcal{W}$ is even more than an algebra, as seen in the following well-known theorem due to N. Wiener.

**Theorem 5.15.** *Let $f \in \mathcal{W}$ and suppose that $f(z) \neq 0$ for all $z$ on the unit circle $|z| = 1$. Then $\frac{1}{f} \in \mathcal{W}$ also.*

A proof of this theorem is unfortunately beyond the scope of this book. For the reader who is unwilling to accept this theorem, it is only a small sacrifice to consider the subclass of Laurent series of rational functions which are pole-free on $|z| = 1$, since the Laurent series of special interest are finite or at least have exponential decay.

Let $\phi$ be a scaling function whose two-scale symbol

$$P_\phi(z) = \frac{1}{2} \sum_{k=-\infty}^{\infty} p_k z^k \qquad (5.3.1)$$

is in $\mathcal{W}$. Recall that $P_\phi$ governs the relation $V_0 \subset V_1$ in the sense that

$$\phi(x) = \sum_k p_k \phi(2x - k) \qquad (5.3.2)$$

and $\phi$ "generates" $V_0$. Let us now consider any other $\ell^1$-sequence $\{q_k\}$ and its "symbol"

$$Q(z) = \frac{1}{2} \sum_{k=-\infty}^{\infty} q_k z^k, \qquad (5.3.3)$$

(which is halved to match the two-scale symbol $P$). Then $Q$ is also in $\mathcal{W}$ and defines a function

$$\psi(x) := \sum_k q_k \phi(2x - k) \tag{5.3.4}$$

in $V_1$. This function $\psi$ also generates a closed subspace $W_0$ in the same manner as $\phi$ generates $V_0$, namely:

$$W_0 := \text{clos}_{L^2(\mathbb{R})} \langle \psi(\cdot - k) : k \in \mathbb{Z} \rangle. \tag{5.3.5}$$

Hence, analogous to what

$$P := P_\phi \tag{5.3.6}$$

does, the symbol $Q$ governs the relation $W_0 \subset V_1$ in the sense that (5.3.4) and (5.3.5) are satisfied.

Of course the relation between the two subspaces $V_0$ and $W_0$ of $V_1$ must depend on the relation between the two symbols $P$ and $Q$. Our main concern in the construction of wavelets is at least to ensure that $V_0$ and $W_0$ are complementary subspaces of $V_1$, in the sense that

$$V_0 \cap W_0 = \{0\} \quad \text{and} \quad V_1 = V_0 + W_0. \tag{5.3.7}$$

As in (1.4.4), the two properties in (5.3.7) together are referred to by saying that $V_1$ is the "*direct sum*" of $V_0$ and $W_0$, and the notation

$$V_1 = V_0 \dotplus W_0 \tag{5.3.8}$$

is used in place of (5.3.7). In the following, we will see that the matrix

$$M_{P,Q}(z) := \begin{bmatrix} P(z) & Q(z) \\ P(-z) & Q(-z) \end{bmatrix} \tag{5.3.9}$$

plays an essential role in characterizing (5.3.8). Hence, we must consider the determinant

$$\Delta_{P,Q}(z) := \det M_{P,Q}(z) \tag{5.3.10}$$

of the matrix in (5.3.9). Since $P$ and $Q$ are in $\mathcal{W}$ and $\mathcal{W}$ is an algebra, we have

$$\Delta_{P,Q} \in \mathcal{W}$$

also. In addition, if $\Delta_{P,Q}(z) \neq 0$ on $|z| = 1$, then by Theorem 5.15, we also have

$$\frac{1}{\Delta_{P,Q}} \in \mathcal{W}.$$

So, under the condition $\Delta_{P,Q} \neq 0$ on $|z| = 1$, the two functions

$$\begin{cases} G(z) := \dfrac{Q(-z)}{\Delta_{P,Q}(z)}; \\[2mm] H(z) := \dfrac{-P(-z)}{\Delta_{P,Q}(z)} \end{cases} \tag{5.3.11}$$

are both in the Wiener Class $\mathcal{W}$. The reason for considering the functions $G$ and $H$ in (5.3.11) is that the transpose $M_{G,H}^T$ of $M_{G,H}$ is the inverse of $M_{P,Q}$, namely:

$$\begin{cases} M_{P,Q}(z)M_{G,H}^T(z) = \begin{bmatrix} 1 & 0 \\ 0 & 1 \end{bmatrix}; \\ M_{G,H}^T(z)M_{P,Q}(z) = \begin{bmatrix} 1 & 0 \\ 0 & 1 \end{bmatrix}, \quad |z| = 1. \end{cases} \tag{5.3.12}$$

The first identity in (5.3.12) is equivalent to the pair of identities

$$\begin{cases} P(z)G(z) + Q(z)H(z) = 1; \\ P(z)G(-z) + Q(z)H(-z) = 0, \quad |z| = 1, \end{cases} \tag{5.3.13}$$

while the second identity in (5.3.12) is equivalent to the following set of four identities:

$$\begin{cases} P(z)G(z) + P(-z)G(-z) = 1; \\ P(z)H(z) + P(-z)H(-z) = 0; \\ G(z)Q(z) + G(-z)Q(-z) = 0; \\ Q(z)H(z) + Q(-z)H(-z) = 1, \quad |z| = 1. \end{cases} \tag{5.3.14}$$

For $L^2(\mathbb{R})$ decomposition, we do not need the identities in (5.3.14). However, this set of identities will be crucial to our discussion of "duality" in the next section.

Since $G, H \in \mathcal{W}$, we may write

$$\begin{cases} G(z) = \dfrac{1}{2} \displaystyle\sum_{n=-\infty}^{\infty} g_n z^n; \\ H(z) = \dfrac{1}{2} \displaystyle\sum_{n=-\infty}^{\infty} h_n z^n, \end{cases} \tag{5.3.15}$$

where $\{g_n\}, \{h_n\} \in \ell^1$, whenever $\Delta_{P,Q}(z) \neq 0$ on the unit circle. We are now ready to formulate the following decomposition result.

**Theorem 5.16.** *A necessary and sufficient condition for the direct-sum decomposition (5.3.8) to hold is that the (continuous) function $\Delta_{P,Q}$ never vanishes on the unit circle $|z| = 1$. Furthermore, if $\Delta_{P,Q}(z) \neq 0$ for all $|z| = 1$, then the family $\{\psi(\cdot - k): k \in \mathbb{Z}\}$, governed by $Q(z)$ as in (5.3.4), is a Riesz basis of $W_0$, and the "decomposition relation"*

$$\phi(2x - \ell) = \frac{1}{2} \sum_{k=-\infty}^{\infty} \{g_{2k-\ell}\phi(x - k) + h_{2k-\ell}\psi(x - k)\}, \quad \ell \in \mathbb{Z}, \tag{5.3.16}$$

*holds for all $x \in \mathbb{R}$.*

**Proof in one direction.** We will only take care of the important direction. Hence, in the following, it is assumed that $\Delta_{P,Q}(z) \neq 0$ for all $z$ satisfying

$|z| = 1$. As a consequence, all sequences to be considered are in $\ell^1$ and there will be no danger in interchanging the orders of summation.

Observe that as an equivalent formulation of (5.3.13), we have

$$\begin{cases} P(z)(G(z) + G(-z)) + Q(z)(H(z) + H(-z)) = 1; \\ P(z)(G(z) - G(-z)) + Q(z)(H(z) - H(-z)) = 1, \quad |z| = 1, \end{cases} \quad (5.3.17)$$

which, in view of (5.3.15), may be written as

$$\begin{cases} P(z) \sum_k g_{2k} z^{2k} + Q(z) \sum_k h_{2k} z^{2k} = 1; \\ P(z) \sum_k g_{2k-1} z^{2k-1} + Q(z) \sum_k h_{2k-1} z^{2k-1} = 1, \quad |z| = 1. \end{cases} \quad (5.3.18)$$

Hence, by setting $z = e^{-i\omega/2}$ and multiplying the two identities in (5.3.18) by $\hat{\phi}(\frac{\omega}{2})$ and $z\hat{\phi}(\frac{\omega}{2})$, consecutively, we have

$$\begin{cases} \hat{\phi}\left(\frac{\omega}{2}\right) = \sum_k \left( g_{2k} z^{2k} P(z) \hat{\phi}\left(\frac{\omega}{2}\right) + h_{2k} z^{2k} Q(z) \hat{\phi}\left(\frac{\omega}{2}\right) \right); \\ \hat{\phi}\left(\frac{\omega}{2}\right) e^{-i\frac{\omega}{2}} = \sum_k \left( g_{2k-1} z^{2k} P(z) \hat{\phi}\left(\frac{\omega}{2}\right) + h_{2k-1} z^{2k} Q(z) \hat{\phi}\left(\frac{\omega}{2}\right) \right), \end{cases}$$

which is equivalent to

$$\begin{cases} \hat{\phi}\left(\frac{\omega}{2}\right) = \sum_k \left( g_{2k} z^{2k} \hat{\phi}(\omega) + h_{2k} z^{2k} \widehat{\psi}(\omega) \right); \\ \hat{\phi}\left(\frac{\omega}{2}\right) e^{-i\frac{\omega}{2}} = \sum_k \left( g_{2k-1} z^{2k} \hat{\phi}(\omega) + h_{2k-1} z^{2k} \widehat{\psi}(\omega) \right), \end{cases} \quad (5.3.19)$$

where the Fourier transform formulations of (5.3.2) and (5.3.4) have been used. Consequently, by taking the inverse Fourier transform on both sides in (5.3.19), we obtain

$$\begin{cases} 2\phi(2x) = \sum_k \left( g_{2k}\phi(x - k) + h_{2k}\psi(x - k) \right); \\ 2\phi(2x - 1) = \sum_k \left( g_{2k-1}\phi(x - k) + h_{2k-1}\psi(x - k) \right). \end{cases} \quad (5.3.20)$$

It is clear that (5.3.20) is equivalent to (5.3.16). As a consequence, since $\{g_k\}$ and $\{h_k\}$ are in $\ell^1$, and since

$$V_1 = \text{clos}_{L^2(\mathbb{R})} \langle \phi(2 \cdot -k) \colon k \in \mathbb{Z} \rangle,$$

we have now shown that $V_1 \subset V_0 + W_0$, so that

$$V_1 = V_0 + W_0.$$

To prove that this is a direct sum, we consider

$$\sum_k a_k \phi(x-k) + \sum_k b_k \psi(x-k) = 0, \qquad (5.3.21)$$

where $\{a_k\}$ and $\{b_k\}$ are in $\ell^2$. Then by applying the two-scale relations in (5.3.2) and (5.3.4), we obtain

$$\sum_\ell \left( \sum_k a_k p_{\ell-2k} + \sum_k b_k q_{\ell-2k} \right) \phi(2x-\ell) = 0,$$

so that

$$\sum_k a_k p_{\ell-2k} + \sum_k b_k q_{\ell-2k} = 0, \quad \ell \in \mathbb{Z}, \qquad (5.3.22)$$

by referring to the fact that $\{\phi(2 \cdot -\ell) \colon \ell \in \mathbb{Z}\}$ is a Riesz basis of $V_1$. Now, taking the symbols (or "$z$-transforms") of both sides of (5.3.22), we have

$$A(z^2)P(z) + B(z^2)Q(z) = 0, \qquad (5.3.23)$$

where $A$ and $B$ denote the symbols of $\{a_k\}$ and $\{b_k\}$, respectively. So, if $z$ is also replaced by $-z$, then (5.3.23) gives rise to the linear equations:

$$\begin{cases} P(z)A(z^2) + Q(z)B(z^2) = 0; \\ P(-z)A(z^2) + Q(-z)B(z^2) = 0, \end{cases}$$

with two unknowns $A(z^2)$ and $B(z^2)$, where the coefficient matrix is $M_{P,Q}(z)$, which is nonsingular for all $z$ on $|z| = 1$. Hence $A(z^2)$ and $B(z^2)$ must be zero, and the $\ell^2$-sequences $\{a_k\}$ and $\{b_k\}$ in (5.3.21) are trivial. This proves that $V_0 \cap W_0 = \{0\}$.

To prove that the family $\{\psi(\cdot - k) \colon k \in \mathbb{Z}\}$ is a Riesz basis of $W_0$, we will rely on Theorem 3.24. In particular, since $\{\phi(\cdot - k) \colon k \in \mathbb{Z}\}$ is a Riesz basis of $V_0$, we have

$$0 < A \le \sum_k |\hat{\phi}(\omega + 2\pi k)|^2 \le B < \infty, \quad \omega \in \mathbb{R}. \qquad (5.3.24)$$

Also, it follows from the Fourier transform formulation of (5.3.4) that

$$\sum_k |\hat{\psi}(\omega + 2\pi k)|^2 = \sum_k \left| Q(e^{-i(\frac{\omega}{2} + \pi k)}) \right|^2 \left| \hat{\phi}\left(\frac{\omega}{2} + \pi k\right) \right|^2$$

$$= |Q(z)|^2 \sum_k \left| \hat{\phi}\left(\frac{\omega}{2} + 2\pi k\right) \right|^2$$

$$+ |Q(-z)|^2 \sum_k \left| \hat{\phi}\left(\frac{\omega}{2} + \pi + 2\pi k\right) \right|^2,$$

where $z = e^{-i\omega/2}$, so that an application of (5.3.24) yields

$$A\{|Q(z)|^2 + |Q(-z)|^2\} \leq \sum_k |\hat{\psi}(\omega + 2\pi k)|^2 \tag{5.3.25}$$

$$\leq B\{|Q(z)|^2 + |Q(-z)|^2\}.$$

Since $Q \in \mathcal{W}$, it is continuous on $|z| = 1$, and we have

$$B' := 2 \max_{|z|=1} |Q(z)| < \infty. \tag{5.3.26}$$

On the other hand, in view of

$$\Delta_{P,Q}(z) = \det \begin{bmatrix} P(z) & Q(z) \\ P(-z) & Q(-z) \end{bmatrix} \neq 0, \quad |z| = 1,$$

we see that not both $Q(z)$ and $Q(-z)$ can vanish at the same $z$ on the unit circle, and so, again by the continuity of $Q$ on $|z| = 1$, we have

$$A' := \min_{|z|=1} (|Q(z)|^2 + |Q(-z)|^2) > 0. \tag{5.3.27}$$

Hence, it follows from (5.3.25), (5.3.26), and (5.3.27) that

$$AA' \leq \sum_k |\hat{\psi}(\omega + 2\pi k)|^2 \leq BB', \quad \omega \in \mathbb{R}, \tag{5.3.28}$$

or $\{\psi(\cdot - k): k \in \mathbb{Z}\}$ is a Riesz basis of $W_0$. $\blacksquare$

We must now pause for a moment and comment on the decomposition of $L^2(\mathbb{R})$ via Theorem 5.16.

**Remark 5.17.** Let $\Delta_{P,Q}(z) \neq 0$ for all $z$ on the unit circle and define

$$W_j := \text{clos}_{L^2(\mathbb{R})} \langle \psi(2^j \cdot - k): k \in \mathbb{Z} \rangle, \quad j \in \mathbb{Z}. \tag{5.3.29}$$

Then in view of the definition of $V_j$, $j \in \mathbb{Z}$, and the assertion $V_1 = V_0 \dot{+} W_0$ in Theorem 5.16, we have

$$V_{j+1} = V_j \dot{+} W_j, \quad j \in \mathbb{Z}. \tag{5.3.30}$$

Hence, since $\{V_j\}$ is an MRA of $L^2(\mathbb{R})$, it follows that the family $\{W_j\}$ constitutes a direct-sum decomposition of $L^2(\mathbb{R})$, namely:

$$L^2(\mathbb{R}) = \cdots \dot{+} W_{-1} \dot{+} W_0 \dot{+} \cdots . \tag{5.3.31}$$

Furthermore, the decomposition relation in (5.3.16) gives rise to the decomposition algorithm described by (1.6.9) with $a_k = g_{-k}$ and $b_k = h_{-k}$, and the pair of two-scale relations in (5.3.2) and (5.3.4) gives rise to the reconstruction algorithm described by (1.6.10). (Derivation of this and other details concerning these decomposition and reconstruction algorithms will be given in the next section.) However, from the most modest assumption that $\Delta_{P,Q}(z) \neq 0$ for $|z| = 1$, it is not possible to draw any conclusion on time-frequency analysis.

(i) From (5.3.3), it follows that

$$\int_{-\infty}^{\infty} \psi(x)dx = \sum_{k=-\infty}^{\infty} q_k \int_{-\infty}^{\infty} \phi(2x - k)dx \qquad (5.3.32)$$

$$= \frac{1}{2} \sum_{k=-\infty}^{\infty} q_k \hat{\phi}(0) = Q(1).$$

As usual, let

$$\psi_{j,k}(x) := 2^{j/2} \psi(2^j x - k).$$

Then Theorem 5.16 says that for each $j \in \mathbb{Z}$, the family $\{\psi_{j,k} \colon k \in \mathbb{Z}\}$ is a Riesz basis of $W_j$. However, the whole family $\{\psi_{j,k} \colon j, k \in \mathbb{Z}\}$ is not necessarily a Riesz basis of $L^2(\mathbb{R})$. Indeed, as shown in Chapter 3, for a function $\psi$ to generate a Riesz basis of $L^2(\mathbb{R})$ such that $\hat{\psi}$ is continuous, its integral over $(-\infty, \infty)$ must be zero, and so, in view of (5.3.32), a necessary condition is that

$$Q(1) = 0. \qquad (5.3.33)$$

(ii) Even if $\psi$ would generate a Riesz basis of $L^2(\mathbb{R})$, $\psi$ may not be a wavelet (or more precisely, an $\mathcal{R}$-wavelet), since the existence of the dual $\tilde{\psi}$ of $\psi$ still has to be investigated. (See Definition 1.5 and the example in (1.4.1) of an $\mathcal{R}$-function which does not have a dual.) Recall that in any series representation

$$f(x) = \sum_{j,k} c_{j,k} \psi_{j,k}(x), \quad f \in L^2(\mathbb{R}),$$

it requires a dual $\tilde{\psi}$ of $\psi$ to extract any time-frequency information of $f$ from the coefficients $c_{j,k}$ (see Section 1.4 and Theorem 3.27).

## 5.4. Wavelets and their duals

We continue our discussion of the decomposition of $L^2(\mathbb{R})$ and extend our effort to ensure that the decompositions are "*wavelet decompositions*". As observed in Remark 5.17, this requires the function $\psi$, governed by the Laurent series $Q \in \mathcal{W}$ according to (5.3.4), to be a wavelet with some dual wavelet $\tilde{\psi}$. In particular, $Q$ must satisfy (5.3.33). Recall that the two-scale symbol $P = P_\phi \in \mathcal{W}$ must also satisfy the conditions in (5.1.12) and (5.1.13). Hence, $P$ and $Q$ necessarily satisfy the conditions

$$\begin{cases} P(1) = 1 \quad \text{and} \quad P(-1) = 0; \\ Q(1) = 0. \end{cases} \qquad (5.4.1)$$

Let $G$ and $H$ be the Laurent series defined by (5.3.11). Then we have $G, H \in \mathcal{W}$, and the four Laurent series $P, Q, G, H$ satisfy the identities in (5.3.13). Therefore it follows from this set of identities and (5.4.1) that $G$ must also satisfy the conditions

$$G^*(1) = 1 \quad \text{and} \quad G^*(-1) = 0, \tag{5.4.2}$$

where the notation

$$G^*(z) := \overline{G(z)} = \overline{G}\left(\frac{1}{z}\right), \quad |z| = 1, \tag{5.4.3}$$

is used to facilitate our forthcoming presentation. The similarity between $P$ and $G^*$, as described by (5.4.1) and (5.4.2), suggests that

$$G^*(z) = \frac{1}{2} \sum_{n=-\infty}^{\infty} \bar{g}_{-n} z^n, \quad |z| = 1, \tag{5.4.4}$$

(see (5.3.15)) should also be chosen as the two-scale symbol of some scaling function that generates a possibly different MRA of $L^2(\mathbb{R})$.

This motivates the following strategy for constructing wavelets and their duals. We will start from two admissible two-scale symbols $P = P_\phi$ and $G^* = G^*_{\tilde{\phi}}$, such that both

$$\begin{cases} \hat{\phi}(\omega) = \prod_{k=1}^{\infty} P(e^{-i\omega/2^k}) \quad \text{and} \\ \hat{\tilde{\phi}}(\omega) = \prod_{k=1}^{\infty} G^*(e^{-i\omega/2^k}) \end{cases} \tag{5.4.5}$$

are in $L^2(\mathbb{R})$ (see Definition 5.4 and Theorem 5.5). Moreover, we assume that $\phi$ generates an MRA $\{V_j\}$ and $\tilde{\phi}$ generates an MRA $\{\tilde{V}_j\}$ of $L^2(\mathbb{R})$. Then according to Theorem 5.16, selecting any two arbitrary Laurent series $Q$ and $H$ that satisfy

$$\Delta_{P,Q}(z) \neq 0 \quad \text{and} \quad \Delta_{G,H}(z) \neq 0, \quad |z| = 1, \tag{5.4.6}$$

will result in two totally unrelated direct-sum decompositions of $L^2(\mathbb{R})$. In view of the discussion in the previous section, we will make use of the first identity in (5.3.14) to make a connection between these two decompositions.

**Definition 5.18.** *The two-scale symbols $P = P_\phi$ and $G^* = G^*_{\tilde{\phi}}$ are said to be "duals" of each other if they satisfy the identity*

$$P(z)G(z) + P(-z)G(-z) = 1, \quad |z| = 1, \tag{5.4.7}$$

*(see (5.4.3) for the relation between $G^*$ and $G$).*

Hence, if the two Laurent series $Q$ and $H$ are so chosen that the two nonsingular matrices $M_{P,Q}(z)$ and $M_{G,H}^T(z)$ are inverses of each other on $|z| = 1$, that is,

$$M_{P,Q}(z)M_{G,H}^T(z) = M_{G,H}^T(z)M_{P,Q}(z) = \begin{bmatrix} 1 & 0 \\ 0 & 1 \end{bmatrix}, \quad |z| = 1, \qquad (5.4.8)$$

then by (5.3.14) and the equivalence between this identity (see also (5.3.12)), we have

$$\begin{cases} P(z)H(z) + P(-z)H(-z) = 0; \\ G(z)Q(z) + G(-z)Q(-z) = 0; \\ Q(z)H(z) + Q(-z)H(-z) = 1, \quad |z| = 1. \end{cases} \qquad (5.4.9)$$

Of course, (5.4.8) is also equivalent to

$$\begin{cases} P(z)G(z) + Q(z)H(z) = 1; \\ P(-z)G(z) + Q(-z)H(z) = 0, \quad |z| = 1, \end{cases} \qquad (5.4.10)$$

(see (5.3.13)). In this regard, we have the following.

**Theorem 5.19.** *Let $P$ and $G^*$ be dual two-scale symbols as in Definition 5.18. Then the Laurent series $Q$ and $H$ in $\mathcal{W}$ satisfy (5.4.8) if and only if they are chosen from the class:*

$$\begin{cases} Q(z) = z^{-1}G(-z)K(z^2) \quad \text{and} \quad H(z) = zP(-z)K^{-1}(z^2), \\ \text{where } K \in \mathcal{W}, \text{ with } K(z) \neq 0 \text{ on } |z| = 1. \end{cases} \qquad (5.4.11)$$

**Proof.** It is easy to verify that every pair of $Q$ and $H$ from (5.4.11) satisfies (5.4.8). To derive the converse, we rely on the equivalence between (5.4.8) and (5.4.10). So, by applying Cramer's rule, we may express $G$ and $H$ in terms of $P$ and $Q$, namely:

$$G(z) = \frac{Q(-z)}{\Delta_{P,Q}(z)} \quad \text{and} \quad H(z) = \frac{-P(-z)}{\Delta_{P,Q}(z)}, \quad |z| = 1 \qquad (5.4.12)$$

(see (5.3.11)), where $\Delta_{P,Q}(z) = P(z)Q(-z) - P(-z)Q(z) \neq 0$ for $|z| = 1$. Since $\Delta_{P,Q}(-z) = -\Delta_{P,Q}(z)$, we may define

$$K(z^2) := z\Delta_{P,Q}(-z), \quad |z| = 1, \qquad (5.4.13)$$

so that $K \in \mathcal{W}$ by Theorem 5.15, and $K(z) \neq 0$ for $|z| = 1$. Now, (5.4.11) follows from (5.4.12) and (5.4.13). ∎

We remark that by (5.4.2) and the first identity in (5.4.9), the pair $(G^*, H^*)$ satisfies the condition

$$\begin{cases} G^*(1) = 1 \quad \text{and} \quad G^*(-1) = 0; \\ H^*(1) = 0, \end{cases} \qquad (5.4.14)$$

which is the same set of conditions as in (5.4.1) for the pair $(P, Q)$. In addition, in our strategy for constructing wavelets and dual wavelets through $Q$ and $H$ of the class described by (5.4.11), the two-scale symbols $P = P_\phi$ and $G^* = G^*_{\tilde{\phi}}$ play the same leading roles. Hence, the two pairs $(P, Q)$ and $(G^*, H^*)$ are interchangeable. This is called the "*duality principle*" to be discussed at greater length later in this section.

It is therefore important to study the two admissible two-scale symbols $P$ and $G^*$ in more detail. According to Definition 5.4, we can write

$$
\begin{cases}
P(z) = \left(\dfrac{1+z}{2}\right)^N S(z); \\
G^*(z) = \left(\dfrac{1+z}{2}\right)^{\tilde{N}} \tilde{S}(z), \quad |z| = 1,
\end{cases}
\tag{5.4.15}
$$

where $N$ and $\tilde{N}$ are positive integers, $S(1) = \tilde{S}(1) = 1$, and the $L^\infty(0, 2\pi)$ moduli of continuity of both $S(e^{-i\omega})$ and $\tilde{S}(e^{-i\omega})$ are of orders $O(\eta^\alpha)$ and $O(\eta^{\tilde{\alpha}})$, respectively, where $0 < \alpha, \tilde{\alpha} \le 1$. In what follows, we will require the factors $S$ and $\tilde{S}$ in (5.4.15) to satisfy, in addition,

$$
\begin{cases}
B := \max_{|z|=1} |S(z)| < 2^{N-\frac{1}{2}}; \\
\tilde{B} := \max_{|z|=1} |\tilde{S}(z)| < 2^{\tilde{N}-\frac{1}{2}},
\end{cases}
\tag{5.4.16}
$$

and employ the standard notation $\chi_A$ for the characteristic function of a set $A$.

**Lemma 5.20.** *Let $P$ and $G^*$ be admissible two-scale symbols as in (5.4.15) that satisfy (5.4.16). Then*

$$
\lim_{n\to\infty} \int_{-\infty}^{\infty} \left| \chi_{[-2^n\pi, 2^n\pi]}(\omega) \prod_{k=1}^{n} P(e^{-i\omega/2^k}) G(e^{-i\omega/2^k}) \right.
\tag{5.4.17}
$$
$$
\left. - \prod_{k=1}^{\infty} P(e^{-i\omega/2^k}) G(e^{-i\omega/2^k}) \right| d\omega = 0.
$$

**Proof.** Since some of the required estimates are quite similar to those in the proof of Theorem 5.5, we will not elaborate on them. Let us first show that

$$
\left| \prod_{k=1}^{\infty} P(e^{-i\omega/2^k}) G(e^{-i\omega/2^k}) \right| \le \frac{C}{(1+|\omega|)^{1+\eta}}, \quad \omega \in \mathbb{R},
\tag{5.4.18}
$$

for some $\eta > 0$. This estimate is quite straightforward. Indeed, for any positive integer $n_0$, and all $\omega$ with $2^{n_0} < |\omega|/\pi \le 2^{n_0+1}$, it follows from the first assumption in (5.4.16) that

$$\left| \prod_{k=1}^{n_0} S(e^{-i\omega/2^k}) \right| \leq B^{n_0} \leq C_1 \left( 1 + \frac{|\omega|}{\pi} \right)^{\log_2 B}$$

$$\leq C_1'(1 + |\omega|)^{\log_2 B} \leq C_1'(1 + |\omega|)^{N - \frac{1}{2} - \eta_1}.$$

for some $\eta_1 > 0$. In addition, for any $K > n_0$, we have, by the admissibility condition,

$$\left| \prod_{k=n_0+1}^{K} S(e^{-i\omega/2^k}) \right| = \prod_{k=n_0+1}^{K} |1 + (S(e^{-i\omega/2^k}) - 1)|$$

$$\leq \prod_{k=n_0+1}^{K} \left| 1 + O\left( \frac{|\omega|^\alpha}{2^{k\alpha}} \right) \right| \leq C_2.$$

Hence, we have

$$\left| \prod_{k=1}^{\infty} S(e^{-i\omega/2^k}) \right| \leq C_1 C_2 (1 + |\omega|)^{N - \frac{1}{2} - \eta_1},$$

so that

$$\left| \prod_{k=1}^{\infty} P(e^{-i\omega/2^k}) \right| \leq C \left| \frac{\sin(\omega/2)}{\omega/2} \right|^N (1 + |\omega|)^{N - \frac{1}{2} - \eta_1}$$

$$\leq C(1 + |\omega|)^{-\frac{1}{2} - \eta_1}.$$

Since the same estimate applies to $G$, by using $\eta_2 > 0$, say, we obtain (5.4.18) with $\eta = \eta_1 + \eta_2$.

Next, we will derive that $C > 0$ and $\eta > 0$ exist, such that for any sufficiently large positive integer $n$, and $|\omega| \leq 2^n \pi$, we have

$$\left| \prod_{k=1}^{n} P(e^{-i\omega/2^k}) G(e^{-i\omega/2^k}) \right| \leq \frac{C}{(1 + |\omega|)^{1+\eta}}, \tag{5.4.19}$$

for some $\eta > 0$. To arrive at (5.4.19), we continue with the argument outlined above to get, for $|\omega| \leq 2^n \pi$,

$$\left| \prod_{k=1}^{n} P(e^{-i\omega/2^k}) \right| = \left| \frac{2\sin(\omega/2)}{2^{n+1}\sin(\omega/2^{n+1})} \right|^N \left| \prod_{k=1}^{n} S(e^{-i\omega/2^k}) \right|$$

$$\leq C(1 + |\omega|)^{-\frac{1}{2} - \eta_1},$$

where the inequalities, $\frac{2}{\pi}|\omega| \leq |\sin \omega| \leq |\omega|$, which hold for $|\omega| \leq \frac{\pi}{2}$ are used. Again the same estimate applies to $G$.

We now proceed to establish (5.4.17). First observe that by (5.4.18), the function

$$\prod_{k=1}^{\infty} P(e^{-i\omega/2^k})G(e^{-i\omega/2^k})$$

is in $L^1(\mathbb{R})$. Let $\varepsilon > 0$ be arbitrarily given. Choose $M > 0$ such that

$$\int_{|\omega|>M} (1+|\omega|)^{-1-2\eta}d\omega < \varepsilon.$$

Then we break the integral in (5.4.17) into the sum of two integrals. For the integral over $|\omega| \leq M$, the integrand converges uniformly to zero; and for the integral over $|\omega| > M$, we bound it by the sum of two integrals, one of which may be estimated by (5.4.18), while the other may be estimated by (5.4.19). This completes the proof of the lemma. ∎

Recall that the two admissible two-scale symbols $P$ and $G^*$ give rise to two scaling functions $\phi$ and $\tilde{\phi}$, as in (5.4.5). Although $\phi$ and $\tilde{\phi}$ might generate two different MRA's of $L^2(\mathbb{R})$, they could still be related in the following sense.

**Definition 5.21.** *Two scaling functions $\phi$ and $\tilde{\phi}$, generating possibly different MRA's $\{V_j\}$ and $\{\tilde{V}_j\}$, respectively, of $L^2(\mathbb{R})$, are said to be "dual scaling functions", if they satisfy the condition*

$$\langle \phi(\cdot - j), \tilde{\phi}(\cdot - k)\rangle = \int_{-\infty}^{\infty} \phi(x-j)\overline{\tilde{\phi}(x-k)}dx = \delta_{j,k}, \quad j,k \in \mathbb{Z}. \quad (5.4.20)$$

In the following, we shall give the connection between dual scaling functions and admissible two-scale symbols that are dual to each other.

**Theorem 5.22.** *Let $P = P_\phi$ and $G^* = G^*_{\tilde{\phi}}$ be two admissible two-scale symbols as defined in (5.4.15). Also, let $\phi$ and $\tilde{\phi}$ be the corresponding scaling functions whose Fourier transforms are given by (5.4.5). If $\phi$ and $\tilde{\phi}$ are dual scaling functions as in Definition 5.21, then $P$ and $G^*$ are dual to each other in the sense of (5.4.7). Conversely, if $P$ and $G^*$ are dual to each other and satisfy (5.4.16), then $\phi$ and $\tilde{\phi}$ are dual scaling functions.*

**Proof.** Let $\phi$ and $\tilde{\phi}$ be dual scaling functions. Then for each $n \in \mathbb{Z}$, we have

$$\delta_{n,0} = \langle \phi, \tilde{\phi}(\cdot - n)\rangle = \frac{1}{2\pi} \int_{-\infty}^{\infty} \hat{\phi}(\omega)\overline{\hat{\tilde{\phi}}(\omega)} \, e^{in\omega}d\omega$$

$$= \sum_{k=-\infty}^{\infty} \frac{1}{2\pi} \int_{2\pi k}^{2\pi(k+1)} \hat{\phi}(\omega)\overline{\hat{\tilde{\phi}}(\omega)} \, e^{in\omega}d\omega$$

$$= \frac{1}{2\pi} \int_0^{2\pi} \left( \sum_{k=-\infty}^{\infty} \hat{\phi}(\omega+2\pi k)\overline{\hat{\tilde{\phi}}(\omega+2\pi k)} \right) e^{in\omega}d\omega,$$

so that

$$\sum_{k=-\infty}^{\infty} \hat{\phi}(\omega + 2\pi k)\overline{\hat{\tilde{\phi}}(\omega + 2\pi k)} = 1, \quad \text{a.e.} \tag{5.4.21}$$

Hence, setting $z = e^{-i\omega/2}$ and applying (5.4.21), we obtain

$$\begin{aligned}
\delta_{n,0} &= \frac{1}{2\pi} \int_{-\infty}^{\infty} P(z)\overline{G^*(z)}\, \hat{\phi}\left(\frac{\omega}{2}\right) \overline{\hat{\tilde{\phi}}\left(\frac{\omega}{2}\right)} e^{in\omega}\, d\omega \\
&= \frac{1}{2\pi} \int_0^{2\pi} \sum_k \left[ P(z)G(z)\hat{\phi}\left(\frac{\omega}{2} + 2\pi k\right) \overline{\hat{\tilde{\phi}}\left(\frac{\omega}{2} + 2\pi k\right)} \right. \\
&\qquad \left. + P(-z)G(-z)\hat{\phi}\left(\frac{\omega}{2} + \pi + 2\pi k\right) \overline{\hat{\tilde{\phi}}\left(\frac{\omega}{2} + \pi + 2\pi k\right)} \right] e^{in\omega}\, d\omega \\
&= \frac{1}{2\pi} \int_0^{2\pi} [P(z)G(z) + P(-z)G(-z)] e^{in\omega}\, d\omega,
\end{aligned}$$

so that, by the continuity of $P$ and $G$ on $|z| = 1$, we have

$$P(z)G(z) + P(-z)G(-z) = 1, \quad |z| = 1.$$

That is, $P_\phi$ and $G_\phi^*$ are dual to each other.

To prove the converse, we fix a $j \in \mathbb{Z}$ and consider, for any positive integer $n$,

$$I_n := \frac{1}{2\pi} \int_{-2^n\pi}^{2^n\pi} \left( \prod_{k=1}^{n} P(e^{-i\omega/2^k})G(e^{-i\omega/2^k}) \right) e^{ij\omega}\, d\omega. \tag{5.4.22}$$

Then by a change of variable $x = 2^{-n}\omega$, we have

$$\begin{aligned}
I_n &= 2^n \frac{1}{2\pi} \int_{-\pi}^{\pi} \left( \prod_{k=1}^{n} P(e^{-i2^{n-k}x})G(e^{-i2^{n-k}x}) \right) e^{ij2^n x}\, dx \tag{5.4.23} \\
&= 2^n \frac{1}{2\pi} \int_0^{\pi} \left( \prod_{k=1}^{n-1} P(e^{-i2^{n-k}x})G(e^{-i2^{n-k}x}) \right) \\
&\qquad \times [P(e^{-ix})G(e^{-ix}) + P(-e^{-ix})G(-e^{-ix})] e^{ij2^n x}\, dx.
\end{aligned}$$

Now, by invoking the duality between $P$ and $G^*$ and making another change of variable $y = 2x$, it follows that

$$\begin{aligned}
I_n &= 2^{n-1} \frac{1}{2\pi} \int_0^{2\pi} \left( \prod_{k=1}^{n-1} P(e^{-i2^{n-k-1}y})G(e^{-i2^{n-k-1}y}) \right) e^{ij2^{n-1}y}\, dy \tag{5.4.24} \\
&= 2^{n-1} \frac{1}{2\pi} \int_{-\pi}^{\pi} \left( \prod_{k=1}^{n-1} P(e^{-i2^{(n-1)-k}y})G(e^{-i2^{(n-1)-k}y}) \right) e^{ij2^{n-1}y}\, dy.
\end{aligned}$$

Hence, comparing (5.4.24) with (5.4.23), we have $I_n = I_{n-1}$. Since this conclusion is valid for any positive integer $n$, we obtain

$$I_n = I_{n-1} = \cdots = I_0 = \frac{1}{2\pi} \int_{-\pi}^{\pi} e^{ij\omega} d\omega = \delta_{j,0}. \tag{5.4.25}$$

Finally, by applying Lemma 5.20, the result in (5.4.25) yields

$$\langle \phi, \tilde{\phi}(\cdot - j)\rangle = \frac{1}{2\pi} \int_{-\infty}^{\infty} \hat{\phi}(\omega)\overline{\hat{\tilde{\phi}}(\omega)}\, e^{ij\omega} d\omega$$

$$= \frac{1}{2\pi} \int_{-\infty}^{\infty} \left( \prod_{k=1}^{\infty} P(e^{-i\omega/2^k})G(e^{-i\omega/2^k}) \right) e^{ij\omega} d\omega$$

$$= \lim_{n \to \infty} I_n = I_0 = \delta_{j,0}.$$

This completes the proof of the theorem. ∎

Let us now select any $Q$ and $H$ from the class of functions in (5.4.11). By Theorem 5.19, the matrices $M_{P,Q}$ and $M_{G^*,H^*}$ are invertible on $|z| = 1$, and so Theorem 5.16 applies. In particular, by considering the functions

$$\begin{cases} \psi(x) := \sum_k q_k \phi(2x - k); \\ \tilde{\psi}(x) := \sum_k \bar{h}_{-k}\tilde{\phi}(2x - k), \end{cases} \tag{5.4.26}$$

where

$$\begin{cases} Q(z) := \frac{1}{2} \sum_k q_k z^k; \\ H^*(z) := \frac{1}{2} \sum_k \bar{h}_{-k} z^k, \end{cases} \tag{5.4.27}$$

(see (5.4.3) for the analogous formulation of $G^*$), and setting

$$\begin{cases} \psi_{j,k} := 2^{j/2}\psi(2^j \cdot -k); \\ \tilde{\psi}_{j,k} := 2^{j/2}\tilde{\psi}(2^j \cdot -k), \end{cases} \tag{5.4.28}$$

as well as

$$\begin{cases} W_j := \text{clos}_{L^2(\mathbb{R})}\langle \psi_{j,k} \colon k \in \mathbb{Z}\rangle; \\ \widetilde{W}_j := \text{clos}_{L^2(\mathbb{R})}\langle \tilde{\psi}_{j,k} \colon k \in \mathbb{Z}\rangle, \end{cases} \tag{5.4.29}$$

we have

$$\begin{cases} V_{j+1} = V_j \dot{+} W_j; \\ \widetilde{V}_{j+1} = \widetilde{V}_j \dot{+} \widetilde{W}_j, \quad j \in \mathbb{Z}. \end{cases} \tag{5.4.30}$$

Here, as usual, we set

$$\begin{cases} V_j := \mathrm{clos}_{L^2(\mathbb{R})} \langle \phi_{j,k} \colon \ k \in \mathbb{Z} \rangle; \\ \tilde{V}_j := \mathrm{clos}_{L^2(\mathbb{R})} \langle \tilde{\phi}_{j,k} \colon \ k \in \mathbb{Z} \rangle, \end{cases} \tag{5.4.31}$$

where

$$\begin{cases} \phi_{j,k} = 2^{j/2} \phi(2^j \cdot -k); \\ \tilde{\phi}_{j,k} := 2^{j/2} \tilde{\phi}(2^j \cdot -k), \end{cases} \tag{5.4.32}$$

with $\phi$ and $\tilde{\phi}$ being the scaling functions whose two-scale symbols are $P = P_\phi$ and $G^* = G^*_{\tilde{\phi}}$, respectively.

We shall next show that if the admissible two-scale symbols $P$ and $G^*$ are dual to each other in the sense that the identity

$$P(z)G(z) + P(-z)G(-z) = 1, \quad |z| = 1,$$

is satisfied, then not only are $\{\psi_{j,k}\}$ and $\{\tilde{\psi}_{j,k}\}$ dual to each other, but additional orthogonality properties are achieved as well.

**Theorem 5.23.** *Let $P = P_\phi$ and $G^* = G^*_{\tilde{\phi}}$ be two admissible two-scale symbols which satisfy (5.4.16) and are dual to each other. Then for any $Q, H \in \mathcal{W}$ chosen from the class (5.4.11), the functions $\phi, \tilde{\phi}, \psi,$ and $\tilde{\psi}$ defined as in (5.4.5) and (5.4.26) satisfy*

$$\langle \psi_{j,k}, \tilde{\psi}_{\ell,m} \rangle = \delta_{j,\ell} \delta_{k,m}, \quad j, k, \ell, m \in \mathbb{Z}; \tag{5.4.33}$$

*and*

$$\begin{cases} \langle \phi_{j,k}, \tilde{\psi}_{j,\ell} \rangle = 0; \\ \langle \tilde{\phi}_{j,k}, \psi_{j,\ell} \rangle = 0, \quad j, k, \ell \in \mathbb{Z}, \end{cases} \tag{5.4.34}$$

*that is, $V_j \perp \widetilde{W}_j$ and $\tilde{V}_j \perp W_j$ for all $j \in \mathbb{Z}$.*

**Proof.** Let us first consider the case $j = \ell$ in (5.4.33). In this case, by the third identity in (5.4.9) and (5.4.21) we have, again using the notation $z = e^{-i\omega/2}$,

$$\begin{aligned} \langle \psi_{j,k}, \tilde{\psi}_{j,m} \rangle &= \frac{1}{2\pi} \int_{-\infty}^{\infty} \hat{\psi}(\omega) \overline{\hat{\tilde{\psi}}(\omega)} \, e^{-i(k-m)\omega} d\omega \\ &= \frac{1}{2\pi} \int_{-\infty}^{\infty} Q(z) \overline{H^*(z)} \hat{\phi}\left(\frac{\omega}{2}\right) \overline{\hat{\tilde{\phi}}\left(\frac{\omega}{2}\right)} \, e^{-i(k-m)\omega} d\omega \\ &= \frac{1}{2\pi} \sum_{\ell} \int_0^{2\pi} \left[ Q(e^{-i(\frac{\omega}{2} + \pi\ell)}) H(e^{i(\frac{\omega}{2} + \pi\ell)}) \right. \\ &\qquad \left. \times \hat{\phi}\left(\frac{\omega}{2} + \pi\ell\right) \overline{\hat{\tilde{\phi}}\left(\frac{\omega}{2} + \pi\ell\right)} \right] e^{-i(k-m)\omega} d\omega \end{aligned} \tag{5.4.35}$$

$$= \frac{1}{2\pi} \int_0^{2\pi} \sum_\ell \left[ Q(z)H(z)\hat{\phi}\left(\frac{\omega}{2} + 2\pi\ell\right) \overline{\hat{\widetilde{\phi}}\left(\frac{\omega}{2} + 2\pi\ell\right)} \right.$$

$$\left. + Q(-z)H(-z)\hat{\phi}\left(\frac{\omega}{2} + \pi + 2\pi\ell\right) \overline{\hat{\widetilde{\phi}}\left(\frac{\omega}{2} + \pi + 2\pi\ell\right)} \right] e^{-i(k-m)\omega} d\omega$$

$$= \frac{1}{2\pi} \int_0^{2\pi} [Q(z)H(z) + Q(-z)H(-z)]e^{-i(k-m)\omega} d\omega$$

$$= \frac{1}{2\pi} \int_0^{2\pi} e^{-i(k-m)\omega} d\omega = \delta_{k,m}.$$

Proceeding to the general case, we observe, by applying the first two identities in (5.4.9) instead, that the same derivation given above also yields (5.4.34), so that

$$V_j \perp \widetilde{W}_j \quad \text{and} \quad \widetilde{V}_j \perp W_j, \quad j \in \mathbb{Z}. \tag{5.4.36}$$

Hence, if $j < \ell$, then

$$\psi_{j,k} \in W_j \subset V_{j+1} \subset V_\ell,$$

and by the first assertion in (5.4.36), we have

$$\langle \psi_{j,k}, \widetilde{\psi}_{\ell,m} \rangle = 0, \quad k, m \in \mathbb{Z}.$$

For $j > \ell$, the same conclusion can be drawn by applying the second assertion in (5.4.36). This completes the proof of the theorem. ∎

As a consequence of the biorthogonality property in (5.4.33), both families $\{\psi_{j,k}\}$ and $\{\widetilde{\psi}_{j,k}\}$ are $\ell^2$-linearly independent. Therefore, since

$$L^2(\mathbb{R}) = \cdots + W_{-1} \dot{+} W_0 \dot{+} W_1 \dot{+} \cdots \tag{5.4.37}$$
$$= \cdots + \widetilde{W}_{-1} \dot{+} \widetilde{W}_0 \dot{+} \widetilde{W}_1 \dot{+} \cdots,$$

both $\{\psi_{j,k}\}$ and $\{\widetilde{\psi}_{j,k}\}$ are bases of $L^2(\mathbb{R})$. In fact, under the hypotheses of Theorem 5.23, it follows that both $\{\psi_{j,k}\}$ and $\{\widetilde{\psi}_{j,k}\}$ are frames of $L^2(\mathbb{R})$ also. We do not intend to give a proof of this fact, since no simple derivation seems to be available. By an application of Theorem 3.20, we may now conclude that $\{\psi_{j,k}\}$ and $\{\widetilde{\psi}_{j,k}\}$ are actually Riesz bases of $L^2(\mathbb{R})$. That is, we have the following result.

**Theorem 5.24.** *Under the hypotheses of Theorem 5.23, the two functions $\psi \in W_0$ and $\widetilde{\psi} \in \widetilde{W}_0$ are wavelets which are dual to each other.*

Consequently, as stated in Theorem 3.27, every function $f \in L^2(\mathbb{R})$ has two (unique) wavelet series representation:

$$\begin{cases} f(x) = \sum_{j,k} \langle f, \widetilde{\psi}_{j,k} \rangle \psi_{j,k}(x); \\ f(x) = \sum_{j,k} \langle f, \psi_{j,k} \rangle \widetilde{\psi}_{j,k}(x), \end{cases} \tag{5.4.38}$$

where the coefficients are values of the IWT of $f$, relative to the basic wavelets $\tilde{\psi}$ and $\psi$ respectively, evaluated at the time-scale positions

$$(b, a) = \left( \frac{k}{2^j}, \frac{1}{2^j} \right)$$

(see Section 1.4 and Theorem 3.27).

It is therefore very important to derive efficient algorithms for finding these IWT values from $f$ and for reconstructing $f$ from these IWT values. It turns out that the two-scale sequences $\{\bar{g}_{-n}\}$ and $\{\bar{h}_{-n}\}$ (whose two-scale symbols are $G^* = G^*_{\phi}$ and $H^*$ as given by (5.4.4) and (5.4.27)) can be used for obtaining the IWT values $\langle f, \tilde{\psi}_{j,k} \rangle$. This computational scheme, called "decomposition algorithm" is a consequence of the decomposition relation (5.3.16) in Theorem 5.16. On the other hand, the two-scale sequences $\{p_n\}$ and $\{q_n\}$ (whose two-scale symbols are $P = P_{\phi}$ and $Q$ as given by (5.3.1) and (5.3.3)) can be used for reconstructing $f$ from its IWT values $\langle f, \tilde{\psi}_{j,k} \rangle$. This computational scheme, called the "reconstruction algorithm" is a consequence of the two-scale relations (5.3.2) and (5.3.4). If we wish to use $\psi$, instead of $\tilde{\psi}$, as the basic wavelet, then the two-scale sequences $\{p_n\}$ and $\{q_n\}$ are used in the decomposition algorithm, while the two-scale sequences $\{\bar{g}_{-n}\}$ and $\{\bar{h}_{-n}\}$ are used in the reconstruction algorithm.

In other words, the roles of the pairs

$$(\{\bar{g}_{-n}\}, \{\bar{h}_{-n}\}) \quad \text{and} \quad (\{p_n\}, \{q_n\})$$

for decomposition and reconstruction purposes are interchanged, if the IWT information

$$\left\{ (W_{\tilde{\psi}} f) \left( \frac{k}{2^j}, \frac{1}{2^j} \right) : j, k \in \mathbb{Z} \right\} \tag{5.4.39}$$

is replaced by the IWT information

$$\left\{ (W_{\psi} f) \left( \frac{k}{2^j}, \frac{1}{2^j} \right) : j, k \in \mathbb{Z} \right\}.$$

This is called the "duality principle" in wavelet decomposition-reconstruction. As a result, there is no need to describe both situations.

In what follows, we only discuss the IWT in (5.4.39) using $\tilde{\psi}$ as the basic wavelet. For any $f \in L^2(\mathbb{R})$, let $f_N$ be some approximant of $f$ from $V_N$ for a fixed $N \in \mathbb{Z}$. Note that this approximation does not have to be the $L^2(\mathbb{R})$ orthogonal projection. We may consider $V_N$ as the "sample space" and $f_N$ the "data" (or measurement) of $f$ on $V_N$. Since

$$V_N = W_{N-1} \dotplus V_{N-1} \tag{5.4.40}$$
$$= \dots = W_{N-1} \dotplus \dots \dotplus W_{N-M} \dotplus V_{N-M}$$

for any positive integer $M$, $f_N$ has a *unique* decomposition:

$$f_N(x) = g_{N-1}(x) + g_{N-2}(x) + \cdots + g_{N-M}(x) + f_{N-M}(x), \qquad (5.4.41)$$

where

$$\begin{cases} g_j(x) \in W_j, \quad j = N - M, \ldots, N - 1; \\ f_{N-M}(x) \in V_{N-M}. \end{cases} \qquad (5.4.42)$$

Let us write

$$\begin{cases} f_j(x) = \sum_k c_k^j \phi(2^j x - k) \in V_j, \text{ with} \\ \mathbf{c}^j := \{c_k^j\}, \quad k \in \mathbb{Z}, \end{cases} \qquad (5.4.43)$$

and

$$\begin{cases} g_j(x) = \sum_k d_k^j \psi(2^j x - k), \text{ with} \\ \mathbf{d}^j := \{d_k^j\}, \quad k \in \mathbb{Z}. \end{cases} \qquad (5.4.44)$$

Then the decomposition in (5.4.41) is uniquely determined by the sequences $\mathbf{c}^j$ and $\mathbf{d}^j$ in (5.4.43) and (5.4.44). It is important to note that

$$d_k^j = (W_{\tilde{\psi}} f_N)\left(\frac{k}{2^j}, \frac{1}{2^j}\right), \quad j, k \in \mathbb{Z}, \qquad (5.4.45)$$

are the values of the IWT of $f_N$, using $\tilde{\psi}$ as the basic wavelet. Observe that the decomposition in (5.4.41) is data-dependent. In the wavelet decomposition and reconstruction schemes to be discussed below, we will use the "*digital representations*" $\mathbf{c}^j, \mathbf{d}^j$ of $f_j(x)$ and $g_j(x)$, respectively.

To facilitate our discussion (and to avoid any possible confusion), we introduce the notation

$$\begin{cases} a_n := \frac{1}{2} g_{-n}; \\ b_n := \frac{1}{2} h_{-n}, \end{cases} \qquad (5.4.46)$$

where $\{\bar{g}_{-n}\}$ and $\{\bar{h}_{-n}\}$ are the two-scale sequences corresponding to the two-scale symbols $G^* = G_\phi^*$ and $H^*$, respectively (see (5.4.4) and (5.4.27)). Hence, the decomposition relation (5.3.16) in Theorem 5.16 now becomes

$$\phi(2x - \ell) = \sum_{k=-\infty}^{\infty} \{a_{\ell-2k}\phi(x-k) + b_{\ell-2k}\psi(x-k)\}, \quad \ell \in \mathbb{Z}. \qquad (5.4.47)$$

Let us now derive the decomposition and reconstruction algorithms stated in (1.6.9) and (1.6.10) in Chapter 1.

(i) *Decomposition algorithm*

$$\begin{cases} c_k^{j-1} = \sum_{\ell} a_{\ell-2k} c_\ell^j; \\ d_k^{j-1} = \sum_{\ell} b_{\ell-2k} c_\ell^j. \end{cases} \tag{5.4.48}$$

$$\begin{array}{ccccccccc} & \mathbf{d}^{N-1} & & \mathbf{d}^{N-2} & & & & \mathbf{d}^{N-M} & \\ & \nearrow & & \nearrow & & \nearrow & & \nearrow & \\ \mathbf{c}^N & \longrightarrow & \mathbf{c}^{N-1} & \longrightarrow & \mathbf{c}^{N-2} & \longrightarrow & \cdots & \longrightarrow & \mathbf{c}^{N-M} \end{array}$$

**Proof.** By applying the decomposition relation (5.4.47), we have

$$f_j(x) = \sum_{\ell} c_\ell^j \phi(2^j x - \ell)$$

$$= \sum_{\ell} c_\ell^j \left[ \sum_k \{ a_{\ell-2k} \phi(2^{j-1}x - k) + b_{\ell-2k} \psi(2^{j-1}x - k) \} \right]$$

$$= \sum_k \left\{ \sum_{\ell} a_{\ell-2k} c_\ell^j \right\} \phi(2^{j-1}x - k)$$

$$+ \sum_k \left\{ \sum_{\ell} b_{\ell-2k} c_\ell^j \right\} \psi(2^{j-1}x - k).$$

Hence, from the decomposition $f_j(x) = f_{j-1}(x) + g_{j-1}(x)$, where $f_{j-1}(x)$ and $g_{j-1}(x)$ are given as in (5.4.43) and (5.4.44) with $j$ replaced by $j-1$, it follows that

$$\sum_k \left\{ \sum_{\ell} a_{\ell-2k} c_\ell^j - c_k^{j-1} \right\} \phi(2^{j-1}x - k)$$

$$+ \sum_k \left\{ \sum_{\ell} b_{\ell-2k} c_\ell^j - d_k^{j-1} \right\} \psi(2^{j-1}x - k) = 0,$$

so that (5.4.48) follows by invoking the $\ell^2$-linear independence of $\{ \phi_{j-1,k} : k \in \mathbb{Z} \}$ and $\{ \psi_{j-1,k} : k \in \mathbb{Z} \}$ and the fact that $V_{j-1} \cap W_{j-1} = \{0\}$. ∎

(ii) *Reconstruction algorithm*

$$c_k^j = \sum_{\ell} [p_{k-2\ell} c_\ell^{j-1} + q_{k-2\ell} d_\ell^{j-1}]. \tag{5.4.49}$$

$$\begin{array}{ccccccccc} \mathbf{d}^{N-M} & & \mathbf{d}^{N-M+1} & & & & \mathbf{d}^{N-1} & & \\ & \searrow & & \searrow & & & & \searrow & \\ \mathbf{c}^{N-M} & \longrightarrow & \mathbf{c}^{N-M+1} & \longrightarrow & \cdots & \mathbf{c}^{N-1} & \longrightarrow & \mathbf{c}^N \end{array}$$

**Proof.** By applying the two-scale relations (5.3.2) and (5.3.4), we have

$$f_{j-1}(x) + g_{j-1}(x) = \sum_{\ell}[c_\ell^{j-1}\phi(2^{j-1}x - \ell) + d_\ell^{j-1}\psi(2^{j-1}x - \ell)]$$

$$= \sum_{\ell}\left[c_\ell^{j-1}\sum_{k}p_k\phi(2^jx - 2\ell - k)\right.$$

$$\left. + d_\ell^{j-1}\sum_{k}q_k\phi(2^jx - 2\ell - k)\right]$$

$$= \sum_{\ell}\sum_{k}(c_\ell^{j-1}p_{k-2\ell} + d_\ell^{j-1}q_{k-2\ell})\phi(2^jx - k)$$

$$= \sum_{k}\left\{\sum_{\ell}[p_{k-2\ell}c_\ell^{j-1} + q_{k-2\ell}d_\ell^{j-1}]\right\}\phi(2^jx - k).$$

Since $f_{j-1}(x) + g_{j-1}(x) = f_j(x)$, we obtain (5.4.49) by referring to the representation formula (5.4.43) of $f_j(x)$ and the $\ell^2$-linear independence of $\{\phi_{j,k}:$ $k \in \mathbb{Z}\}$. ∎

Observe that both the decomposition and the reconstruction algorithms are moving average (MA) schemes, except that "downsampling" is required in decomposition and "upsampling" is required in reconstruction. To downsample, we simply keep every other term of the output sequence. More precisely, in (5.4.48), only the terms with even indices are kept, and the (even) indices of this output sequence are halved. To upsample, a zero is placed in between every two consecutive terms of the input sequences before the MA schemes are applied. More precisely, in (5.4.49), the indices of the input sequences $\{c_\ell^{j-1}\}$ and $\{d_\ell^{j-1}\}$ are multiplied by 2, and zeros are used as the terms with odd indices in the new input sequences (see (4.3.11) in Algorithm 4.7 in Chapter 4).

## 5.5. Linear-phase filtering

Scaling functions and wavelets can be considered as filter functions. If the space $L^2(\mathbb{R})$ represents the space of all analog signals with finite energy, and $\{V_j\}$ is an MRA of $L^2(\mathbb{R})$, then sampling an analog signal $f \in L^2(\mathbb{R})$ is accomplished by approximation (which may or may not be interpolation) from some "*sample space*" $V_N$, where $N$ should be chosen large enough to avoid undersampling. It must be emphasized that even if a digital sampling procedure is applied, the sampled signal $f_N \in V_N$ is still an analog signal, although $f_N$ has a series representation in terms of a scaling function, as described by (5.4.43), where the coefficient sequence $\mathbf{c}^N = \{c_k^N\}$ is formulated in terms of the digital samples. For instance, if the $m^{\text{th}}$ order cardinal spline space $V_N^m$ with knot sequence $2^{-N}\mathbb{Z}$ is used as the sampling space $V_N$, then the coefficient sequence $\mathbf{c}^N$ can be obtained by applying a finite moving average procedure to give a quasi-interpolant or interpolant $f_N$ of $f$ as studied in Section 4.5 and Section 4.6, respectively, in the previous chapter. In any case, the analog sample $f_N \in V_N$ of $f$ can now be decomposed as in (5.4.41), where for each $j = N - M, \ldots, N - 1$, $g_j(x)$ gives localized time-frequency information of $f_N$ in the $j^{\text{th}}$ octave (or frequency band). The importance of this filter-bank method is that the details of the sampled signal $f_N$ are sorted out and stored in different subspaces $W_j$ of $V_N$ for better analysis. For instance, in data compression, by simply applying thresholding at each octave we may obtain substantial saving in general. What we really mean by this is that after deleting information of very small magnitudes in each subspace $W_j$, much less data information has to be stored or transmitted, and the reconstruction algorithm can be applied later to give a good approximation of the original signal. Of course there are many more important applications of similar nature. However, since each component $g_j$ has been altered, we no longer have perfect reconstructions, so that special attention must be paid to possible distortion. The reconstructed signal is nothing but a wavelet series, which in turn means that it is a result of linear filtering. Therefore, distortion can be avoided if the filter has linear, or at least generalized linear, phase.

**Definition 5.25.** *Let $f \in L^2(\mathbb{R})$. Then $f$ is said to have "linear phase" if its Fourier transform satisfies*

$$\hat{f}(\omega) = \pm|\hat{f}(\omega)|e^{-ia\omega}, \quad a.e., \tag{5.5.1}$$

*where $a$ is some real constant and the $+$ or $-$ sign is independent of $\omega$. Also, $f$ is said to have "generalized linear phase" if*

$$\hat{f}(\omega) = F(\omega)e^{-i(a\omega+b)}, \quad a.e., \tag{5.5.2}$$

*where $F(\omega)$ is a real-valued function and $a, b$ are real constants. The constant $a$ in both (5.5.1) and (5.5.2) is called the phase of $\hat{f}$.*

**Example 5.26.** The Fourier transform of the $m^{\text{th}}$ order cardinal $B$-spline $N_m$ is given by

$$\widehat{N}_m(\omega) = \left(\frac{\sin(\omega/2)}{\omega/2}\right)^m e^{-im\omega/2},$$

and hence, $N_m$ has linear phase, and the phase of $\widehat{N}_m$ is $m/2$.

**Definition 5.27.** *Let $\{a_n\} \in \ell^1$ and $A(e^{-i\omega})$ be its discrete Fourier transform (or Fourier series). Then $\{a_n\}$ is said to have "linear phase" if*

$$A(e^{-i\omega}) = \pm|A(e^{-i\omega})|e^{-in_0\omega}, \quad \omega \in \mathbb{R}, \tag{5.5.3}$$

*where $n_0 \in \frac{1}{2}\mathbb{Z}$ and the $+$ or $-$ sign is independent of $\omega$. Also, $\{a_n\}$ is said to have "generalized linear phase" if*

$$A(e^{-i\omega}) = F(\omega)e^{-i(n_0\omega+b)}, \quad \omega \in \mathbb{R}, \tag{5.5.4}$$

*for some real-valued function $F(\omega)$, $n_0 \in \frac{1}{2}\mathbb{Z}$ and $b \in \mathbb{R}$. The value $n_0$ in both (5.5.3) and (5.5.4) is called the phase of the symbol of $\{a_n\}$.*

Let us first give a characterization of both functions and sequences with generalized linear phases.

**Lemma 5.28.**

(i) *A function $f \in L^2(\mathbb{R})$ has generalized linear phase in the sense of (5.5.2), where $a, b \in \mathbb{R}$, if and only if $e^{ib} f(x)$ is "skew-symmetric" with respect to $a$ in the sense that*

$$e^{ib} f(a+x) = \overline{e^{ib} f(a-x)}, \quad x \in \mathbb{R}. \tag{5.5.5}$$

(ii) *A sequence $\{a_n\} \in \ell^1$ has generalized linear phase in the sense of (5.5.4), where $n_0 \in \frac{1}{2}\mathbb{Z}$ and $b \in \mathbb{R}$, if and only if $\{e^{ib} a_n\}$ is "skew-symmetric" with respect to $n_0$ in the sense that*

$$e^{ib} a_n = \overline{e^{ib} a_{2n_0-n}}, \quad n \in \mathbb{Z}. \tag{5.5.6}$$

**Proof.** (i) Suppose that $f \in L^2(\mathbb{R})$ satisfies (5.5.2). Then

$$f(x) = \frac{1}{2\pi} \int_{-\infty}^{\infty} F(\omega) e^{-i(a\omega+b)} e^{ix\omega} d\omega,$$

or equivalently,

$$e^{ib} f(a-x) = \frac{1}{2\pi} \int_{-\infty}^{\infty} F(\omega) e^{-ix\omega} d\omega. \tag{5.5.7}$$

Since $F(\omega)$ is real, assertion (5.5.5) follows by equating the complex conjugate of the expression in (5.5.7) with itself.

Conversely, if (5.5.5) is satisfied, then taking the Fourier transform of both sides of (5.5.5) yields

$$e^{ib} \hat{f}(\omega) e^{ia\omega} = e^{-ib} \int_{-\infty}^{\infty} \overline{f(a-x)} \, e^{-i\omega x} dx$$

$$= e^{-ib} \overline{\int_{-\infty}^{\infty} f(a-x) e^{i\omega x} dx} = \overline{e^{ib} \hat{f}(\omega) e^{ia\omega}}.$$

Hence this quantity is real, and (5.5.2) follows by setting this real-valued function to be $F(\omega)$.

(ii) Suppose that $\{a_n\} \in \ell^1$ satisfies (5.5.4). Then we have

$$e^{i(n_0\omega+b)} A(e^{-i\omega}) = F(\omega) = \overline{F(\omega)} = e^{-i(n_0\omega+b)} \overline{A(e^{-i\omega})},$$

or equivalently,

$$e^{i2n_0\omega} e^{ib} A(e^{-i\omega}) = \overline{e^{ib} A(e^{-i\omega})}. \tag{5.5.8}$$

Hence, assertion (5.5.6) follows by comparing the coefficients of $e^{in\omega}$ in (5.5.8). Conversely, if (5.5.6) holds, then we have (5.5.8), and consequently

$$e^{i(n_0\omega+b)} A(e^{-i\omega}) = \overline{e^{i(n_0\omega+b)} A(e^{-i\omega})},$$

and we define this real-valued expression to be $F(\omega)$. This yields (5.5.4). ∎

**Remark.** The notion of *"skew-symmetry"* in (5.5.5) and (5.5.6) is not very satisfying because of the necessary complex conjugation. When $f(x)$ is real-valued, however, it is clear that for (5.5.5) to hold, $e^{i2b}$ must also be real, or $b = \frac{1}{2}\pi k$ where $k \in \mathbb{Z}$. That is, (5.5.5) becomes

$(1°)$ $f(a + x) = f(a - x)$, $x \in \mathbb{R}$ (symmetry),

or

$(2°)$ $f(a + x) = -f(a - x)$, $x \in \mathbb{R}$ (antisymmetry).

Of course, an analogous conclusion can be drawn for real $\ell^1$ sequences.

**Theorem 5.29.**

(i) *A real-valued function $f \in L^2(\mathbb{R})$ has generalized linear phase if and only if it is either symmetric or antisymmetric (with respect to the phase of $\hat{f}$).*

(ii) *A real-valued sequence $\{a_n\} \in \ell^1$ has generalized linear phase if and only if it is either symmetric or antisymmetric (with respect to the phase of the symbol of $\{a_n\}$).*

Characterization of linear phase is a little harder. However, in view of the foregoing discussion (see particularly the remark), we will only consider real-valued functions and sequences. Also, since the phase property of a two-scale sequence directly influences that of the corresponding scaling function, we give the following characterization of linear-phase sequences.

**Lemma 5.30.** *A real-valued $\ell^1$-sequence $\{a_n\}$, with symbol $A(e^{-i\omega})$, has linear phase if and only if there is some $n_0 \in \frac{1}{2}\mathbb{Z}$, such that $A(e^{-i\omega})e^{in_0\omega}$ is real-valued, even, and has no sign changes.*

The proof of this result is easy. However, if the sequence is finite, then we can say a little more.

**Lemma 5.31.** *A real-valued finite sequence $\{a_n\}$ with support $[0, N]$ has linear phase if and only if the following statements hold:*

(i) $a_{N-n} = a_n$, $n \in \mathbb{Z}$; *and*

(ii) *the symbol*

$$A(z) = \sum_{n=0}^{N} a_n z^n$$

*has only zeros of even order on the unit circle.*

**Proof.** By Lemma 5.30, the real-valued finite sequence $\{a_n\}$, $n = 0, \ldots, N$, has linear phase if and only if there exists some $n_0 \in \frac{1}{2}\mathbb{Z}$, such that the function

$$F(\omega) := A(e^{-i\omega})e^{in_0\omega}$$

is real-valued, even, and has no sign changes. On the other hand, it is clear that $F(\omega) = F(-\omega)$ is equivalent to

$$\sum_{n=0}^{N} a_n e^{in\omega} = A(e^{i\omega}) = e^{i2n_0\omega} A(e^{-i\omega})$$

$$= \sum_{n=0}^{N} a_n e^{i(2n_0-n)\omega} = \sum_{n=2n_0-N}^{2n_0} a_{2n_0-n} e^{in\omega},$$

which, in turn, is also equivalent to $n_0 = \frac{1}{2}N$ and $a_{N-n} = a_n$ for all $n \in \mathbb{Z}$. Of course, the real-valued function $F(\omega)$ has no sign changes if and only if its real zeros (if any) are of even order; and this, in turn, is equivalent to the statement that $A(z)$ has only zeros of even order on the unit circle. ∎

We now turn to a study of the phase properties of scaling functions.

**Theorem 5.32.** *Let $\phi$ be a scaling function with two-scale sequence $\{p_n\} \in \ell^1$. Also, let $P = P_\phi$ denote the two-scale symbol of $\phi$. Then*
  (i) *$\phi$ has generalized linear phase if and only if*

$$\overline{P(z)} = z^{-2n_0} P(z), \quad |z| = 1, \tag{5.5.9}$$

  *for some $n_0 \in \frac{1}{2}\mathbb{Z}$, and*
  (ii) *$\phi$ has linear phase if and only if*

$$P(e^{-i\omega}) = |P(e^{-i\omega})| e^{-in_0\omega}, \tag{5.5.10}$$

  *where $n_0 \in \frac{1}{2}\mathbb{Z}$.*

**Proof.** If $\phi$ has generalized linear phase, then by Definition 5.25, we have

$$\hat{\phi}(\omega) = F(\omega) e^{-i(a\omega+b)}, \quad \text{a.e.}$$

for some real-valued function $F(\omega)$ and some $a, b \in \mathbb{R}$. Hence, $\overline{\hat{\phi}(\omega)} = F(\omega) e^{i(a\omega+b)}$ and therefore

$$\overline{P(e^{-i\omega/2})} = \frac{\overline{\hat{\phi}(\omega)}}{\overline{\hat{\phi}(\omega/2)}} = e^{ia\omega/2} \frac{F(\omega)}{F(\omega/2)}$$

$$= e^{ia\omega} \frac{\hat{\phi}(\omega)}{\hat{\phi}(\omega/2)} = e^{ia\omega} P(e^{-i\omega/2}),$$

for almost all $\omega \in \mathbb{R}$. This implies that

$$n_0 := a \in \frac{1}{2}\mathbb{Z},$$

and (5.5.9) holds. If, in addition, $\phi$ has linear phase, then by Definition 5.25, we have $b = 0$ and $F(\omega)$ has no sign changes. Consequently,

$$P(e^{-i\omega/2}) = \frac{\hat{\phi}(\omega)}{\hat{\phi}(\omega/2)} = e^{-ia\omega/2}\frac{F(\omega)}{F(\omega/2)}$$

$$= e^{-ia\omega/2}\left|\frac{F(\omega)}{F(\omega/2)}\right| = e^{-in_0\omega/2}|P(e^{-i\omega/2})|,$$

which agrees with (5.5.10).

Conversely, if (5.5.9) holds, then we have

$$\hat{\phi}(\omega) = \prod_{k=1}^{\infty} P(e^{-i\omega/2^k}) \tag{5.5.11}$$

$$= \left\{\prod_{k=1}^{\infty} \overline{P(e^{-i\omega/2^k})}\right\} e^{-i2n_0\omega}$$

$$= \overline{\hat{\phi}(\omega)}\, e^{-i2n_0\omega}.$$

Hence, the function

$$F(\omega) := e^{in_0\omega}\hat{\phi}(\omega)$$

is real-valued, and since

$$\hat{\phi}(\omega) = F(\omega)e^{-in_0\omega},$$

$\phi$ has generalized linear phase. If the hypothesis (5.5.10) is assumed, then we have

$$\hat{\phi}(\omega) = \prod_{k=1}^{\infty} P(e^{-i\omega/2^k}) = \left|\prod_{k=1}^{\infty} P(e^{-i\omega/2^k})\right| e^{-in_0\omega} \tag{5.5.12}$$

$$= |\hat{\phi}(\omega)|e^{-in_0\omega},$$

so that $\phi$ has linear phase.   ∎.

**Remark.** It follows from Lemma 5.28, (i), and the above argument, that for a scaling function $\phi$ to have generalized linear phase, it is necessary and sufficient that $\phi$ is "skew-symmetric" with respect to some $n_0 \in \frac{1}{2}\mathbb{Z}$, in the sense that

$$\phi(n_0 + x) = \overline{\phi(n_0 - x)}, \text{ a.e.} \tag{5.5.13}$$

Indeed, for $\phi$ to have generalized linear phase, we have (5.5.9) and consequently (5.5.11), so that $\hat{\phi}(\omega) = F(\omega)e^{-in_0\omega}$ for some real-valued function $F(\omega)$ and some $n_0 \in \frac{1}{2}\mathbb{Z}$. Hence, (5.5.13) follows from Lemma 5.28, (i). The converse is trivial.   ∎

If the two-scale sequence $\{p_k\}$ is real-valued and finite, then by applying Theorem 5.32 and Lemma 5.31, we can say a little more, as in the following.

**Theorem 5.33.** *Let $\phi$ be a real-valued scaling function whose two-scale sequence $\{p_n\}$ is a finite real sequence with support $[0, N]$. Then*

    (i) *$\phi$ has generalized linear phase if and only if $p_{N-n} = p_n$ for all $n \in \mathbb{Z}$;*
        *and*

    (ii) *$\phi$ has linear phase if and only if $p_{N-n} = p_n$ for all $n$ and all the zeros of the two-scale symbol $P_\phi$ that lie on the circle, if any, have even multiplicities.*

To investigate the phase properties of a wavelet, one has to have some knowledge of its two-scale symbol $Q$. For instance, if the scaling function $\phi$ has generalized linear phase, then by the two-scale relation $\widehat{\psi}(\omega) = Q(e^{-i\omega/2})\hat{\phi}(\omega)$ and Definitions 5.25 and 5.27, it follows that $\psi$ also has generalized linear phase provided the sequence $\{q_k\}$ has generalized linear phase, and an analogous conclusion can be made concerning the property of linear phase. Of course, more can be said by a more careful analysis as in the study of $\phi$.

**Example 5.34.** Consider the first order cardinal $B$-spline $N_1$ and its corresponding Haar wavelet $\psi_1(x) = \psi_H(x) := N_1(2x) - N_1(2x - 1)$ (see (1.5.7), (1.1.16), and Example 3.2). From Example 5.26, we see that $N_1$ has linear phase. Since the two-scale symbol $Q$ for $\psi_1$ is

$$Q(z) = \frac{1}{2}(1 - z) = \left(\sin \frac{\omega}{4}\right) e^{-i\left(\frac{1}{4}\omega - \frac{\pi}{2}\right)}, \tag{5.5.14}$$

where $z = e^{-i\omega/2}$, we also see that

$$\widehat{\psi_1}(\omega) = Q(z)\widehat{N_1}\left(\frac{\omega}{2}\right) = \frac{(\sin \frac{\omega}{4})^2}{\omega/4} e^{-i\left(\frac{1}{2}\omega - \frac{\pi}{2}\right)}, \tag{5.5.15}$$

has generalized linear phase, but does not have linear phase.

Observe that the Haar wavelet $\psi_1 = \psi_H$ is a compactly supported orthogonal wavelet (see (1.1.16)). From the following result, we may conclude that $\psi_1$ is the *only* o.n. wavelet with compact support such that its corresponding scaling function has generalized linear phase. (The relation between an o.n. wavelet $\psi$ and its corresponding o.n. scaling function $\phi$ in general will be discussed in the next section as well as in Chapter 7.)

**Theorem 5.35.** *Let $\phi$ be a scaling function governed by a finite two-scale relation*

$$\phi(x) = \sum_{k=0}^{N} p_k \phi(2x - k), \quad p_0, p_N \neq 0, \tag{5.5.16}$$

*as in (5.2.2) with $N = N_\phi$ and $p_k = p_k^\phi$. Suppose that $\{\phi(\cdot - k)\colon k \in \mathbb{Z}\}$ is an orthonormal family that constitutes a partition of unity, and $\phi$ is skew-symmetric in the sense that*

$$\phi(a + x) = \overline{\phi(a - x)}, \quad x \in \mathbb{R}, \tag{5.5.17}$$

for some $a \in \mathbb{R}$ (see (5.5.5) in Lemma 5.28). Then $\phi$ must be the first order cardinal B-spline.

**Proof.** As usual, let

$$P(z) = \frac{1}{2} \sum_{k=0}^{N} p_k z^k$$

and $z = e^{-i\omega/2}$. Then (5.5.16) is equivalent to

$$\hat{\phi}(\omega) = P(z)\hat{\phi}\left(\frac{\omega}{2}\right), \quad \omega \in \mathbb{R}. \tag{5.5.18}$$

On the other hand, assertion (5.5.17) is equivalent to

$$\hat{\phi}(\omega)e^{ia\omega} = \overline{\hat{\phi}(\omega)e^{ia\omega}}, \quad \omega \in \mathbb{R}. \tag{5.5.19}$$

Hence, from (5.5.18) and (5.5.19), we have

$$P(z) = \frac{\hat{\phi}(\omega)}{\hat{\phi}(\frac{\omega}{2})} = e^{-ia\omega}\frac{\overline{\hat{\phi}(\omega)}}{\overline{\hat{\phi}(\frac{\omega}{2})}} = z^{2a}\overline{P(z)},$$

for $|z| = 1$. Since $P(z)$ is a polynomial of degree $N$ with nonzero leading coefficient and nonzero constant term, it follows that $2a = N$, so that

$$\overline{P(z)} = z^{-N}P(z), \quad |z| = 1. \tag{5.5.20}$$

Let us now consider the hypothesis that $\{\phi(\cdot - k): k \in \mathbb{Z}\}$ is an orthonormal family. By Theorem 3.23, this hypothesis is equivalent to

$$\sum_{k=-\infty}^{\infty} |\hat{\phi}(\omega + 2\pi k)|^2 = 1, \tag{5.5.21}$$

so that an application of (5.5.18) yields

$$|P(z)|^2 + |P(-z)|^2 = 1, \quad |z| = 1. \tag{5.5.22}$$

So, by substituting (5.5.20) into (5.5.22), we have

$$(P(z))^2 + (-1)^N(P(-z))^2 = z^N, \quad |z| = 1. \tag{5.5.23}$$

Recall from Section 5.2 that, for $\phi$ to be a function, it is necessary that $N > 0$. Hence, from (5.5.23) and the hypothesis $p_N \neq 0$, we see that $N$ must be an odd integer. Now, by setting

$$\begin{cases} P_e(z) = \dfrac{1}{2}\sum_{k} p_{2k}z^k; \\[2mm] P_o(z) = \dfrac{1}{2}\sum_{k} p_{2k+1}z^k, \end{cases} \tag{5.5.24}$$

we can write

$$P(z) = P_e(z^2) + zP_o(z^2), \qquad (5.5.25)$$

and the identity (5.5.22) yields

$$|P_e(z)|^2 + |P_o(z)|^2 = \frac{1}{2}, \quad |z| = 1. \qquad (5.5.26)$$

Hence, by applying (5.5.20) and (5.5.25), we obtain

$$\begin{aligned} P_e(z^2) + zP_o(z^2) = P(z) &= z^N \overline{P(z)} \\ &= z^N \overline{[P_e(z^2) + zP_o(z^2)]} \\ &= z^N \overline{P_e(z^2)} + z^{N-1} \overline{P_o(z^2)}, \quad |z| = 1. \end{aligned}$$

Since $N$ is odd, equating the odd and even parts gives

$$\begin{cases} P_e(z^2) = z^{N-1} \overline{P_o(z^2)}; \\ P_o(z^2) = z^{N-1} \overline{P_e(z^2)}, \quad |z| = 1, \end{cases}$$

so that

$$|P_e(z)|^2 = |P_o(z)|^2, \quad |z| = 1.$$

Applying this identity to (5.5.26) gives rise to

$$|P_e(z)|^2 = |P_o(z)|^2 = \frac{1}{4}, \quad |z| = 1. \qquad (5.5.27)$$

This is not possible unless $P_e$ and $P_o$ are monomials, or

$$P(z) = \frac{1}{2}(p_0 + p_N z^N), \quad N \text{ odd}.$$

(To verify this claim, we may simply multiply out the polynomials in (5.5.27) and compare coefficients of equal powers of $z$.) Since $\{\phi(\cdot - k) \colon k \in \mathbb{Z}\}$ is a partition of unity, we have $P(1) = 1$ and $P(-1) = 0$ (see (5.1.12) and (5.1.13)), so that

$$P(z) = \frac{1 + z^N}{2};$$

and hence,

$$\hat{\phi}(\omega) = \prod_{k=1}^{\infty} \left( \frac{1 + e^{-i\omega N/2^k}}{2} \right) = \frac{1 - e^{-i\omega N}}{i\omega N}$$

(see Example 5.3). Consequently, by (4.2.9), we obtain

$$\begin{aligned} \sum_k |\hat{\phi}(\omega + 2\pi k)|^2 &= \frac{4 \sin^2(\frac{\omega N}{2})}{N^2} \sum_{k=-\infty}^{\infty} \frac{1}{(\omega + 2\pi k)^2} \qquad (5.5.28) \\ &= \frac{\sin^2(\omega N/2)}{N^2 \sin^2(\omega/2)}. \end{aligned}$$

It now follows from (5.5.21) that $N = 1$. That is, $P_\phi(z) = P(z) = (1 + z)/2$, or $\phi$ is the first order cardinal $B$-spline $N_1$ (see Section 5.2).   ∎

**Remark.** Although $\phi$ is only assumed to be skew-symmetric in (5.5.17), we have shown that it is real-valued, and hence, $\phi$ is actually symmetric. Theorem 5.35 says that any compactly supported scaling function $\phi$ that generates a partition of unity and an orthonormal family $\{\phi(\cdot - k):\ k \in \mathbb{Z}\}$ cannot have generalized linear phase unless it is almost everywhere equal to the characteristic function of an interval $[k, k + 1)$ for some integer $k$.

## 5.6. Compactly supported wavelets

The objective of this section is to investigate the structure of wavelets with compact supports. Motivated by the need of linear-phase filtering in signal analysis as discussed in the previous section, we are particularly interested in skew-symmetric wavelets. (Recall from Section 5.5 that for any real-valued function $f$, $e^{ib}f(x)$ is skew-symmetric for some $b \in \mathbb{R}$ if and only if $f$ is symmetric or antisymmetric.) Following the strategy for constructing wavelets as developed in Section 5.4, we will consider a pair of admissible two-scale symbols $P = P_\phi$ and $G^* = G_{\tilde\phi}^*$ which are dual to each other in the sense that

$$P(z)G(z) + P(-z)G(-z) = 1, \quad |z| = 1, \tag{5.6.1}$$

where $G^*(z) = \overline{G(z)}$, $|z| = 1$ (see Definition 5.18, (5.4.3), and (5.4.4)). Then the wavelet $\psi$ and its dual $\tilde\psi$ have two-scale symbols in the sense that

$$\begin{cases} \widehat{\psi}(\omega) = Q(e^{-i\omega/2})\hat\phi\left(\dfrac{\omega}{2}\right); \\[2mm] \widehat{\tilde\psi}(\omega) = H^*(e^{-i\omega/2})\widehat{\tilde\phi}\left(\dfrac{\omega}{2}\right), \end{cases} \tag{5.6.2}$$

where $H^*(z) = \overline{H(z)}$, $|z| = 1$, and $Q$ and $H$ are arbitrarily, but necessarily, selected from the class in (5.4.11), namely:

$$\begin{cases} Q(z) = z^{-1}G(-z)K(z^2); \\[2mm] H(z) = zP(-z)K^{-1}(z^2), \quad |z| = 1, \end{cases} \tag{5.6.3}$$

where $K$ is in Wiener's class $\mathcal{W}$ with $K(z) \neq 0$ for $|z| = 1$. Also recall that the spaces $\{V_j\}$, $\{W_j\}$, $\{\tilde V_j\}$, and $\{\widetilde W_j\}$ generated by $\phi$, $\psi$, $\tilde\phi$, and $\tilde\psi$, respectively, satisfy

$$\begin{cases} \cdots \subset V_{-1} \subset V_0 \subset V_1 \subset \cdots \\[2mm] V_{j+1} = V_j \dot+ W_j, \quad j \in \mathbb{Z}, \end{cases} \tag{5.6.4}$$

$$\begin{cases} \cdots \subset \tilde V_{-1} \subset \tilde V_0 \subset \tilde V_1 \subset \cdots \\[2mm] \tilde V_{j+1} = \tilde V_j \dot+ \widetilde W_j, \quad j \in \mathbb{Z}, \end{cases} \tag{5.6.5}$$

and

$$\begin{cases} V_j \perp \widetilde W_j, \quad j \in \mathbb{Z}; \\[2mm] \tilde V_j \perp W_j, \quad j \in \mathbb{Z}. \end{cases} \tag{5.6.6}$$

In addition, the pairs $(\phi, \tilde{\phi})$ and $(\psi, \tilde{\psi})$ are dual pairs, in the sense that

$$\begin{cases} \langle (\phi \cdot -k), \tilde{\phi}(\cdot - m) \rangle = \delta_{k,m}, & k, m \in \mathbb{Z}; \\ \langle \psi_{j,k}, \tilde{\psi}_{\ell,m} \rangle = \delta_{j,\ell} \delta_{k,m}, & j, k, \ell, m \in \mathbb{Z}. \end{cases} \tag{5.6.7}$$

Details have been discussed in Section 5.4.

Let us first study the structure of semi-orthogonal (s.o.) wavelets, and particularly orthogonal (o.n.) wavelets (see Definition 3.22). It is clear from Theorem 3.25 that the spaces $W_j$ and $\widetilde{W}_j$, $j \in \mathbb{Z}$, generated by any s.o. wavelet $\psi$ and its dual $\tilde{\psi}$ are identical, namely: $W_j = \widetilde{W}_j$ for all $j \in \mathbb{Z}$. Hence, from (5.1.4), we also have $V_j = \widetilde{V}_j$ for all $j \in \mathbb{Z}$, so that the scaling function $\phi$ and its dual $\tilde{\phi}$ generate the same MRA. In fact, going back to the proof of Theorem 5.22 (see (5.4.21)), we observe that the (unique) dual $\tilde{\phi}$ of $\phi$ is given by

$$\hat{\tilde{\phi}}(\omega) = \frac{\hat{\phi}(\omega)}{\sum\limits_{k=-\infty}^{\infty} |\hat{\phi}(\omega + 2\pi k)|^2}, \tag{5.6.8}$$

(see (3.6.13) for dual s.o. wavelets). Now, let us restrict our attention to scaling functions $\phi$ with finite two-scale sequences $\{p_n\}$, namely:

$$\phi(x) = \sum_{n=0}^{N} p_n \phi(2x - n), \quad p_0, p_N \neq 0,$$

where $p_n = p_n^\phi$ and $N = N_\phi$ as studied in Section 5.2. Recall, from (5.2.24), (5.2.25), and (5.2.30), that the generalized Euler-Frobenius Laurent polynomial

$$E(z) = E_\phi(z) := \sum_{k=-\infty}^{\infty} \left\{ \int_{-\infty}^{\infty} \phi(k + y) \overline{\phi(y)} \, dy \right\} z^k \tag{5.6.9}$$

$$= \sum_{k=-\infty}^{\infty} \left| \hat{\phi} \left( \frac{\omega}{2} + 2\pi k \right) \right|^2,$$

relative to $\phi$, where $z = e^{-i\omega/2}$, is zero-free and pole-free on $|z| = 1$. Hence, it follows from (5.6.8) that

$$\hat{\tilde{\phi}}(\omega) = \frac{1}{E(z^2)} \hat{\phi}(\omega), \tag{5.6.10}$$

and consequently the dual $\tilde{\phi}$ of the compactly supported scaling function $\phi$ does *not* have compact support unless $E(z)$ is a positive constant, although it is of exponential decay. (Of course, the s.o. wavelets $\psi$, relative to $\phi$, may still have compact supports.) First, we must find the two-scale symbol $G^* = G^*_{\tilde{\phi}}$

of $\tilde{\phi}$. This is easily done by applying (5.6.10) and the two-scale relation of $\phi$ as follows:

$$\hat{\tilde{\phi}}(\omega) = \frac{1}{E(z^2)}\hat{\phi}(\omega) = \frac{1}{E(z^2)}P(z)\hat{\phi}\left(\frac{\omega}{2}\right)$$

$$= \frac{E(z)}{E(z^2)}P(z)\hat{\tilde{\phi}}\left(\frac{\omega}{2}\right),$$

so that

$$G^*(z) = \frac{E(z)}{E(z^2)}P(z), \quad z = e^{-i\omega/2}. \tag{5.6.11}$$

It is also easy to verify that for this $G^*$, the dual relation (5.6.1) is equivalent to the identity (5.2.31) for generalized Euler-Frobenius Laurent polynomials. Now, by Theorem 5.19 and (5.6.11), the two-scale symbol $Q$ for any wavelet $\psi$ relative to the scaling function $\phi$ is given by

$$Q(z) = z^{-1}G(-z)K(z^2) \tag{5.6.12}$$

$$= z^{-1}\frac{E(-z)\overline{P(-z)}}{E(z^2)}K(z^2),$$

where $K \in \mathcal{W}$ with $K(z) \neq 0$ for $|z| = 1$ (see (5.6.3)). Thus, we have some freedom in the choice of $\psi$. In particular, the wavelet $\psi$ with minimum support is obtained by selecting the admissible $K \in \mathcal{W}$ (that is, $K(z) \neq 0$ for $|z| = 1$) such that $Q(z)$ is a polynomial with lowest degree. We do not intend to pursue this in the general setting any further, but only mention that the results from Theorems 5.11 and 5.12 concerning two-scale symbols of minimally supported scaling functions would be useful for this study. A detailed investigation for scaling functions which are cardinal $B$-splines will be given in Section 6.2 in the next chapter.

Since $E$ is a Laurent polynomial which is zero-free and pole-free on $|z| = 1$, we may choose $K(z) = -zE(z)$ in (5.6.12), so that the two-scale symbols $Q$ and $H^*$ for a compactly supported s.o. wavelet $\psi$ and its dual $\tilde{\psi}$, respectively, are given by

$$\begin{cases} Q(z) = -zE(-z)\overline{P(-z)}; \\ H(z) = -z^{-1}\dfrac{P(-z)}{E(z^2)}. \end{cases} \tag{5.6.13}$$

(See (5.6.3). The reason for this choice of $K$, instead of simply $K = E$, is that this normalization is consistent with the formulation of the Haar function.) Observe that if $E$ is not a constant, then again the dual wavelet $\tilde{\psi}$ does not have compact support, although it decays exponentially. One advantage of the choice of $Q$ in (5.6.13) is that it is very easy to determine whether or not the wavelet $\psi$ has generalized linear phase. Indeed, since the finite coefficient sequence of the Laurent polynomial $E(z)$ is skew-symmetric (see (5.2.26) or (5.6.9)), it is clear that the coefficient sequence of the polynomial $Q(z)$ in (5.6.13) is also skew-symmetric provided that the two-scale sequence $\{p_n\}$ has this property. Furthermore, in view of (5.6.11) and (5.6.13), the same conclusion holds for both $G^*$ and $H^*$. In particular, for real sequences, by applying Theorems 5.32 (i) and 5.33, we have the following result.

**Theorem 5.36.** *Let $\{p_n\}$ be a finite, symmetric, real-valued two-scale sequence of a scaling function $\phi$. Also, let $\psi$, $\tilde{\phi}$, and $\tilde{\psi}$ be the s.o. wavelet, dual scaling function, and dual wavelet with two-scale symbols $Q$, $G^*$, and $H^*$, respectively, as given by (5.6.13) and (5.6.11). Then $\phi$ and its corresponding s.o. wavelet $\psi$ have compact supports, $\tilde{\phi}$ and $\tilde{\psi}$ are of exponential decay, and all of $\phi, \psi, \tilde{\phi}, \tilde{\psi}$ have generalized linear phases.*

Next, let us consider orthogonal (o.n.) wavelets $\psi$ relative to compactly supported scaling functions $\phi$. The general approach is to construct $\phi$ which is orthonormal (o.n.), in the sense that

$$\{\phi(\cdot - k): \ k \in \mathbb{Z}\}$$

is an o.n. family. For such a $\phi$, it follows from (5.6.9) that the generalized Euler-Frobenius polynomial $E(z)$ is the constant 1. Hence, by (5.6.13), we have

$$Q(z) = -z\overline{P(-z)}, \quad |z| = 1.$$

Consequently, if the two-scale relation of $\phi$ is given by

$$\phi(x) = \sum_{n=0}^{N} p_n \phi(2x - n), \quad p_0, p_N \neq 0,$$

then the two-scale relation of the o.n. wavelet $\psi$ is given by

$$\psi(x) = \sum_{n=-N+1}^{1} (-1)^n \bar{p}_{1-n} \phi(2x - n). \tag{5.6.14}$$

Observe that since the coefficient sequences $\{p_n\}$ and $\{(-1)^n \bar{p}_{1-n}\}$ for $\phi$ and $\psi$ have similar phase properties, one expects, in view of Theorem 5.35, that the compactly supported o.n. wavelets also fail to have generalized phases. Let us reformulate Theorem 5.35 as the following.

**Theorem 5.37.** *Let $\psi$ be a compactly supported o.n. wavelet as given by (5.6.14) whose corresponding o.n. scaling function $\phi$ generates a partition of unity. Suppose that $\phi$ is skew-symmetric in the sense of (5.5.17). Then $\psi$ must be the Haar function $\psi_H$.*

Hence, for a compactly supported continuous wavelet function $\psi$ to have generalized linear phase, there seem to be only two alternatives. First, we may settle for semi-orthogonality. This certainly works, provided that we are willing to accept duals which are only of exponential decay. In the following, we will totally give up orthogonality and look for compactly supported $\psi$ and $\tilde{\psi}$ with generalized linear phases. Following the strategy formulated in Section 5.4, we will start with two admissible two-scale (polynomial) symbols $P = P_\phi$ and $G^* = G^*_{\tilde{\phi}}$ that are dual to each other. (For $Q$ and $H^*$ to be Laurent polynomials as well, it suffices to choose a monomial $K$.)

Recall from Theorem 5.32 that $\phi$ has generalized linear phase if and only if its two-scale symbol $P$ satisfies

$$P(z) = z^m \overline{P(z)}, \quad |z| = 1, \tag{5.6.15}$$

for some integer $m$ (see (5.5.9), with $m = 2n_0$, $n_0 \in \frac{1}{2}\mathbb{Z}$). In order to be able to construct a compactly supported dual $\tilde{\phi}$ of $\phi$ that also has generalized linear phase, we have to come up with an admissible two-scale polynomial symbol of $\tilde{\phi}$ that also satisfies (5.6.15) for some integer power of $z$. In this direction, we have the following result.

**Theorem 5.38.** *Let $P = P_\phi$ and $G^* = G^*_{\tilde{\phi}}$ be admissible two-scale Laurent polynomial symbols that are dual to each other, such that $P$ satisfies (5.6.15). Then*

$$G_1(z) := \frac{1}{2}\{G(z) + z^{-m}G^*(z)\} \tag{5.6.16}$$

*satisfies the same duality relation*

$$P(z)G_1(z) + P(-z)G_1(-z) = 1, \quad |z| = 1, \tag{5.6.17}$$

*as $G(z)$, and moreover,*

$$G_1(z) = z^{-m}\overline{G_1(z)}, \quad |z| = 1, \tag{5.6.18}$$

*or equivalently,*

$$G^*_1(z) = z^m \overline{G^*_1(z)}, \quad |z| = 1. \tag{5.6.19}$$

**Proof.** It is clear that $G_1$ satisfies (5.6.18). Indeed, for $|z| = 1$, we have

$$G^*_1(z) = \overline{G_1(z)} = \frac{1}{2}\{\overline{G(z)} + z^m\overline{G^*(z)}\}$$

$$= \frac{1}{2}\{G^*(z) + z^m G(z)\} = \frac{1}{2}\{G(z) + z^{-m}G^*(z)\}z^m$$

$$= z^m G_1(z) = z^m\overline{G^*_1(z)}.$$

To verify (5.6.17), we simply apply (5.6.15) and (5.6.1), and obtain

$$P(z)G_1(z) + P(-z)G_1(-z)$$

$$= \frac{1}{2}\{P(z)[G(z) + z^{-m}G^*(z)] + P(-z)[G(-z) + (-z)^{-m}G^*(-z)]\}$$

$$= \frac{1}{2}\{[P(z)G(z) + P(-z)G(-z)] + [\overline{P(z)}G^*(z) + \overline{P(-z)}G^*(-z)]\}$$

$$= \frac{1}{2}\{[P(z)G(z) + P(-z)G(-z)] + [\overline{P(z)G(z) + P(-z)G(-z)}]\}$$

$$= \frac{1}{2}(1 + 1) = 1, \quad |z| = 1.$$

This completes the proof of the theorem. ■

In the following, we will only consider finite two-scale sequences that are real-valued. For such sequences the generalized linear-phase property (5.6.15) becomes

$$P(e^{-i\omega}) = e^{-im\omega}P(e^{i\omega}), \quad \omega \in \mathbb{R}. \tag{5.6.20}$$

**Lemma 5.39.** *Let $P$ be a Laurent polynomial with real coefficients that satisfy (5.6.20) for some $m \in \mathbb{Z}$. Then there is another polynomial $P_1$ (depending on $m$) with real coefficients such that*

$$P(e^{-i\omega}) = \begin{cases} e^{-im\omega/2}P_1(\cos\omega), & \text{for even } m; \\ e^{-im\omega/2}\left(\cos\frac{\omega}{2}\right)P_1(\cos\omega), & \text{for odd } m. \end{cases} \tag{5.6.21}$$

**Proof.** From the assumption (5.6.20), we see that $e^{im\omega/2}P(e^{-i\omega})$ is an even function of $\omega$. Thus, if $m$ is an even integer, then $e^{im\omega/2}P(e^{-i\omega})$ is a $2\pi$-periodic function and is therefore a real polynomial in $\cos\omega$. This gives (5.6.21) for even $m$. On the other hand, if $m$ is an odd integer, then by selecting $\omega = \pi$ in (5.6.20), we have $P(-1) = -P(-1)$, so that $P(-1) = 0$. Therefore, we can write

$$P(z) = \left(\frac{1+z}{2}\right)P_0(z), \tag{5.6.22}$$

for some polynomial $P_0$ with real coefficients. Substituting (5.6.22) into (5.6.20) yields

$$P_0(e^{-i\omega}) = e^{-i(m-1)\omega}P_0(e^{i\omega}).$$

Now, since $m - 1$ is even, we have

$$P_0(e^{-i\omega}) = e^{-i(m-1)\omega/2}P_1(\cos\omega), \tag{5.6.23}$$

for some polynomial $P_1$ with real coefficients. Hence, by putting (5.6.23) into (5.6.22), we obtain (5.6.21) for odd $m$. ■

In addition to the result in Lemma 5.39, we recall that, as a two-scale symbol, $P$ can be written as

$$P(z) = \left(\frac{1+z}{2}\right)^\ell P_2(z), \tag{5.6.24}$$

where $P_2$ is a Laurent polynomial with real coefficients satisfying

$$P_2(1) = 1 \quad \text{and} \quad P_2(-1) \neq 0, \tag{5.6.25}$$

and $\ell$ is some positive integer.

**Lemma 5.40.** *Let $P$ be a Laurent polynomial with real coefficients that satisfies (5.6.20) and (5.6.24)-(5.6.25). Then $(m - \ell)$ must be an even integer, and*

$$P(e^{-i\omega}) = e^{-im\omega/2} \left(\cos\frac{\omega}{2}\right)^\ell S(\cos\omega), \qquad (5.6.26)$$

*where $S$ is polynomial with real coefficients that satisfies*

$$S(1) = 1 \quad \text{and} \quad S(-1) \neq 0. \qquad (5.6.27)$$

**Proof.** From (5.6.24), we have

$$P(e^{-i\omega}) = e^{-i\ell\omega/2} \left(\cos\frac{\omega}{2}\right)^\ell P_2(e^{-i\omega}). \qquad (5.6.28)$$

Hence, by (5.6.20), we obtain

$$P_2(e^{-i\omega}) = e^{-i(m-\ell)\omega} P_2(e^{i\omega}). \qquad (5.6.29)$$

We will first show that $(m - \ell)$ is an even integer. Suppose, on the contrary, that $(m - \ell)$ is odd. Then as before, we see, by applying (5.6.29) with $\omega = \pi$, that $P_2(-1) = 0$. This is a contradiction to (5.6.25). Now, since $(m - \ell)$ is even, we may apply Lemma 5.39 to write

$$P_2(e^{-i\omega}) = e^{-i(m-\ell)\omega/2} S(\cos\omega) \qquad (5.6.30)$$

for some polynomial $S$ with real coefficients. Hence, assertion (5.6.26) is established by substituting (5.6.30) into (5.6.28). In addition, by (5.6.25), it is clear that the polynomial $S$ in (5.6.30) satisfies (5.6.27).    ∎

From Lemma 5.40, we see that any two-scale (Laurent) polynomial symbol with real coefficients and generalized linear phase (i.e. satisfying (5.6.20)) takes on the representation (5.6.26) with $S$ being a polynomial with real coefficients satisfying (5.6.27). In view of Theorem 5.38 and Lemma 5.40, we will look for the dual symbol $G^*(z) = \overline{G(z)}$, $|z| = 1$, of the form

$$G(e^{-i\omega}) = e^{im\omega/2} \left(\cos\frac{\omega}{2}\right)^{\tilde\ell} \widetilde{S}(\cos\omega), \qquad (5.6.31)$$

where $\widetilde{S}$ is a real polynomial satisfying $\widetilde{S}(1) = 1$ and $\ell$ is some positive integer such that

$$N := (\tilde\ell + \ell)/2$$

is also a positive integer. By applying (5.6.26) and (5.6.31), the dual identity in (5.6.1) becomes

$$\left(\cos\frac{\omega}{2}\right)^{2N} S(\cos\omega)\widetilde{S}(\cos\omega) + \left(\sin\frac{\omega}{2}\right)^{2N} S(-\cos\omega)\widetilde{S}(-\cos\omega) = 1. \quad (5.6.32)$$

Also, by setting $x = \sin^2(\omega/2)$, we have $\cos\omega = 1 - 2x$. Hence, if we define

$$R(x) := S(1 - 2x)\widetilde{S}(1 - 2x),$$

then (5.6.32) becomes

$$(1-x)^N R(x) + x^N R(1-x) = 1. \tag{5.6.33}$$

So, our problem now is to give a characterization of the real polynomial $R(x)$.

By the Euclidean algoirthm, there exists two polynomials $A$ and $B$ such that

$$x^N A(x) + (1-x)^N B(x) = 1. \tag{5.6.34}$$

Let us write

$$A(x) = C(x)(1-x)^N + A_1(x),$$

where $\deg A_1 \leq N - 1$; and set

$$B_1(x) = B(x) + C(x)x^N.$$

Since

$$x^N \{C(x)(1-x)^N + A_1(x)\} + (1-x)^N B(x) = 1,$$

we have

$$x^N A_1(x) = 1 - (1-x)^N B_1(x)$$

This implies that $\deg B_1 \leq N - 1$ also. That is, there exist polynomials $A_1$ and $B_1$ such that

$$\begin{cases} x^N A_1(x) + (1-x)^N B_1(x) = 1; \\ \deg A_1 \leq N - 1 \quad \text{and} \quad \deg B_1 \leq N - 1. \end{cases} \tag{5.6.35}$$

The polynomials $A_1(x)$ and $B_1(x)$ in (5.6.35) are actually unique. Indeed, if there is another solution pair $(\tilde{A}_1, \tilde{B}_1)$, then the difference $(A - \tilde{A}_1, B - \tilde{B}_1)$ satisfies

$$x^N (A_1(x) - \tilde{A}_1(x)) + (1-x)^N (B(x) - \tilde{B}_1(x)) = 0,$$

where $\deg(A_1 - \tilde{A}_1)$, $\deg(B_1 - \tilde{B}_1) \leq N - 1$, and hence, $A_1 - \tilde{A}_1 = 0$ and $B_1 - \tilde{B}_1 = 0$, due to the fact that $x^N$ and $(1-x)^N$ are relatively prime. Now, by interchanging $x$ and $(1-x)$ in (5.6.35), the uniqueness of $A_1$ and $B_1$ implies that $A_1(x) = B_1(1-x)$. That is, there is indeed a unique algebraic polynomial $R_0$ with $\deg R_0 \leq N - 1$ that solves (5.6.33). To determine $R_0$, we multiply (5.6.33) throughout by $(1-x)^{-N}$ and expand it in terms of $x$, yielding

$$R_0(x) = (1-x)^{-N} \{1 - x^N R_0(1-x)\}$$

$$= \sum_{k=0}^{\infty} \binom{N+k-1}{k} x^k \{1 - x^N R_0(1-x)\}$$

$$= \sum_{k=0}^{N-1} \binom{N+k-1}{k} x^k + \widetilde{R}_0(x),$$

where

$$\tilde{R}_0(x) = \sum_{k=N}^{\infty} \binom{N+k-1}{k} x^k - R_0(1-x) \sum_{k=0}^{\infty} \binom{N+k-1}{k} x^{N+k}.$$

Since $\tilde{R}_0(x)$ consists of powers of $x$ of order $N$ or higher, while $R_0(x)$ and the other finite sum are polynomials of degree $\leq N-1$, we have $\tilde{R}_0(x) \equiv 0$, or

$$R_0(x) = \sum_{k=0}^{N-1} \binom{N+k-1}{k} x^k. \tag{5.6.36}$$

This is a "particular solution" of (5.6.33). The general solution must be the sum of $R_0(x)$ and a term which is divisible by $x^N$. Let us call this term $x^N T(x)$. Since this function solves the "homogeneous equation" (with 1 replaced by 0 in (5.6.33)), it follows that $T$ satisfies $T(1-x) = -T(x)$. That is, setting

$$\begin{cases} T_0(y) := T\left(\dfrac{1-y}{2}\right); \\ y := 1 - 2x = \cos\omega, \end{cases} \tag{5.6.37}$$

we have

$$R\left(\frac{1-y}{2}\right) = R_0\left(\frac{1-y}{2}\right) + \left(\frac{1-y}{2}\right)^N T_0(y), \tag{5.6.38}$$

where $T_0(-y) = -T_0(y)$. Now, returning to (5.6.32), we arrive at the general formulation of $S(y)\tilde{S}(y) = S(\cos\omega)\tilde{S}(\cos\omega)$, namely:

$$\begin{cases} S(y)\tilde{S}(y) = \displaystyle\sum_{k=0}^{N-1} \binom{N+k-1}{k} \left(\dfrac{1-y}{2}\right)^k + \left(\dfrac{1-y}{2}\right)^N T_0(y) \\ T_0(-y) = -T_0(y). \end{cases} \tag{5.6.39}$$

We end this chapter by writing down the skew-symmetric two-scale polynomial symbol $G_1^*$ for a (compactly supported) dual of the $\ell^{\text{th}}$ order cardinal B-spline.

**Example 5.41.** The skew-symmetric two-scale symbols $G_1^*$ which are dual to

$$P_\ell(z) = \left(\frac{1+z}{2}\right)^\ell$$

are given by $G_{\ell,\tilde{\ell}}^*(z) = \overline{G_{\ell,\tilde{\ell}}(z)}$, for $|z| = 1$, where

$$G_{\ell,\tilde{\ell}}(e^{-i\omega}) = e^{i\ell\omega/2} \left(\cos\frac{\omega}{2}\right)^{\tilde{\ell}} \tilde{S}(\cos\omega) = e^{i\ell\omega/2}\left(\cos\frac{\omega}{2}\right)^{\tilde{\ell}}$$

$$\times \left\{ \sum_{k=0}^{N-1} \binom{N+k-1}{k}\left(\sin\frac{\omega}{2}\right)^{2k} + \left(\sin\frac{\omega}{2}\right)^{2N} T\left(\sin^2\frac{\omega}{2}\right) \right\},$$

with $N := (\ell + \tilde{\ell})/2$. This is obtained by applying (5.6.39) and (5.6.26) with $S = 1$ and $m = \ell$.  ∎

# 6 Cardinal Spline-Wavelets

A very general framework for the study of scaling functions and wavelets along with their duals has been established in the previous chapter. One of the main ingredients in this approach is the notion of multiresolution analysis (MRA), which is not only essential in the construction scheme, but also necessary for the formulation of the wavelet decomposition and reconstruction algorithms. In applications such as real-time signal analysis, for instance, a finite-energy signal (i.e., a function in $L^2(\mathbb{R})$) has to be mapped into some sample space $V_N$ belonging to the nested sequence $\{V_j\}$ that constitutes an MRA, before it can be separated into wavelet components by applying the decomposition algorithm. In this regard, the sequence $\{V_j^m\}$, $j \in \mathbb{Z}$, of cardinal spline spaces of an arbitrary order $m$ is a very attractive MRA of $L^2(\mathbb{R})$, in that spline and finite element methods are available for constructing the projection operators. (If we are satisfied with real-time optimal-order approximation, then the quasi-interpolation and interpolation algorithms developed in Sections 4.5 and 4.6, are readily applicable.) In addition, the structure of cardinal splines, though very simple, consists of many desirable properties as studied in Sections 4.2–4.4. This singles them out uniquely as a prime candidate for (nonparametric) modeling of arbitrary functions such as signals.

The objective of this chapter is to formulate wavelets in terms of cardinal $B$-splines and to study the structure of these "*spline-wavelets*". Special emphasis will be given on semi-orthogonal (s.o.) spline-wavelets, since their explicit expressions facilitate not only our study of these special features, but also hardware and software implementations.

## 6.1. Interpolatory spline-wavelets

The only wavelet we are very familiar with so far, at least in explicit formulation, is the Haar wavelet $\psi_1 = \psi_H$. On the one hand, its companion scaling function is the first order cardinal $B$-spline $N_1$, namely:

$$\psi_H(x) = N_1(2x) - N_1(2x - 1), \qquad (6.1.1)$$

while on the other hand, it is interesting to note that $\psi_H$ is also related to the derivative of the second order cardinal $B$-spline $N_2$, in the sense that

$$\psi_H(x) = N_2'(2x). \qquad (6.1.2)$$

It is therefore natural to ask to what extent the observation in (6.1.2) would generalize. To answer this question, let us first remark that the second order cardinal $B$-spline $N_2$ can be viewed as a fundamental cardinal spline, introduced in Section 4.6. In fact, the second order fundamental cardinal spline function $L_2$, defined as in (4.6.2)–(4.6.3), is given by

$$L_2(x) = N_2(x+1).$$

Hence, an equivalent statement of assertion (6.1.2) is

$$\psi_H(x) = L_2'(2x-1). \tag{6.1.3}$$

If we follow this point of view, then we can get spline-wavelets of arbitrary orders. To be precise, let $\{V_j^m\}$ be the MRA of $L^2(\mathbb{R})$ generated by the $m^{\text{th}}$ order cardinal $B$-spline as introduced in Section 4.1, and let $\{W_j^m\}$, $j \in \mathbb{Z}$, denote the sequence of orthogonal complementary (wavelet) spaces, in the sense that

$$V_{j+1}^m = V_j^m \oplus W_j^m, \quad j \in \mathbb{Z}, \tag{6.1.4}$$

where it should be recalled that the circle around the plus sign indicates orthogonal summation (see (1.4.8) and (1.5.9)). In the following, for each positive integer $m$, $L_m$ denotes the $m^{\text{th}}$ order fundamental cardinal spline function introduced in (4.6.2)-(4.6.3).

**Theorem 6.1.** *Let $m$ be any positive integer, and define*

$$\psi_{I,m}(x) = L_{2m}^{(m)}(2x-1), \tag{6.1.5}$$

*where $L_{2m}$ is the $(2m)^{\text{th}}$ order fundamental cardinal spline. Then $\psi_{I,m}$ generates the (wavelet) spaces $W_j^m$, $j \in \mathbb{Z}$, in the sense that*

$$W_j^m = clos_{L^2(\mathbb{R})}\langle 2^{j/2}\psi_{I,m}(2^j x - k) \colon k \in \mathbb{Z}\rangle, \quad j \in \mathbb{Z}. \tag{6.1.6}$$

**Proof.** Let us first verify that $\psi_{I,m}$ is in $W_0^m$. For every $n \in \mathbb{Z}$, by applying successive integration by parts and noting that the $m^{\text{th}}$ derivative of the $m^{\text{th}}$ order cardinal $B$-spline $N_m$ is a finite linear combination of integer translates of the delta distribution, we have

$$\langle N_m(\cdot - n), \psi_{m,I}\rangle = \int_{-\infty}^{\infty} N_m(x-n)L_{2m}^{(m)}(2x-1)dx$$

$$= \frac{(-1)^m}{2^m}\int_{-\infty}^{\infty} L_{2m}(2x-1)N_m^{(m)}(x-n)dx$$

$$= \sum_{k=0}^{m}\frac{1}{2^m}(-1)^{m-k}\binom{m}{k}\int_{-\infty}^{\infty} L_{2m}(2x-1)\delta(x-n-k)dx$$

$$= \sum_{k=0}^{m}\frac{1}{2^m}(-1)^{m-k}\binom{m}{k}L_{2m}(2n+2k-1) = 0,$$

since $L_{2m}(\ell) = \delta_{\ell,0}$, $\ell \in \mathbb{Z}$. Hence, $\psi_{I,m} \in W_0^m$.

Next, let us investigate the two-scale relation of $\psi_{I,m}$ with respect to $N_m(2x - k)$, $k \in \mathbb{Z}$. That is, we are interested in studying the $\ell^2$ sequence $\{q_k\}$ for which

$$\psi_{I,m}(x) = L_{2m}^{(m)}(2x - 1) = \sum_{k=-\infty}^{\infty} q_k N_m(2x - k). \tag{6.1.7}$$

Keeping the same notation as in (4.6.2), we write

$$L_{2m}(x) = \sum_{k=-\infty}^{\infty} c_k^{(2m)} N_{2m}(x + m - k). \tag{6.1.8}$$

On the other hand, by applying the cardinal $B$-spline identity (vii) in Theorem 4.3 repeatedly, it follows that

$$N_{2m}^{(m)}(x) = (\Delta N_{2m-1}^{(m-1)})(x) \tag{6.1.9}$$
$$= \cdots = (\Delta^m N_m)(x)$$
$$= \sum_{k=0}^{m}(-1)^k \binom{m}{k} N_m(x - k),$$

where $\Delta$ denotes the backward difference operator introduced in (4.1.9). Hence, we obtain, from (6.1.5), (6.1.8) and (6.1.9),

$$\psi_{I,m}(x) = L_{2m}^{(m)}(2x - 1) = \sum_{k=-\infty}^{\infty} c_k^{(2m)} N_{2m}^{(m)}(2x - 1 + m - k)$$

$$= \sum_{k=-\infty}^{\infty} c_k^{(2m)} \sum_{\ell=0}^{m}(-1)^\ell \binom{m}{\ell} N_m(2x - 1 + m - k - \ell)$$

$$= \sum_{n=-\infty}^{\infty} q_n N_m(2x - n),$$

with

$$q_n := \sum_{\ell=0}^{m}(-1)^\ell \binom{m}{\ell} c_{m+n-1-\ell}^{(2m)}. \tag{6.1.10}$$

The two-scale symbol $Q$ corresponding to the two-scale sequence $\{q_k\}$ in (6.1.7), as given by (6.1.10), is now

$$Q(z) = \frac{1}{2} \sum_{n=-\infty}^{\infty} \left( \sum_{\ell=0}^{m}(-1)^\ell \binom{m}{\ell} c_{m+n-1-\ell}^{(2m)} \right) z^n \tag{6.1.11}$$

$$= \frac{z^{-m+1}}{2}(1 - z)^m \sum_{n=-\infty}^{\infty} c_n^{(2m)} z^n,$$

where it follows from (6.1.8) and the interpolatory property $L_{2m}(\ell) = \delta_{\ell,0}$, that

$$\sum_{n=-\infty}^{\infty} c_n^{(2m)} z^n = \frac{1}{F_m(z)}, \tag{6.1.12}$$

with

$$F_m(z) := E_{N_m}(z) = \sum_{k=-m+1}^{m-1} N_{2m}(m+k)z^k \tag{6.1.13}$$

$$= \sum_{k=-\infty}^{\infty} \left\{ \int_{-\infty}^{\infty} N_m(k+x)N_m(x)dx \right\} z^k,$$

being the generalized Euler-Frobenius Laurent polynomial relative to the $m^{\text{th}}$ order cardinal $B$-spline $N_m$ (see (4.2.14), (5.2.24) and (5.2.25)). In spline theory, where algebraic polynomials with integer coefficients are very desirable, the Euler-Frobenius polynomials of order $2m - 1$ are defined, as in (4.2.18), by

$$E_{2m-1}(z) := (2m - 1)!z^{m-1}F_m(z).$$

Thus, substituting (6.1.12) into (6.1.11), we have found the formula for the two-scale symbol $Q$, namely:

$$Q(z) = \frac{z^{-m+1}}{2}(1-z)^m \frac{1}{F_m(z)}. \tag{6.1.14}$$

We remark that $F_m$ never vanishes on the unit circle because of (i) in Theorem 5.10.

Now, since the two-scale symbol of $N_m$ is given by

$$P(z) = P_{N_m}(z) = \left(\frac{1+z}{2}\right)^m,$$

(see (4.3.3)), we can compute the determinant $\Delta_{P,Q}$, defined in (5.3.9) and (5.3.10), as follows.

$$\Delta_{P,Q}(z) = \det \begin{bmatrix} P(z) & Q(z) \\ P(-z) & Q(-z) \end{bmatrix} \tag{6.1.15}$$

$$= \frac{(-z)^{-m+1}(1+z)^{2m}}{2^{m+1}F_m(-z)} - \frac{z^{-m+1}(1-z)^{2m}}{2^{m+1}F_m(z)}$$

$$= 2^{m-1}\left[\frac{(P(z))^2}{\Pi_m(-z)} - \frac{(P(-z))^2}{\Pi_m(z)}\right]$$

$$= 2^{m-1}\frac{z\Pi_m(z^2)}{\Pi_m(-z)\Pi_m(z)} = (-2)^{m-1}\frac{zF_m(z^2)}{F_m(-z)F_m(z)},$$

where we have applied the identity (5.2.32) in Theorem 5.10, with $N_\phi = m$, $k_\phi = m - 1$, $\Pi_\phi = \Pi_m(z) = z^{m-1}F_m(z)$, and $P_\phi^r = P_\phi = P$. Since $F_m$ never vanishes on $|z| = 1$, we have shown that

$$\Delta_{P,Q}(z) \neq 0, \quad |z| = 1.$$

Hence, an appeal to Theorem 5.16 shows that we have indeed the proof of Theorem 6.1. ∎

In view of the preceding result, let us consider the subspace

$$V_1^{2m,0} := \{s \in V_1^{2m}: s(k) = 0, \quad k \in \mathbb{Z}\} \tag{6.1.16}$$

of cardinal splines of order $2m$ with knot sequence $\frac{1}{2}\mathbb{Z}$, and vanishing at all the integers. It is clear that the function

$$\Psi_{2m}(x) := \frac{1}{2^m}L_{2m}(2x - 1) \tag{6.1.17}$$

is in $V_1^{2m,0}$ and so are all its integer translates. In fact, we have the following result.

**Theorem 6.2.** *For each $m$, the family*

$$\{\Psi_{2m}(\cdot - k): k \in \mathbb{Z}\} \tag{6.1.18}$$

*is a Riesz basis of $V_1^{2m,0}$.*

**Proof.** To show that the linear span of the family in (6.1.18) is dense in $V_1^{2m,0}$, we let $G \in V_1^{2m,0}$ be chosen arbitrarily. Then by applying Theorem 4.3, (ii), we have

$$\int_{-\infty}^{\infty} G^{(m)}(x)N_m(x - \ell)dx = \sum_{k=0}^{m}(-1)^{m-k}\binom{m}{k}G(k + \ell) = 0 \tag{6.1.19}$$

for all $\ell \in \mathbb{Z}$. Consequently, since it is clear that $G^{(m)}$ is in $V_1^m$, the derivation in (6.1.19) shows that $G^{(m)}$ is orthogonal to the subspace $V_0^m$ of $V_1^m$, and hence lies in $W_1^m$. By Theorem 6.1, we have

$$G^{(m)}(x) = \sum_{n=-\infty}^{\infty} a_n\psi_{I,m}(x - n)$$

for some sequence $\{a_n\} \in \ell^2$. Also, observe that from the definition (6.1.17) of $\Psi_{2m}$, we have

$$\Psi_{2m}^{(m)}(x) = \psi_{I,m}(x), \tag{6.1.20}$$

so that

$$D^m\left\{G(x) - \sum_{n=-\infty}^{\infty} a_k\Psi_{2m}(x - n)\right\} = 0, \quad x \in \mathbb{R},$$

where $D^m$ denotes the $m^{\text{th}}$ order differential operator introduced in (4.5.2). Since $V_1^{2m,0}$ consists only of functions in $L^2(\mathbb{R})$ that vanish at $\mathbb{Z}$, we have

$$G(x) = \sum_{n=-\infty}^{\infty} a_k \Psi_{2m}(x - k), \quad \{a_k\} \in \ell^2.$$

To show that the basis in (6.1.18) is unconditional (or is a Riesz basis), we simply note that since

$$\widehat{\Psi}_{2m}(\omega) = 2^{-m-1}\hat{L}_{2m}\left(\frac{\omega}{2}\right) e^{-i\omega/2}$$

$$= 2^{-m-1} e^{i(m-1)\omega/2} \left( \sum_k c_k^{(2m)} e^{-ik\omega/2} \right) \widehat{N}_{2m}\left(\frac{\omega}{2}\right),$$

it follows from (4.2.16) and (6.1.12) that

$$\sum_{k=-\infty}^{\infty} |\widehat{\Psi}_{2m}(\omega + 2\pi k)|^2 = 2^{-2m-2} \left\{ \frac{F_{2m}(z)}{(F_m(z))^2} + \frac{F_{2m}(-z)}{(F_m(-z))^2} \right\}, \qquad (6.1.21)$$

where $z = e^{-i\omega/2}$ and

$$F_{2m}(z) = E_{N_{2m}}(z) = \frac{1}{(4m - 1)!} z^{2m-1} E_{4m-1}(z), \qquad (6.1.22)$$

with $E_{4m-1}$ being the Euler-Frobenius polynomial of order $4m - 1$ (or degree $4m - 2$). Hence, Theorem 5.10, (i), applies, and the proof is complete. ∎

As a consequence of Theorems 6.1 and 6.2, we have the following result.

**Theorem 6.3.** *For each positive integer $m$, the $m^{\text{th}}$ order differential operator $D^m$ maps the spline space $V_1^{2m,0}$ one-to-one onto the wavelet space $W_0^m$. In addition, the Riesz basis $\{\Psi_{2m}(\cdot - k): k \in \mathbb{Z}\}$ of $V_1^{2m,0}$ corresponds to the Riesz basis $\{\psi_{I,m}(\cdot - k): k \in \mathbb{Z}\}$ of $W_0^m$ via the relation $\psi_{I,m} = D^m \Psi_{2m}$.*

## 6.2. Compactly supported spline-wavelets

The interpolatory wavelets $\psi_{I,m}$ introduced in (6.1.5) have exponential decay but are not compactly supported. From the development in Chapter 5 (see Section 5.6), however, we already know that semi-orthogonal (s.o.) wavelets with compact supports always exist, provided that the two-scale sequence of the corresponding scaling function $\phi$ is finite. Indeed, if $P = P_\phi$ denotes, as usual, the two-scale symbol of $\phi$, then by considering the dual $\tilde{\phi}$ of $\phi$ to be in the same space $V_0$ as $\phi$, the two-scale symbol $G^* = G_{\tilde{\phi}}^*$ of $\tilde{\phi}$, where $G^*(z) = \overline{G(z)}$ for $|z| = 1$, is given by

$$G(z) = \frac{E_\phi(z)}{E_\phi(z^2)} \overline{P(z)}, \quad |z| = 1, \qquad (6.2.1)$$

where $E_\phi$ is the generalized Euler-Frobenius Laurent polynomial relative to $\phi$ (see (5.6.11)). As a result, the two-scale symbol $Q$ of any s.o. wavelet $\psi$ corresponding to $\phi$ is given by

$$Q(z) = z^{-1}G(-z)K(z^2) \tag{6.2.2}$$

$$= z^{-1}E_\phi(-z)\overline{P(-z)}\frac{K(z^2)}{E_\phi(z^2)}, \quad |z| = 1,$$

for any $K \in \mathcal{W}$ with $K(z) \neq 0$ on $|z| = 1$ (see (5.6.12)), so that by simply choosing $K(z) = -zE_\phi(z)$, we have the two-scale polynomial symbol for a compactly supported s.o. wavelet $\psi$. In the cardinal spline setting, it follows from Theorem 5.19 that the general formulation of the two-scale symbols for the class of all possible s.o. wavelets relative to $N_m$ is given by

$$\begin{cases} Q(z) = -z^{-2}\left(\dfrac{1-z}{2}\right)^m E_{2m-1}(-z)\dfrac{K(z^2)}{E_{2m-1}(z^2)}, \\ K \in \mathcal{W}, \text{ with } K(z) \neq 0 \text{ on } |z| = 1, \end{cases} \tag{6.2.3}$$

where $E_{2m-1}$ is the Euler-Frobenius polynomial of order $2m-1$ (or degree $2m-2$) as defined in (4.2.18). So, to ensure that the wavelet $\psi$ has compact support, the Laurent series $K$ in (6.2.3) must be so chosen that $Q$ is a polynomial. Now, since $E_{2m-1}$ never vanishes on $|z| = 1$, the only way to reduce the degree of the polynomial $(1-z)^m E_{2m-1}(-z)$ is for $E_{2m-1}(z^2)$ to have a common zero with $E_{2m-1}(-z)$, while, at the same time, when a common factor is cancelled out, the remaining factor of $E_{2m-1}(z^2)$ must still be in powers of $z^2$ for it to be cancelled out by $K(z^2)$ in (6.2.3). This is not possible since $E_{2m-1}(z)$ does not have any symmetric zeros and in fact, as we will see in Section 6.4, all the zeros of $E_{2m-1}(z)$ are negative (see also (4.2.18)-(4.2.19)). Hence, for $Q$ in (6.2.3) to be the two-scale symbol of an s.o. cardinal spline-wavelet $\psi$ with minimum support, it is necessary and sufficient that

$$K(z) = c_0 z^{n_0} E_{2m-1}(z), \tag{6.2.4}$$

where $c_0$ is any nonzero constant and $n_0$ is any integer. In other words, up to multiplication by a nonzero constant and shift by any integer, the compactly supported s.o. wavelet $\psi_m$ with minimum support that corresponds to the $m^{\text{th}}$ order cardinal B-spline $N_m$ is unique, and is given by

$$\begin{cases} \psi_m(x) := \displaystyle\sum_n q_n N_m(2x - n); \\ Q_m(z) = \dfrac{1}{2}\displaystyle\sum_n q_n z^n := \left(\dfrac{1-z}{2}\right)^m \displaystyle\sum_{k=0}^{2m-2} N_{2m}(k+1)(-z)^k, \end{cases} \tag{6.2.5}$$

where we have chosen $c_0 = -[(2m-1)!]^{-1}$ and $n_0 = 1$ in (6.2.4) in order to have $q_0 \neq 0$ and $q_n = 0$ for $n < 0$. In view of the fact that, analogous to $N_m$,

$\psi_m$ also has minimum support, we will call $\psi_m$ the $m^{\text{th}}$ order "*B-wavelet*" for convenience. Since $Q_m$ in (6.2.5) is the product of two polynomial symbols, the sequence $\{q_n\}$ is the convolution of these two polynomial coefficient sequences. That is,

$$q_n = \frac{(-1)^n}{2^{m-1}} \sum_{\ell=0}^{m} \binom{m}{\ell} N_{2m}(n+1-\ell), \quad n = 0, \ldots, 3m-2. \tag{6.2.6}$$

Let us summarize these findings in the following.

**Theorem 6.4.** *Let $m$ be any positive integer. Also, let $N_m$ be the $m^{\text{th}}$ order cardinal B-spline and $\psi_m$ the B-wavelet as defined in (6.2.5) with coefficients given by (6.2.6). Then*

$$\{\psi_m(\cdot - k) \colon k \in \mathbb{Z}\} \tag{6.2.7}$$

*is a Riesz basis of $W_0$. Furthermore, $\psi_m$ has compact support with*

$$\text{supp } \psi_m = [0, 2m-1]. \tag{6.2.8}$$

*It is the "unique" wavelet in $W_0$ with minimum support in the sense that if $\eta \in W_0$ generates $W_0$ as $\psi_m$ does and the support of $\eta$ is an interval with length not exceeding $2m - 1$, then $\eta(x) = c_0 \psi_m(x - n_0)$ for some constant $c_0 \neq 0$ and $n_0 \in \mathbb{Z}$.*

We now turn to the dual $\widetilde{\psi}_m$ of $\psi_m$. By (5.4.11), the two-scale symbol $H^*$ of $\widetilde{\psi}$ relative to $\widetilde{\phi}$ is given by $H(z) = zP(-z)K^{-1}(z^2)$, where $H(z) = \overline{H^*(z)}$ for $|z| = 1$. Hence, in view of the dual relation of $\widetilde{\phi}$ and $\phi$ (see (5.6.10)), we have, for $z = e^{-i\omega/2}$,

$$\widehat{\widetilde{\psi}}_m(\omega) = H^*(z)\widehat{\widetilde{N}}_m\left(\frac{\omega}{2}\right) = H^*(z)\frac{1}{E_{N_m}(z)}\widehat{N}_m\left(\frac{\omega}{2}\right).$$

Hence, if we try to relate the dual wavelet $\widetilde{\psi}_m$ with the interpolatory wavelet $\psi_{I,m}$ introduced in Section 6.1, then using the two-scale symbol $Q(z)$ in (6.1.14) with $F_m(z) = E_{N_m}(z)$ (see (6.1.13)), we have

$$\widehat{\widetilde{\psi}}_m(\omega) = H^*(z)\left(\frac{2}{z^{-m+1}(1-z)^m}\right)\widehat{\psi}_{I,m}(\omega)$$

$$= \left(\overline{zP(-z)K^{-1}(z^2)}\right)\frac{2z^{m-1}}{(1-z)^m}\widehat{\psi}_{I,m}(\omega)$$

$$= \frac{z^{-1}\left(1-\frac{1}{z}\right)^m}{2^m K(z^2)}\frac{2z^{m-1}}{(1-z)^m}\widehat{\psi}_{I,m}(\omega)$$

$$= (-1)^m 2^{-m+1} z^{-2}\frac{1}{K(z^2)}\widehat{\psi}_{I,m}(\omega).$$

Recall that in our normalization of $\psi_m$ in (6.2.5), we have chosen $c_0 = -[(2m-1)!]^{-1}$ and $n_0 = 1$ in (6.2.4) for $K(z)$. This gives

$$\widehat{\widetilde{\psi}}_m(\omega) = (-1)^{m+1} 2^{-m+1}[(2m-1)!]z^{-4}\frac{1}{E_{2m-1}(z^2)}\widehat{\psi}_{I,m}(\omega) \quad (6.2.9)$$

$$= \frac{(-1)^{m+1}}{2^{m-1}}z^{-2(m+1)}\frac{1}{F_m(z^2)}\widehat{\psi}_{I,m}(\omega),$$

where $F_m$ is the reciprocal of the symbol of the $B$-spline coefficient sequence $\{c_k^{(2m)}\}$ in the definition of the fundamental cardinal spline $L_{2m}$. Hence, noting that $z^2 = e^{-i\omega}$ in (6.2.9), we have

$$\widetilde{\psi}_m(x) = \frac{(-1)^{m+1}}{2^{m-1}}\sum_{k=-\infty}^{\infty}c_k^{(2m)}\psi_{I,m}(x+m+1-k). \quad (6.2.10)$$

We end this section with a discussion of the phase properties and an investigation of the time-frequency windows of the $B$-wavelet $\psi_m$ and its dual $\widetilde{\psi}_m$. First observe that the cardinal $B$-spline $N_m$ and the fundamental cardinal spline $L_{2m}$ are symmetric for any $m$ (see Theorem 4.3, (ix)). Hence, the sequence $\{c_k^{(2m)}\}$ is also symmetric. So, from the definition of $\psi_{I,m}$, we see that $\psi_{I,m}$ must be symmetric for even $m$ and antisymmetric for odd $m$, and the same conclusion holds for $\widetilde{\psi}_m$ in view of (6.2.10). Similarly, from (6.2.6), since it is clear that the sequence $\{q_n\}$ is also symmetric for even order $m$ and antisymmetric for odd order $m$, we may draw the same conclusion for the cardinal $B$-wavelets $\psi_m$.

**Theorem 6.5.** *All the wavelets $\psi_m$, $\widetilde{\psi}_m$, and $\psi_{I,m}$ are symmetric for even $m$ and antisymmetric for odd $m$. Consequently, they all have generalized linear phases.*

The graphs of the $B$-wavelets $\psi_m$ are particularly interesting. For $m \geq 3$, the even order $\psi_m$'s match almost exactly with

$$Re\, G_{b,\omega}^{\alpha}(t) = (\cos \omega t)g_\alpha(t-b) \quad (6.2.11)$$

and the odd order ones with

$$Im\, G_{b,\omega}^{\alpha}(t) = (\sin \omega t)g_\alpha(t-b), \quad (6.2.12)$$

for certain values of $\alpha, b, \omega$, where $g_\alpha$ is the Gaussian function with parameter $\alpha$ (see (3.1.10)). In Figures 6.2.1 and 6.2.2, we show the graphs of $\psi_4$ and $\psi_3$, respectively. Observe the resemblance between these graphs and those of the corresponding Gaussian functions in Figures 3.1.1 and 3.1.2. The error curves are shown in Figures 6.2.3 and 6.2.4.

Recall from Chapter 3 that when a wavelet $\psi$ is used as a basic wavelet in the IWT, the area of the window is given by $4\Delta_\psi\Delta_{\widehat{\psi}}$, and the smaller the value of $\Delta_\psi \Delta_{\widehat{\psi}}$, the better the wavelet is for applications to time-frequency localization. Since the Gaussian function $g_\alpha$ cannot be used as a basic wavelet, the Uncertainly Principle (see Theorem 3.5) says that $\Delta_\psi \Delta_{\widehat{\psi}} > \frac{1}{2}$, no matter what basic wavelet is considered. In Table 6.2.1, we give the values of $\Delta_{\psi_m} \Delta_{\widehat{\psi_m}}$ for $m = 2, 3, 4, 5, 6$. Observe how close the cardinal $B$-wavelets $\psi_m$ are to being optimal for larger values of $m$.

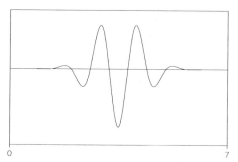

**Figure 6.2.1.** Cubic spline-wavelet $\psi_4$.

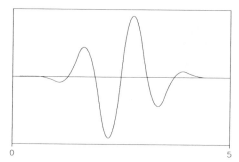

**Figure 6.2.2.** Quadratic spline-wavelet $\psi_3$.

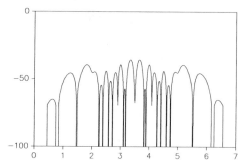

**Figure 6.2.3.** $|\psi_4 - Re\, G^\alpha_{3.5, 2\pi}|^2$ in dB, $\alpha = 0.2925$.

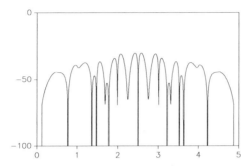

**Figure 6.2.4.** $|\psi_3 - Im\, G^{\alpha}_{2.5,2\pi}|^2$ in dB, $\alpha = 0.2300$.

| $m$ | Products of $\Delta_{\psi_m}$ and $\Delta_{\widehat{\psi}_m}$ |
|---|---|
| 2 | 0.971715 |
| 3 | 0.535070 |
| 4 | 0.504839 |
| 5 | 0.500929 |
| 6 | 0.500367 |

**Table 6.2.1.** Values of $\Delta_{\psi_m}\Delta_{\widehat{\psi}_m}$.

## 6.3. Computation of cardinal spline-wavelets

This section is devoted to a discussion of certain computational schemes for the compactly supported cardinal spline-wavelets (or $B$-wavelets) $\psi_m$ in (6.2.5)-(6.2.6). In Section 5.2, a method for computing any compactly supported scaling function was offered. It consists of two main steps: the first step is to find the eigenvector $[\phi(1)\ldots\phi(N_\phi - 1)]^T$ of the matrix $[p^\phi_{2j-k}]$, $1 \leq j,k \leq N_\phi - 1$, corresponding to the eigenvalue $\lambda = 1$ and subject to the normalization condition $\phi(1) + \cdots + \phi(N_\phi - 1) = 1$, while the second step is to apply the Interpolatory Graphical Display Algorithm described in Section 4.3 using $j_0 = 0$, $a^{(0)}_\ell = \delta_{\ell,0}$, $w_{m,k} = \phi(k)$ and replacing $p_{m,k}$ by $p^\phi_k$. This method yields the values of

$$\phi\left(k + \frac{\ell}{2^{j-1}}\right), \quad k \in \mathbb{Z}, \quad \ell = 0,\ldots,2^{j-1} - 1, \qquad (6.3.1)$$

for any desirable positive integer $j$. Now from the two-scale relation

$$\psi(x) = \sum_k q_k \phi(2x - k)$$

for the wavelet $\psi$, we can easily formulate the following scheme for computing $\psi$ at the dyadic points, namely:

$$
\begin{cases}
\psi\left(n + \dfrac{\ell}{2^j}\right) = \displaystyle\sum_k q_{2n-k}\phi\left(k + \dfrac{\ell}{2^{j-1}}\right); \\[2mm]
n \in \mathbb{Z}, \quad \ell = 0,\ldots 2^j - 1, \text{ and } \phi\left(k + \dfrac{\ell}{2^{j-1}}\right) \text{ from (6.3.1).}
\end{cases}
\tag{6.3.2}
$$

Note that again this scheme consists of taking moving averages followed by upsampling.

Observe that if the values of $p_k^\phi$ and $q_k$ are exact, then the computational algorithm described above gives precise values of $\psi$ at the dyadic points. For the setting of cardinal splines of order $m$, recall from (4.3.3) and (6.2.6) that

$$
p_k^\phi = p_{m,k} = \begin{cases} 2^{-m+1}\binom{m}{k} & \text{for } 0 \le k \le m; \\ 0 & \text{otherwise,} \end{cases}
\tag{6.3.3}
$$

and

$$
q_{m,k} := q_k = \begin{cases} \dfrac{(-1)^k}{2^{m-1}} \displaystyle\sum_{\ell=0}^m \binom{m}{\ell} N_{2m}(k+1-\ell) & \text{for } 0 \le k \le 3m - 2; \\ 0 & \text{otherwise,} \end{cases}
\tag{6.3.4}
$$

where the values of $N_{2m}(k)$, $k \in \mathbb{Z}$, can be computed by applying the recursive scheme

$$
\begin{cases}
N_2(k) = \delta_{k,1}, \quad k \in \mathbb{Z}; \text{ and} \\[2mm]
N_{n+1}(k) = \dfrac{k}{n}N_n(k) + \dfrac{n-k+1}{n}N_n(k-1), \\[2mm]
\text{where } k = 1,\ldots,n \text{ and } n = 2,\ldots,2m-1,
\end{cases}
\tag{6.3.5}
$$

put forth in (4.2.15). Note that the procedure in (6.3.5) is more efficient than finding the eigenvectors of $[p_{m,2j-k}]$, $1 \le j,k \le m-1$, and $[p_{2m,2j-k}]$, $1 \le j,k \le 2m-1$, corresponding to the eigenvalue $\lambda = 1$, as described above and in (5.2.14)-(5.2.17).

The computational advantage of cardinal $B$-splines over other scaling functions is more than just the recursive scheme in (6.3.5). The first goal in this section is to introduce a Pascal triangular algorithm (PTA) for computing the two-scale sequences $\{q_{m,k}\}$, $k \in \mathbb{Z}$, $m = 1,2,\ldots$, more directly. As we will see, this algorithm gives not only the two-scale sequences, but also the coefficient sequences $\{q_{m,k}^{(r)}\}$ of the cardinal $B$-spline series representations of the $r^{\text{th}}$ derivatives

$$
\psi_m^{(r)}(x) = \sum_{k=0}^{3m-2+r} q_{m,k}^{(r)} N_{m-r}(2x - k)
\tag{6.3.6}
$$

of the compactly supported $m^{\text{th}}$ order $B$-wavelet $\psi_m$. The structure of this PTA will be discussed in Section 6.6 in conjunction with a discussion of "total positivity", "complete oscillation", and "zero-crossings". For this purpose, we describe the PTA in somewhat more generality.

In the following, we will use the notation

$$\mathbb{Z}_+ = \{0, 1, 2, \ldots\} \tag{6.3.7}$$

for the set of nonnegative integers, and as in (4.5.9), $\tilde{A}(z)$ will denote the symbol of a sequence $\{a_n\} \in \ell^2$. To facilitate our presentation, we will also need the notation

$$\mathcal{L}^n := \{\{a_k\} \colon a_k = 0 \text{ for } k < 0 \text{ or } k > n, \text{ and } a_0, a_n \neq 0\}, \tag{6.3.8}$$

for any $n \in \mathbb{Z}_+$. Hence, $\{\mathcal{L}^n\}$, $n \in \mathbb{Z}_+$, is a mutually disjoint partition of the family

$$\mathcal{L} := \ell^2 \cap \bigcup_{n=0}^{\infty} \mathcal{L}^n \tag{6.3.9}$$

of the class of all causal $\ell^2$-sequences. In addition, let $\tau$ denote the shift operator on $\ell^2$ defined by

$$(\tau \mathbf{a})_{n+1} := a_n, \quad n \in \mathbb{Z}, \quad \mathbf{a} = \{a_n\} \in \ell^2. \tag{6.3.10}$$

**Definition 6.6.** *A Pascal triangular algorithm (PTA) is a map* $\mathcal{P} \colon \mathbb{Z}_+ \to \mathcal{L}$ *that can be formulated as follows:*

$$\begin{cases} \mathcal{P}(0) = \{\delta_{j,0}\}; \text{ and} \\ (\mathcal{P}(n+1))_j = L(n,j)(\mathcal{P}(n))_j + R(n, j-1)(\mathcal{P}(n))_{j-1}, \quad j \in \mathbb{Z}, \quad n \in \mathbb{Z}_+, \\ \text{where } L(n,0) \neq 0 \text{ and } R(n,n) \neq 0. \end{cases}$$
$$\tag{6.3.11}$$

**Definition 6.7.** *A PTA* $\mathcal{P}$*, as described by (6.3.11), is called a linear Pascal triangular algorithm (LPTA), if* $L(n, \cdot)$ *and* $R(n, \cdot)$ *are both linear (in the second "variable"* $\cdot$*), namely:*

$$\begin{cases} L(n,j) = k_L(n)j + b_L(n); \\ R(n,j) = k_R(n)(n-j) + b_R(n), \quad j \in \mathbb{Z}, \end{cases} \tag{6.3.12}$$

*for some* $k_L(n)$*,* $k_R(n)$*,* $b_L(n)$*, and* $b_R(n)$*.*

**Remark.** For any PTA $\mathcal{P}$, it is clear that

$$\mathcal{P}(n) \in \mathcal{L}^n, \quad n \in \mathbb{Z}_+. \tag{6.3.13}$$

If, in addition, $\mathcal{P}$ is an LPTA, then it is necessary that

$$b_L(n) \neq 0 \quad \text{and} \quad b_R(n) \neq 0, \quad n \in \mathbb{Z}_+, \tag{6.3.14}$$

holds, since $b_L(n) = L(n,0) \neq 0$ and $b_R(n) = R(n,n) \neq 0$.

Because of the initial condition $\mathcal{P}(0) = \{\delta_{j,0}\}$ in (6.3.11), a PTA can be realized as a "tree" algorithm for computing the sequences

$$\mathbf{s}^n := \mathcal{P}(n) \in \mathcal{L}^n, \quad n = 1, 2, \dots .$$

More precisely, if the configurations in Figure 6.3.1 are interpreted as

$$\begin{cases} u = at_1 + bt_2, \\ u = bt_2, \quad \text{and} \\ u = at_1, \text{ respectively,} \end{cases} \tag{6.3.15}$$

then the tree configuration of a PTA can be described by Figure 6.3.2, where we have used the notation

$$L_{n,k} := L(n,k), \quad R_{n,k} := R(n,k), \quad s_k^n = (\mathbf{s}^n)_k$$

**Figure 6.3.1.** Details of PTA.

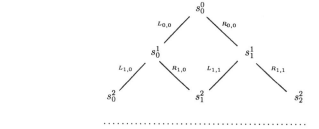

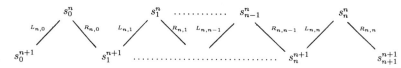

**Figure 6.3.2.** Pascal Triangle.

**Example 6.8.** Consider the LPTA $\mathcal{P}_b$ defined by setting

$$L(n,j) = R(n,j) = 1, \quad n \in \mathbb{Z}_+, \quad j \in \mathbb{Z}; \tag{6.3.16}$$

(that is, $k_L(n) = k_R(n) = 0$ and $b_L(n) = b_R(n) = 1$). Then the Pascal triangle in Figure 6.3.2 is the well-known algorithm for computing the sequence of binomial coefficients

$$\left\{ \binom{n}{j} \right\} = \mathcal{P}_b(n), \quad n \in \mathbb{Z}_+. \tag{6.3.17}$$

**Example 6.9.** Consider the LPTA $\mathcal{P}_e$ defined by setting

$$\begin{cases} L(n,j) = j+1, \text{ and} \\ R(n,j) = (n-j)+1; \end{cases} \tag{6.3.18}$$

(that is, $k_L(n) = k_R(n) = b_L(n) = b_R(n) = 1$). Then the Pascal Triangle in Figure 6.3.2 can be used to calculate the coefficient sequence of the modified Euler-Frobenius polynomials $\widetilde{E}_n$ (see the remark following this example), defined by

$$\widetilde{E}_n(z) := n! \sum_{j=0}^{n-1} N_{n+1}(j+1)z^j \tag{6.3.19}$$

of order $n$ (or degree $n-1$), in the sense that

$$\mathcal{P}_e(n-1) = \{n! N_{n+1}(j+1)\}_{j=0}^{n-1}, \quad n = 1, 2, \dots . \tag{6.3.20}$$

This is a consequence of the recursive scheme (4.2.15) or (6.3.5) for cardinal $B$-splines. It can be shown that this LPTA yields

$$N_{n+1}(j+1) = \frac{1}{n!} \sum_{\nu=0}^{j} (-1)^\nu \binom{n+1}{\nu} (j+1-\nu)^n.$$

In particular, we have listed the first nine modified Euler-Frobenius polynomials in Table 6.3.1.

| $n$ | $\widetilde{E}_n(z)$ |
|---|---|
| 1 | $1$ |
| 2 | $1+z$ |
| 3 | $1 + 4z + z^2$ |
| 4 | $1 + 11z + 11z^2 + z^3$ |
| 5 | $1 + 26z + 66z^2 + 26z^3 + z^4$ |
| 6 | $1 + 57z + 302z^2 + 302z^3 + 57z^4 + z^5$ |
| 7 | $1 + 120z + 1191z^2 + 2416z^3 + 1191z^4 + 120z^5 + z^6$ |
| 8 | $1 + 247z + 4293z^2 + 15619z^3 + 15619z^4 + 4293z^5 + 247z^6 + z^7$ |
| 9 | $1 + 502z + 14608z^2 + 88234z^3 + 156190z^4 + 88234z^5 + 14608z^6 + 502z^7 + z^8$ |

**Table 6.3.1.** Modified Euler-Frobenius polynomials.

**Remark.** Recall from (4.6.6) that the $m^{\text{th}}$ order Euler-Frobenius polynomials $E_m$ are defined by

$$E_m(z) = m! \sum_{k \in \mathbb{Z}} N_{m+1}\left(\frac{m+1}{2} + k\right) z^{k+m/2}.$$

Hence, for odd $m$, we have

$$E_{2n-1}(z) = (2n-1)! \sum_{k \in \mathbb{Z}} N_{2n}(n+k) z^{k+n-1}$$

$$= (2n-1)! \sum_{k=0}^{2n-2} N_{2n}(k+1) z^{k},$$

which agrees with (4.2.18). For even $m$, however, the coefficients of $E_m(z)$ are no longer values of the cardinal $B$-spline $N_{m+1}$ at the integers, and therefore the recursive scheme in (4.2.15) or (6.3.5) does not apply. For this computational reason, the modified Euler-Frobenius polynomials $\widetilde{E}_n(z)$ are introduced in (6.3.19). We have seen that $\widetilde{E}_{2m-1}(z) = E_{2m-1}(z)$. In other applications, the $\widetilde{E}_{2m}$'s are not very useful since interpolation by cardinal splines of odd order (and with knot sequence $\mathbb{Z}$) at the knots $\mathbb{Z}$ is "unstable" (for example, the coefficient matrix is not invertible). This follows from Table 6.1, by observing that

$$\widetilde{E}_{2m}(-1) = 0, \quad m = 1, 2, \ldots .$$

Another reason for introducing $\widetilde{E}_{2m}$ is that again in view of the recursive scheme (6.3.5) for cardinal $B$-splines at the integers, there is a very nice relationship between $\widetilde{E}_n$ and $\widetilde{E}_{n+1}$, namely:

$$\widetilde{E}_{n+1}(z) = (1+nz)\widetilde{E}_n(z) + z(1-z)\widetilde{E}_n'(z), \quad n = 1, 2, \ldots . \tag{6.3.21}$$

The proof of (6.3.21) is a straightforward application of (6.3.5). This identity will be useful in Section 6.5.

We now turn to formulating the PTA for the two-scale sequences $\{q_{m,k}\}$ in (6.3.4) of the $m^{\text{th}}$ order cardinal $B$-wavelets $\psi_m$. As mentioned earlier, in computing $\{q_{m,k}\}$, this same PTA may be applied to yield the coefficient sequences $\{q_{m,k}^{(r)}\}$ of the $B$-spline series of the $r^{\text{th}}$ derivatives $\psi_m^{(r)}$ of $\psi_m$, $r = 0, \ldots, m-1$. In order to avoid any sign changes, let us introduce the notation

$$\begin{cases} \tilde{q}_{m,k}^{(r)} := |q_{m,k}^{(r)}| = (-1)^k q_{m,k}^{(r)} \\ \tilde{q}_{m,k}^{(0)} := |q_{m,k}| = (-1)^k q_{m,k}. \end{cases} \tag{6.3.22}$$

We have the following result.

**Theorem 6.10.** *For every positive integer $m$, let $\mathcal{P}_m$ be the LPTA with*

$$L(n,j) = \begin{cases} j+1 & \text{for} \quad 0 \le n < 2m-2; \\ 1 & \text{for} \quad n \ge 2m-2 \end{cases} \tag{6.3.23}$$

*and*

$$R(n,j) = \begin{cases} (n-j)+1 & \text{for} \quad 0 \le n < 2m-2; \\ 1 & \text{for} \quad n \ge 2m-2. \end{cases} \tag{6.3.24}$$

Then

$$(\mathcal{P}_m(3m - 2 + r))_k = (2m - 1)! 2^{m-r-1} \tilde{q}_{m,k}^{(r)}, \tag{6.3.25}$$

for $r = 0, \ldots, m - 1$.

**Remark.** Observe that in (6.3.23) and (6.3.24), we have $b_L(n) = b_R(n) = 1$ for all $n \in \mathbb{Z}_+$, and

$$k_L(n) = k_R(n) = \begin{cases} 1 & \text{for} \quad 0 \leq n < 2m - 2; \\ 0 & \text{for} \quad n \geq 2m - 2, \end{cases}$$

in the LPTA.

**Proof.** Let us again use the notation

$$\begin{cases} \mathbf{s}^n := \mathcal{P}_m(n); \\ s_j^n := (\mathbf{s}^n)_j; \\ \widetilde{S}^n(z) := \sum_j s_j^n z^j. \end{cases} \tag{6.3.26}$$

Then in view of (6.3.23) and (6.3.24), by applying the results in Examples 6.9 and 6.8 consecutively, we have

$$\widetilde{S}^n(z) = \begin{cases} \widetilde{E}_{n+1}(z) & \text{for} \quad 0 \leq n < 2m - 2; \\ E_{2m-1}(z)(1 + z)^{n-2m+2} & \text{for} \quad n \geq 2m - 2. \end{cases}$$

(Note that $\widetilde{E}_{2m-1} = E_{2m-1}$.) Hence, the symbol of $\mathbf{s}^{3m-2+r}$ is given by

$$\widetilde{S}^{3m-2+r}(z) = E_{2m-1}(z)(1 + z)^{m+r}, \quad r = 0, \ldots, m - 1. \tag{6.3.27}$$

On the other hand, by applying the identity (vii) in Theorem 4.3, we have

$$q_{m,k}^{(r)} = 2^r \sum_{\ell=0}^{r} (-1)^\ell \binom{r}{\ell} q_{m,k-\ell}. \tag{6.3.28}$$

Therefore, since the symbol of $\{q_{m,k}\}$ (which is twice the two-scale symbol $Q$ for the $B$-wavelet $\psi_m$ relative to the scaling function $N_m$) is

$$2Q_m(z) = 2^{-m+1} \frac{1}{(2m - 1)!} E_{2m-1}(-z)(1 - z)^m$$

(see (6.2.5)), it follows from (6.3.28) that the symbol of $\{q_{m,k}^{(r)}\}$ is given by

$$2^{-m+r+1} \frac{1}{(2m - 1)!} E_{2m-1}(-z)(1 - z)^{m+r}. \tag{6.3.29}$$

So, if we multiply the expression in (6.3.29) by $(2m-1)!2^{m-r-1}$ and change $-z$ to $z$, we obtain the symbol $\widetilde{S}^{3m-2+r}$ in (6.3.27) of $\mathcal{P}_m(3m-2+r)$. That is, we have indeed derived (6.3.25).    ∎

Since the $B$-wavelet $\psi_m$ is a cardinal spline of order $m$, its $(m-1)^{\text{st}}$ derivative is a step function $\psi_m^{(m-1)}$ with jumps at $j/2$, $j = 0,\ldots,4m-2$. Hence, to obtain $\psi_m$, we may simply integrate this step function $(m-1)$ times. In general, to generate any $B$-wavelet series

$$g(x) = \sum_j d_j \psi_m(x-j), \qquad (6.3.30)$$

the integrations can be performed on the step function

$$g^{(m-1)}(x) = \sum_j d_j \psi_m^{(m-1)}(x-j).$$

Of course, if the engineer prefers to work with an *"impulse train"* (that is, a series of delta functions), then he or she may use the series

$$g^{(m)}(x) = \sum_j d_j \psi_m^{(m)}(x-j)$$

which necessitates an extra integration.

**Theorem 6.11.** *For each positive integer $m$, let*

$$\mathcal{P}_m(4m-3) = \mathbf{s}^{4m-3} = \{s_j^{4m-3}\}_{j=0}^{4m-3},$$

*and let $\chi_{[0,\frac{1}{2})}$ denote the characteristic function of the interval $[0,\frac{1}{2})$. Then the $m^{\text{th}}$ order $B$-wavelet $\psi_m$ is given by*

$$\psi_m(x) = \frac{1}{(2m-1)!} \sum_{j=0}^{4m-3} (-1)^j s_j^{4m-3} \qquad (6.3.31)$$

$$\times \int_0^x dx_1 \int_0^{x_1} dx_2 \cdots \int_0^{x_{m-2}} \chi_{[0,\frac{1}{2})}\left(x_{m-1} - \frac{j}{2}\right) dx_{m-1}.$$

Another method for computing the $B$-wavelet $\psi_m$ (and any $B$-wavelet series $g(x)$ as in (6.3.30)), is to compute the $B$-net representation of each of its polynomial pieces. This gives the values of $\psi_m(x)$ not only at the dyadic points $x = n + \ell/2^j$, as given by the general computational scheme (6.3.2), but also at every $x \in \mathbb{R}$. The efficient computation may be performed simply by taking a moving average of the sequence $\mathcal{P}_m(3m-2)$ with the sequence of the $B$-nets of the $m^{\text{th}}$ order cardinal $B$-spline $N_m(2x)$ obtained by applying the Cardinal $B$-spline $B$-net Algorithm in Section 4.4. (See Example 4.11 for the $B$-nets of the quadratic, cubic, and quartic cardinal $B$-splines.)

## 6.4. Euler-Frobenius polynomials

We have seen that Euler-Frobenius polynomials $E_m(z)$ play a very important role in cardinal spline interpolation (see Sections 4.6 and 6.1) and the construction and analysis of cardinal spline-wavelets (see Sections 6.1-6.3). In this section, we shall investigate these polynomials in some detail, paying special attention to their zero structures. These structures have already been used in (4.2.18)-(4.2.21) for determining the sharp lower Riesz bounds of $N_m$, and will again play an important role in the error analysis of spline-wavelet decompositions in the next section. Although the properties of the Euler-Frobenius polynomials $E_m$ of even and odd orders are the same and the derivations of these properties are quite similar, in order not to repeat analogous arguments we will only consider the odd order ones, since the even order ones are not used in our study of spline-wavelets.

Analogous to (4.6.8), by applying the Poisson Summation Formula (2.5.8), the Euler-Frobenius polynomial $E_{2m-1}(z)$, with $z = e^{-i\omega}$, can be written as

$$E_{2m-1}(z) := (2m-1)! \sum_{k=0}^{2m-2} N_{2m}(k+1)z^k \qquad (6.4.1)$$

$$= (2m-1)! e^{i\omega} \sum_k \widehat{N}_{2m}(\omega + 2\pi k)$$

$$= (2m-1)! e^{-i(m-1)\omega} \left(2\sin\frac{\omega}{2}\right)^{2m} \sum_{k=-\infty}^{\infty} \frac{1}{(\omega + 2\pi k)^{2m}}$$

(see (3.2.16)). By defining

$$e_n(\omega) := \left(2\sin\frac{\omega}{2}\right)^{n+2} \sum_{k=-\infty}^{\infty} \frac{1}{(\omega + 2\pi k)^{n+2}}, \qquad (6.4.2)$$

it follows from (6.4.1) that

$$e_{2m-2}(\omega) = \frac{1}{(2m-1)!} z^{-m+1} E_{2m-1}(z) \qquad (6.4.3)$$

$$= \sum_{k=-m+1}^{m-1} N_{2m}(m+k)z^k$$

$$= N_{2m}(m) + 2\sum_{k=1}^{m-1} N_{2m}(m+k)\cos k\omega.$$

As a consequence of (6.4.1) and (6.4.3), it is clear that $e_n(\omega)$ is completely characterized by the recursive relation

$$\begin{cases} e_{n+1}(\omega) = \left(\cos\frac{\omega}{2}\right)e_n(\omega) - \dfrac{2}{n+2}\left(\sin\frac{\omega}{2}\right)e_n'(\omega); \\ e_0(\omega) = 1. \end{cases} \qquad (6.4.4)$$

Another formulation of $e_n$ is given by

$$e_n(\omega) = \left(\sin\frac{\omega}{2}\right)^{n+2}\frac{(-1)^{n+1}}{(n+1)!}(D^{n+1}\cot)\left(\frac{\omega}{2}\right),\tag{6.4.5}$$

which follows from (4.2.9), where $D$ is the derivative operator. In view of the formulations in (6.4.4) and (6.4.5), we find it convenient to introduce the new variable

$$x = \cos\frac{\omega}{2},\tag{6.4.6}$$

so that (6.4.4) becomes

$$\begin{cases} U_{n+1}(x) = xU_n(x) + \dfrac{1-x^2}{n+2}U_n'(x); \\[2mm] U_0(x) = 1, \end{cases}\tag{6.4.7}$$

where

$$U_n(x) := e_n(\omega).\tag{6.4.8}$$

**Remark.** From (6.4.7), it is quite easy to compute $U_n$, $n \in \mathbb{Z}_+$. For instance, we have

$$U_1(x) = x,$$
$$U_2(x) = \frac{1}{3}(1 + 2x^2),$$
$$U_3(x) = \frac{1}{3}(2x + x^3),$$
$$U_4(x) = \frac{1}{15}(2 + 11x^2 + 2x^4).$$

Observe from this remark that at least for $0 \leq n \leq 4$, $U_n(x)$ is a polynomial of exact degree $n$, and has a positive leading coefficient. Moreover, $U_n(1) = 1$ and $U_n$ is an even function for even $n$ and an odd function for odd $n$. That the same properties still hold for all $n \in \mathbb{Z}_+$ can be established by mathematical induction. In fact, one can even draw some interesting conclusions concerning the zeros of $U_n$ as follows.

**Lemma 6.12.** *For each $n \in \mathbb{Z}_+$, $U_n(x)$ is a polynomial of exact degree $n$, consisting only of even powers if $n$ is even and only of odd powers if $n$ is odd, such that $U_n^{(n)} > 0$ and $U_n(1) = 1$. Furthermore, all the zeros of $U_n$ are simple and purely imaginary.*

**Proof.** We will only be concerned with the last statement. In addition, since it is easier to consider real zeros, we study

$$u_n(x) = \frac{1}{i^n}U_n(ix), \quad n \in \mathbb{Z}_+\tag{6.4.9}$$

instead of $U_n$; and it is sufficient to show that all the zeros of $u_n$ are simple and real. From (6.4.7), we also have

$$\begin{cases} u_{n+1}(x) = xu_n(x) - \dfrac{1+x^2}{n+2}u'_n(x); \\ u_0(x) = 1, \end{cases} \qquad (6.4.10)$$

and consequently, it is easy to see that $u_n$ is even for even $n$ and odd for odd $n$. In addition, from (6.4.9), it follows that

$$u_n^{(n)}(x) = U_n^{(n)}(ix),$$

which means that the leading coefficient of $u_n$ is also positive. We now proceed by induction.

(i) Suppose that $u_{2k}$ has only simple and real zeros. Then since $u_{2k}$ is an even function, it cannot have a zero at 0 and all its zeros occur in symmetric pairs, say, $\pm\xi_j$, where

$$0 < \xi_1 < \cdots < \xi_k.$$

Now, $u'_{2k}$ cannot vanish at these zeros, and in fact, the sequence

$$\{u'_{2k}(-\xi_k), \ldots, u'_{2k}(-\xi_1), u'_{2k}(\xi_1), \ldots, u'_{2k}(\xi_k)\}$$

must (strictly) alternate in sign. Consequently, since (6.4.10) asserts that

$$u_{2k+1}(\pm\xi_j) = -\frac{1+\xi_j^2}{n+1}u'_{2k}(\pm\xi_j),$$

the sequence

$$\{u_{2k+1}(-\xi_k), \ldots, u_{2k+1}(-\xi_1), u_{2k+1}(\xi_1), \ldots, u_{2k+1}(\xi_k)\} \qquad (6.4.11)$$

must also (strictly) alternate in sign. Being an odd function, $u_{2k+1}$ must vanish at 0, and from the sign pattern in (6.4.11), it has at least one zero between $\xi_j$ and $\xi_{j+1}$, $j = 1, \ldots, k-1$. So, $u_{2k+1}$ has at least $2(k-1)+1 = 2k-1$ zeros in the open interval $(-\xi_k, \xi_k)$. Now, from the fact that $u_{2k}$ has a positive leading coefficient, it follows that $u'_{2k}(\xi_k) > 0$ (since $\xi_k$ is the largest zero). Hence, again from (6.4.10), we have

$$u_{2k+1}(\xi_k) = -\frac{1+\xi_k^2}{n+1}u'_{2k}(\xi_k) < 0.$$

So, since the leading coefficient of $u_{2k+1}$ is also positive, $u_{2k+1}$ must have at least one zero to the right of $\xi_k$; and being odd, it also has one zero to the left of $-\xi_k$. This shows that $u_{2k+1}$ has $2k+1$ simple real zeros.

(ii) Suppose that $u_{2k+1}$ has only simple real roots at $\pm\eta_j$ and 0 where $0 < \eta_1 < \cdots < \eta_k$. Then by the same argument as in (i), we may conclude that $u_{2k+2}$ has roots in each of the intervals $(0, \eta_1), \ldots, (n_{k-1}, n_k)$, $(\eta_k, \infty)$. Hence, being even, $u_{2k+2}$ has $2k + 2$ simple real zeros. ∎

By applying Lemma 6.12, we may conclude, from the identification of $U_n$ with $e_n$ in (6.4.8), that

$$
\begin{cases}
e_{2k}(\omega) = U_{2k}(x) = c_k \prod_{j=1}^{k} (x^2 + \alpha_j^2), \quad 0 < \alpha_1 < \cdots < \alpha_k, \quad c_k > 0; \\
\\
e_{2k+1}(\omega) = U_{2k+1}(x) = d_k x \prod_{j=1}^{k} (x^2 + \beta_j^2), \quad 0 < \beta_1 < \cdots < \beta_k, \quad d_k > 0.
\end{cases}
$$
$$(6.4.12)$$

Next, we must relate the imaginary zeros of $U_n$ to the zeros of the Euler-Frobenius polynomials. Observe from the fact that $z = e^{-i\omega}$ and the relation (6.4.6) between $x$ and $\omega$, that, for any $\alpha > 0$,

$$
\begin{aligned}
x^2 + \alpha^2 &= \left( \frac{e^{i\omega/2} + e^{-i\omega/2}}{2} \right)^2 + \alpha^2 \\
&= \frac{1}{4} \{ z + (2 + 4\alpha^2) + z^{-1} \} \\
&= \frac{z^{-1}}{4} \{ z^2 + (2 + 4\alpha^2) z + 1 \} \\
&= \frac{z^{-1}}{4} (z - a_1)(z - a_2),
\end{aligned}
$$

where

$$
\begin{aligned}
a_1, a_2 &= \frac{-(2 + 4\alpha^2) \pm \sqrt{(2 + 4\alpha^2)^2 - 4}}{2} \\
&= -(1 + 2\alpha^2) \pm 2\alpha \sqrt{1 + \alpha^2}.
\end{aligned}
$$

So, by setting

$$
a_1 = -(1 + 2\alpha^2) - 2\alpha \sqrt{1 + \alpha^2},
$$

we have

$$
a_2 = \frac{1}{a_1} \quad \text{and} \quad -1 < a_2 < 0.
$$

Hence, we conclude that, for any $\alpha > 0$,

$$
\begin{cases}
(x^2 + \alpha^2) = \frac{z^{-1}}{4} (z - \gamma) \left( z - \frac{1}{\gamma} \right), \\
\text{where } -1 < \gamma < 0.
\end{cases}
$$
$$(6.4.13)$$

To apply this result to (6.4.12), let us set

$$
\lambda_j = -(1 + 2\alpha_j^2) + 2\alpha_j \sqrt{1 + \alpha_j^2}.
$$

Then it is clear that $-1 < \lambda_j < 0$ and

$$e_{2k}(\omega) = c_k \frac{1}{4^k} z^{-k} \sum_{j=1}^{k} (z - \lambda_j) \left( z - \frac{1}{\lambda_j} \right), \tag{6.4.14}$$

where $z = e^{-i\omega}$. When (6.4.14) is substituted into (6.4.3), we arrive at the following result.

**Theorem 6.13.** *Let $m$ be any positive integer. Then the Euler-Frobenius polynomial $E_{2m-1}$ of order $2m - 1$ (or degree $2m - 2$) can be written as*

$$E_{2m-1}(z) = \prod_{j=1}^{2m-2} (z - \lambda_{m,j}), \tag{6.4.15}$$

*where*

$$\lambda_{m,2m-2} < \lambda_{m,2m-3} < \cdots < \lambda_{m,m} < -1 < \lambda_{m,m-1} < \cdots < \lambda_{m,1} < 0, \tag{6.4.16}$$

*and*

$$\lambda_{m,1}\lambda_{m,2m-2} = \cdots = \lambda_{m,m-1}\lambda_{m,m} = 1. \tag{6.4.17}$$

## 6.5. Error analysis in spline-wavelet decomposition

In Section 6.2, when the compactly supported cardinal spline-wavelet (or $B$-wavelet) $\psi_m$ was introduced in (6.2.5), the normalization parameters $c_0$ and $n_0$ for $K(z)$ in (6.2.4) were chosen to be $c_0 = -\frac{1}{(2m-1)!}$ and $n_0 = 1$. Hence, the symbols $G(z)$ and $H(z)$ in (5.3.15) that correspond to the decomposition relation (5.3.16) in Theorem 5.16 are given by

$$\begin{cases} G(z) = \dfrac{1}{2} \displaystyle\sum_n g_n z^n = z^{-1} \left( \dfrac{1+z}{2} \right)^m \dfrac{E_{2m-1}(z)}{E_{2m-1}(z^2)}; \\[4mm] H(z) = \dfrac{1}{2} \displaystyle\sum_n h_n z^n = -z^{-1} \left( \dfrac{1-z}{2} \right)^m \dfrac{(2m-1)!}{E_{2m-1}(z^2)} \end{cases} \tag{6.5.1}$$

(see (6.2.1) and (5.4.11)). Now let us recall from (5.4.46) that the sequences for the decomposition algorithm (5.4.48) are chosen to be

$$\begin{cases} a_n = \dfrac{1}{2} g_{-n}; \\[4mm] b_n = \dfrac{1}{2} h_{-n}. \end{cases} \tag{6.5.2}$$

So, for non-Haar spline-wavelets (i.e. $m \geq 2$), the "weight" sequences $\{a_n\}$ and $\{b_n\}$ are infinite sequences, and must be truncated in order to apply the (finite) moving average scheme in (5.4.48).

**Remark.** The weight sequences for the reconstruction algorithm (5.4.49) are the finite sequences

$$\begin{cases} p_n = 2^{-m+1} \dbinom{m}{n}, & n = 0, \dots, m; \\[2mm] q_n = (-1)^n 2^{-m+1} \displaystyle\sum_{\ell=0}^{m} \dbinom{m}{\ell} N_{2m}(n+1-\ell), & n = 0, \dots, 3m-2, \end{cases} \tag{6.5.3}$$

which can be easily computed for any order $m$ by using the linear PTA's as discussed in Examples 6.8 and 6.9. In addition, since $G(z)$ and $H(z)$ are rational functions, it is conceivable that a recursive algorithm could be devised for decomposition without truncation. However, we will not discuss this approach here.

Observe that the only multiplicative factor in (6.5.1) that contributes to the infiniteness of the weight decomposition sequences $\{a_n\}$ and $\{b_n\}$ in (6.5.1)-(6.5.2) is $1/E_{2m-1}(z^2)$. We will now analyze the "errors" that will arise as a result of truncating this factor.

Let us begin by considering the $(2m)^{\text{th}}$ order fundamental cardinal spline

$$L_{2m}(x) = \sum_{k=-\infty}^{\infty} c_k^{(2m)} N_{2m}(x+m-k) \tag{6.5.4}$$

that has already been discussed in Sections 4.6 and 6.1 (see (6.1.8)). Recall from (6.1.12) and (6.1.13) that the symbol of the coefficient sequence $\{c_k^{(2m)}\}$ in (6.5.4) is the reciprocal of the generalized Euler-Frobenius Laurent polynomial $F_m := E_{N_m}$, namely:

$$\widetilde{C}^{(2m)}(z) := \sum_{n=-\infty}^{\infty} c_n^{(2m)} z^n = \frac{1}{F_m(z)} = \frac{(2m-1)! z^{m-1}}{E_{2m-1}(z)}, \tag{6.5.5}$$

where $E_{2m-1}$ is the Euler-Frobenius polynomial of order $2m-1$ (or degree $2m-2$) studied in the previous section. So, truncation of $1/E_{2m-1}(z^2)$ is equivalent to truncating the $B$-spline series representation (6.5.4) of the fundamental cardinal spline $L_{2m}$. For convenience, let us absorb the factor $(2m-1)!$ with $\{c_n^{(2m)}\}$ by setting

$$\alpha_n = \alpha_n^{(m)} := \frac{1}{(2m-1)!} c_n^{(2m)}, \tag{6.5.6}$$

so that (6.5.5) becomes

$$\frac{z^{m-1}}{E_{2m-1}(z)} = \sum_{n=-\infty}^{\infty} \alpha_n z^n. \tag{6.5.7}$$

We now truncate this Laurent series by introducing

$$T_m^N(z) := \sum_{n=-N}^{N} \alpha_n z^n, \tag{6.5.8}$$

for positive integers $N$. With the quantity in (6.5.7) replaced by that in (6.5.8) in the expressions (6.5.1) for $G$ and $H$, we obtain

$$\begin{cases} G_N(z) = \dfrac{1}{2} \sum_n g_{N,n} z^n := \left(\dfrac{1+z}{2}\right)^m z^{-2m+1} E_{2m-1}(z) T_m^N(z^2); \\[4mm] H_N(z) = \dfrac{1}{2} \sum_n h_{N,n} z^n := -(2m-1)! \left(\dfrac{1+z}{2}\right)^m z^{-2m+1} T_m^N(z^2). \end{cases} \tag{6.5.9}$$

The finite "truncated" decomposition sequences are now

$$\begin{cases} a_{N,n} = \dfrac{1}{2} g_{N,-n}; \\[4mm] b_{N,n} = \dfrac{1}{2} h_{N,-n}. \end{cases} \tag{6.5.10}$$

It is clear from (6.5.9) that the supports of the truncated sequences are given by

$$\begin{cases} \sup\{a_{N,n}\} = [-2N - m + 1, 2N + 2m - 1] \cap \mathbb{Z}, \\[2mm] \sup\{b_{N,n}\} = [-2N + m - 1, 2N + 2m + 1] \cap \mathbb{Z}. \end{cases} \tag{6.5.11}$$

When the finite sequences $\{a_{N,n}\}$ and $\{b_{N,n}\}$ are used instead of the original sequences $\{a_n\}$ and $\{b_n\}$ as weights in the decomposition algorithm (5.4.48), there will be some discrepancy. To measure this error, we simply compare the perfect reconstruction of the truncated decomposed components with the original finite-energy sequence. Precisely, let

$$\begin{cases} f_j(x) = \sum_k c_k^j N_m(2^j x - k); \\[2mm] \mathbf{c}^j = \{c_k^j\} \end{cases} \tag{6.5.12}$$

be any cardinal $B$-spline series in $V_j^m$. The sequence $\mathbf{c}^j$ is a finite-energy sequence representation of the signal $f_j$. If $f_j$ is decomposed into

$$f_j = f_{N,j-1} + g_{N,j-1}, \tag{6.5.13}$$

where $f_{N,j-1} \in V_{j-1}^m$ and $g_{N,j-1} \in W_{j-1}^m$, by using the (finite) decomposition algorithm

$$\begin{cases} c_{N,k}^{j-1} = \sum_\ell a_{N,\ell-2k} c_\ell^j; \\[2mm] d_{N,k}^{j-1} = \sum_\ell b_{N,\ell-2k} c_\ell^j, \end{cases} \tag{6.5.14}$$

then we will compare the perfect reconstruction $\mathbf{c}_N^j$ of its decomposed components

$$\begin{cases} \mathbf{c}_N^{j-1} := \{c_{N,k}^{j-1}\}, & k \in \mathbb{Z}; \\ \mathbf{d}_N^{j-1} := \{d_{N,k}^{j-1}\}, & k \in \mathbb{Z}; \end{cases}$$

that is, $\mathbf{c}_N^j = \{\mathbf{c}_{N,k}^j\}$, $k \in \mathbb{Z}$, where

$$c_{N,k}^j = \sum_\ell [p_{k-2\ell} c_{N,\ell}^{j-1} + q_{k-2\ell} d_{N,\ell}^{j-1}], \qquad (6.5.15)$$

with the original sequence $\mathbf{c}^j$, where $\{p_k\}$ and $\{q_k\}$ are given in (6.5.3). By using $\ell^2$-measurement, the truncation error is the quantity:

$$\mathcal{E}_N^{(m)}(\mathbf{c}^j) := \left\{ \sum_{\ell \in \mathbb{Z}} |c_\ell^j - c_{N,\ell}^j|^2 \right\}^{1/2}. \qquad (6.5.16)$$

Since the cardinal B-spline $N_m$ generates a Riesz basis of $V_0^m$, the error measurement in (6.5.16) is equivalent to the measurement

$$\widetilde{\mathcal{E}}_N^{(m)}(f_j) := \|f_j - f_{N,j}\|_2, \qquad (6.5.17)$$

where

$$f_{N,j} := \sum_k c_{N,k}^j N_m(2^j x - k), \qquad (6.5.18)$$

with $\{c_{N,k}^j\}$ given by (6.5.15).

For the error analysis, we need the following expression for the coefficients of the B-spline series of the fundamental cardinal splines.

**Lemma 6.14.** *Let* $\lambda_{m,j}$, $j = 1, \ldots, 2m - 2$, *be the zeros of the* $(2m-1)^{\mathrm{st}}$ *order Euler-Frobenius polynomial* $E_{2m-1}$ *as in Theorem 6.13. Then the coefficients of the Laurent series in (6.5.7) are given by*

$$\alpha_j = \alpha_j^{(m)} = \sum_{k=1}^{m-1} \left( \frac{\lambda_{m,k}^{m-2}}{E_{2m-1}'(\lambda_{m,k})} \right) \lambda_{m,k}^{|j|}, \qquad (6.5.19)$$

*for all* $j \in \mathbb{Z}$.

**Proof.** To establish (6.5.19), we need the identity in (6.3.21) for the modified Euler-Frobenius polynomials $\widetilde{E}_n$ defined in (6.3.19). Recall that

$$E_{2m-1}(z) = \widetilde{E}_{2m-1}(z), \qquad m = 1, 2, \ldots,$$

but $\widetilde{E}_{2m} \neq E_{2m}$ for all $m$. Now by (6.3.21) and (6.4.17), we have

$$\begin{cases} \widetilde{E}_{2m}(\lambda_{m,j}) = \lambda_{m,j}(1 - \lambda_{m,j})E_{2m-1}'(\lambda_{m,j}) \\ \widetilde{E}_{2m}(\lambda_{m,j}^{-1}) = -\dfrac{1}{\lambda_{m,j}^2}(1 - \lambda_{m,j})E_{2m-1}'(\lambda_{m,j}^{-1}). \end{cases} \qquad (6.5.20)$$

Hence, it follows from the simple observation

$$\widetilde{E}_{2m}(\lambda_{m,j}) = \lambda_{m,j}^{2m-1}\widetilde{E}_{2m}(\lambda_{m,j}^{-1}),$$

and (6.5.20), that

$$E'_{2m-1}(\lambda_{m,j}^{-1}) = -\lambda_{m,j}^{-2m+4}E'_{2m-1}(\lambda_{m,j}). \tag{6.5.21}$$

By resorting to a partial fraction decomposition and the use of the relations in (6.5.21) and (6.4.17), consecutively, we obtain

$$\frac{z^{m-1}}{E_{2m-1}(z)} = \sum_{j=1}^{2m-2} \frac{\lambda_{m,j}^{m-1}}{(z - \lambda_{m,j})E'_{2m-1}(\lambda_{m,j})} \tag{6.5.22}$$

$$= \sum_{j=1}^{m-1}\left\{\frac{\lambda_{m,j}^{m-1}}{E'_{2m-1}(\lambda_{m,j})z\left(1 - \frac{\lambda_{m,j}}{z}\right)} + \frac{\lambda_{m,j}^{m-2}}{E'_{2m-1}(\lambda_{m,j})(1 - \lambda_{m,j}z)}\right\}$$

$$= \sum_{j=1}^{m-1} \frac{\lambda_{m,j}^{m-2}}{E'_{2m-1}(\lambda_{m,j})} \sum_{n=-\infty}^{\infty} \lambda_{m,j}^{|n|}z^n, \quad |z| = 1.$$

This establishes (6.5.19). ∎

**Remark.** By setting

$$\alpha_j = \alpha_j^{(m)} = \sum_{k=1}^{m-1} \kappa_k^{(m)}\lambda_{m,k}^{|j|}, \quad j \in \mathbb{Z}, \tag{6.5.23}$$

in (6.5.19), we have, by applying (6.3.21),

$$\kappa_k^{(m)} = \frac{\lambda_{m,k}^{m-2}}{E'_{2m-1}(\lambda_{m,k})} = \lambda_{m,k}^{m-1}(1 - \lambda_{m,k})\widetilde{E}_{2m}^{-1}(\lambda_{m,k}), \tag{6.5.24}$$

where $\widetilde{E}_{2m}$ is the modified Euler-Frobenius polynomial of order $2m$. Thus, the formula (6.5.24) facilitates the computation of $\kappa_k^{(m)}$, and consequently of $\alpha_j^{(m)}$ in (6.5.23), since $\widetilde{E}_{2m}$ is easy to compute (see the LPTA in Example 6.9 and (6.3.21)).

To formulate an estimate of the error measurement in (6.5.16), let us recall the notation

$$F_m(z) = \frac{1}{(2m-1)!}\frac{E_{2m-1}(z)}{z^{m-1}} \tag{6.5.25}$$

as introduced in (6.1.13) and define

$$R_m^N(z) := (2m-1)!F_m(z^2)\sum_{j=1}^{m-1}\kappa_j^{(m)}\left\{\frac{z^{-2N-2}}{1 - \lambda_{m,j}z^{-2}} + \frac{z^{2N+2}}{1 - \lambda_{m,j}z^2}\right\}\lambda_{m,j}^{N+1}. \tag{6.5.26}$$

We have the following results.

**Theorem 6.15.** *For any positive integer* $m$,

$$\mathcal{E}_N^{(m)}(\mathbf{c}) \le \max_{|z|=1} |R_m^N(z)| \|\mathbf{c}\|_{\ell^2} \tag{6.5.27}$$

*for any* $\mathbf{c} \in \ell^2$.

**Theorem 6.16.** *For any positive integer* $m$, *there exists a positive integer* $N_0 = N_0(m)$, *such that*

$$\max_{|z|=1} |R_m^N(z)| = 2((2m-1)!) \left| \sum_{j=1}^{m-1} \frac{\kappa_j^{(m)}}{1 - \lambda_{m,j}} |\lambda_{m,j}|^{N+1} \right|, \tag{6.5.28}$$

*for all* $N \ge N_0$. *Furthermore,* $N_0(m)$ *can be chosen to be 0 for* $m = 2, 3, 4$.

As a consequence of the two theorems stated above, since the terms in the sum in (6.5.28) for $2 \le m \le 4$ alternate in signs and have monotonically decreasing magnitudes, we have the following.

**Corollary 6.17.** *For* $m = 2, 3, 4$, *and any positive integer* $N$,

$$\mathcal{E}_N^{(m)}(\mathbf{c}) \le \sigma_{m-1}^{(m)} |\lambda_{m,m-1}|^{N+1} \|\mathbf{c}\|_{\ell^2}, \tag{6.5.29}$$

*where*

$$\sigma_j^{(m)} := 2((2m-1)!) \frac{\kappa_j^{(m)}}{1 - \lambda_{m,j}}. \tag{6.5.30}$$

**Proof of Theorem 6.15.** Let $\mathbf{c} = \mathbf{c}^j$ and $\mathbf{c}_N = \mathbf{c}_N^j$ as in (6.5.12) and (6.5.15), and denote their symbols by $\widetilde{C}$ and $\widetilde{C}_N$, respectively. Then from (6.5.14), (6.5.15), and (6.5.9), we have

$$\widetilde{C}_N(z) = \{P(z)G_N(z) + Q(z)H_N(z)\}\widetilde{C}(z) \tag{6.5.31}$$
$$+ \{P(z)G_N(-z) + Q(z)H_N(-z)\}\widetilde{C}(-z),$$

where $P$ and $Q$ are the two-scale symbols of the $m^{\text{th}}$ order cardinal B-spline $N_m$ and B-wavelet $\psi_m$. Hence, by applying the identities (5.3.13), we have

$$\widetilde{C}(z) - \widetilde{C}_N(z) = \{P(z)[G(z) - G_N(z)] + Q(z)[H(z) - H_N(z)]\}\widetilde{C}(z)$$
$$+ \{P(z)[G(-z) - G_N(-z)] \tag{6.5.32}$$
$$+ Q(z)[H(-z) - H_N(-z)]\}\widetilde{C}(-z).$$

On the other hand, it follows from (6.5.7), (6.5.8), (6.5.23), (6.5.25), (6.5.26), that

$$R_m^N(z) = 1 - (2m-1)! F_m(z^2) T_m^N(z^2), \tag{6.5.33}$$

and the identity in Theorem 5.10, (iv), translates into

$$(1+z)^{2m} E_{2m-1}(z) - (1-z)^{2m} E_{2m-1}(-z) = 2^{2m} z E_{2m-1}(z^2). \tag{6.5.34}$$

Therefore, by applying (6.5.33) and (6.5.34), the quantity in (6.5.32) can be simplified as

$$
\begin{aligned}
\widetilde{C}(z) - \widetilde{C}_N(z) &= \frac{z^{-2m+1}}{2^{2m}} \Big\{ [(1+z)^{2m} E_{2m-1}(z) - (1-z)^{2m} E_{2m-1}(-z)] \\
&\quad \times \left[ \frac{1}{(2m-1)!} \frac{1}{F_m(z^2)} - T_m^N(z^2) \right] \widetilde{C}(z) \\
&\quad - [(1+z)^m (1-z)^m E_{2m-1}(-z) \\
&\quad - (1-z)^m (1+z)^m E_{2m-1}(-z)] \\
&\quad \times \left[ \frac{1}{(2m-1)!} \frac{1}{F_m(z^2)} - T_m^N(z^2) \right] \widetilde{C}(-z) \Big\} \\
&= F_m(z^2) \left\{ \frac{1}{F_m(z^2)} - (2m-1)! T_m^N(z^2) \right\} \widetilde{C}(z) \\
&= R_m^N(z) \widetilde{C}(z).
\end{aligned}
$$

(6.5.35)

Assertion (6.5.27) now follows from (6.5.35) by applying the Parseval Identity (2.4.18). ∎

**Proof of Theorem 6.16.** By introducing the notion

$$
F_{m,j}(z) := \frac{F_m(z)}{\left(z + \frac{1}{z}\right) - \left(\lambda_{m,j} + \frac{1}{\lambda_{m,j}}\right)},
$$

(6.5.36)

and observing that

$$
E'_{2m-1}(\lambda_{m,j}) = \prod_{k \neq j} (\lambda_{m,j} - \lambda_{m,k}),
$$

we have, from (6.5.24),

$$
\frac{1}{\kappa_j^{(m)}} = (2m-1)! F_{m,j}(\lambda_{m,j}) \left( \lambda_{m,j} - \frac{1}{\lambda_{m,j}} \right).
$$

(6.5.37)

Next, substituting (6.5.37) into (6.5.26) yields

$$
\begin{aligned}
R_m^N(z) &= \sum_{j=1}^{m-1} \frac{\lambda_{m,j}^{N+1} F_m(z^2)}{F_{m,j}(\lambda_{m,j})\left(\lambda_{m,j} - \frac{1}{\lambda_{m,j}}\right)} \frac{1}{\left(z^2 + \frac{1}{z^2}\right) - \left(\lambda_{m,j} + \frac{1}{\lambda_{m,j}}\right)} \\
&\quad \times [(z^{2N} + z^{-2N}) - \lambda_{m,j}^{-1}(z^{2N+2} + z^{-2N-2})] \\
&= \sum_{j=1}^{m-1} \frac{F_{m,j}(z^2)}{F_{m,j}(\lambda_{m,j})} \frac{\lambda_{m,j}^{N+1}}{\lambda_{m,j} - \frac{1}{\lambda_{m,j}}} [(z^{2N} + z^{-2N}) \\
&\quad - \lambda_{m,j}^{-1}(z^{2N+2} + z^{-2N-2})].
\end{aligned}
$$

(6.5.38)

On the other hand, by setting $z = e^{-i\omega/2}$, the expression $F_{m,j}$ as defined in (6.5.36) can be written as

$$F_{m,j}(z^2) = F_{m,j}(e^{-i\omega}) = \sum_{\ell=0}^{m-2} b_{m,j,\ell} \cos \ell\omega. \qquad (6.5.39)$$

Hence, the formula in (6.5.38) becomes

$$R_m^N(z) = 2 \sum_{j=1}^{m-1} \sum_{\ell=0}^{m-2} \frac{\lambda_{m,j}^{N+1} b_{m,j,\ell}}{F_{m,j}(\lambda_{m,j})\left(\lambda_{m,j} - \frac{1}{\lambda_{m,j}}\right)} \qquad (6.5.40)$$
$$\times \cos \ell\omega \left[\cos N\omega + \frac{1}{|\lambda_{m,j}|} \cos(N+1)\omega\right].$$

Since $0 < |\lambda_{m,1}| < \cdots < |\lambda_{m,m-1}| < 1$, it is clear that

$$\lambda_{m,m-1} - \frac{1}{\lambda_{m,m-1}} = \frac{1}{|\lambda_{m,m-1}|} - |\lambda_{m,m-1}| > 0, \qquad (6.5.41)$$

so that

$$F_{m,m-1}(\lambda_{m,m-1}) = \frac{1}{(2m-1)!} \prod_{j=1}^{m-2} \left[\left(\lambda_{m,m-1} + \frac{1}{\lambda_{m,m-1}}\right)\right.$$
$$\left. - \left(\lambda_{m,j} + \frac{1}{\lambda_{m,j}}\right)\right] > 0. \qquad (6.5.42)$$

But since $F_{m,j}(z)$ is a symmetric Laurent polynomial with only negative zeros, the coefficients, $b_{m,j,\ell}$ in (6.5.39) must be strictly positive. So, it follows from (6.5.41) and (6.5.42) that

$$\lim_{N \to \infty} \frac{1}{|\lambda_{m,m-1}|^N} \sum_{j=1}^{m-1} \frac{b_{m,j,\ell} |\lambda_{m,j}|^N}{F_{m,j}(\lambda_{m,j})\left(\lambda_{m,j} - \frac{1}{\lambda_{m,j}}\right)}$$
$$= \frac{b_{m,m-1,\ell}}{F_{m,m-1}(\lambda_{m,m-1})\left(\lambda_{m,m-1} - \frac{1}{\lambda_{m,m-1}}\right)} > 0,$$

for $\ell = 0, \ldots, m-2$; and hence an integer $N_0 = N_0(m)$ exists, such that for all $N \geq N_0$,

$$\gamma_{m,\ell}^{(N)} := 2 \sum_{j=1}^{m-1} \frac{b_{m,j,\ell} |\lambda_{m,j}|^N}{F_{m,j}(\lambda_{m,j})\left(\lambda_{m,j} - \frac{1}{\lambda_{m,j}}\right)} > 0 \qquad (6.5.43)$$

for $\ell = 0, \ldots, m-2$. As a consequence, we see that for all $N \geq N_0$, the cosine polynomial $R_m^N(z)$, which can be formulated as

$$R_m^N(z) = (-1)^{N+1} \sum_{\ell=0}^{m-2} \{\gamma_{m,\ell}^{(N+1)} \cos N\omega + \gamma_{m,\ell}^{(N)} \cos(N+1)\omega\} \cos \ell\omega,$$

has positive coefficients (except for the $(-1)^{N+1}$ factor) and satisfies

$$\max_{|z|=1} |R_m^N(z)| = |R_m^N(1)|, \quad N \geq N_0. \tag{6.5.44}$$

So, since $F_m(1) = \Sigma N_{2m}(j) = 1$, it follows from (6.5.44) and (6.5.26) that

$$\max_{|z|=1} |R_m^N(z)| = |R_m^N(1)| = 2((2m-1)!) \left| \sum_{j=1}^{m-1} \kappa_j^{(m)} \frac{\lambda_{m,j}^{N+1}}{1 - \lambda_{m,j}} \right|, \tag{6.5.45}$$

for all $N \geq N_0$. Observe that (6.5.45) agrees with (6.5.28) in view of the fact that $\{\lambda_{m,j}^{N+1}\}$ is of one sign. For $m = 2$, it is clear that (6.5.45) holds for all $N \geq 0$, and it only takes a little bit more work to show that (6.5.28) also holds for all $N \geq 0$, when $m = 3$ and $4$. ∎

We end this section by mentioning that for linear and cubic spline wavelets, the truncation errors have the following upper bounds

$$\begin{cases} \mathcal{E}_N^{(2)}(\mathbf{c}) \leq 2.7320509 \times (0.26795)^{N+1} \|\mathbf{c}\|_{\ell^2}; \\ \mathcal{E}_N^{(4)}(\mathbf{c}) \leq 7.8373747 \times (0.5352805)^{N+1} \|\mathbf{c}\|_{\ell^2}; \end{cases} \tag{6.5.46}$$

These are crude estimates made by applying Corollary 6.17. Better estimates, particularly for small $N$ and $m = 2, 3, 4$, can be obtained by applying (6.5.28).

## 6.6. Total positivity, complete oscillation, and zero-crossings

As mentioned in Chapter 4, cardinal $B$-splines possess a very special property called "total positivity". This property is the key ingredient that makes the $B$-spline series

$$\sum_k c_k N_m(\cdot - k) \tag{6.6.1}$$

stand out as a unique tool most suitable for "smoothing" purposes. In this section, we shall demonstrate that the corresponding cardinal $B$-wavelet $\psi_m$ has a property which is in some sense a contrast to total positivity. While a $B$-spline series "smooths" any "bumpy" data, a $B$-wavelet series

$$\sum_k d_k \psi_m(\cdot - k) \tag{6.6.2}$$

"detects" such data. The important difference between these two series is that the series in (6.6.1) never oscillates more than its coefficient sequence

$\{c_k\}$, while the series in (6.6.2) always oscillates more than $\{d_k\}$. We will say that $\psi_m$ possesses a special property called "*complete oscillation*", which is in contrast to the property of total positivity of its companion $N_m$.

**Definition 6.18.**

(i) *A matrix $M$ (finite, infinite, or bi-infinite) is said to be "totally positive" (TP), if any square submatrix of $M$ of any finite dimension, formed by deleting arbitrary rows and columns of $A$, has a nonnegative determinant.*

(ii) *A function $F(x, y)$ of two variables is called a TP kernel, if the matrix $[F(x_j, y_k)]$, where $\{x_j\}$ and $\{y_k\}$ are increasing sequences arbitrarily chosen from the domain of definition of $F$, is a TP matrix.*

(iii) *A sequence $\{a_n\}$, $n = 0, 1, \ldots$, (finite or infinite), with $a_0 \neq 0$, is called a TP sequence, or a Pólya frequency (PF) sequence, if the Toeplitz matrix $[a_{-j+k}]$, formulated by setting $a_{-1} = a_{-2} = \cdots = 0$, is a TP matrix.*

For the cardinal $B$-spline $N_m$, we consider $N_m(x - k)$ as a function of two variables $x \in \mathbb{R}$ and $k \in \mathbb{Z}$. So, by saying that $N_m$ is TP, we mean that for any positive integer $n$ and any sequences $\{x_j\}$ and $\{\ell_j\}$, with

$$\begin{cases} x_1 < \cdots < x_n & x_j \in \mathbb{R}; \\ \ell_1 < \cdots < \ell_n, & \ell_j \in \mathbb{Z}, \end{cases} \tag{6.6.3}$$

we have

$$\det[N_m(x_j - \ell_k)] \geq 0. \tag{6.6.4}$$

A proof of this important fact about cardinal $B$-splines will not be included in this book, but we do point out that as a consequence of (6.6.4), the sequence

$$\{N_{2m}(j + 1)\}, \quad j = 0, 1, \ldots, \tag{6.6.5}$$

is a PF sequence. Indeed, by selecting $x_j = j + 1$ and $\ell_k = k$, where $j, k \in \mathbb{Z}_+$, we see that the transpose of the matrix $[N_{2m}(x_j - \ell_k)] = [N_{2m}(j + 1 - k)]$ agrees with the upper triangular Toeplitz matrix whose first row is given by (6.6.5). It is also interesting to note that the symbol of the sequence in (6.6.5), being a $1/(2m - 1)!$ multiple of the Euler-Frobenius polynomial $E_{2m-1}$, can be written as

$$\sum_{j=0}^{\infty} N_{2m}(j + 1)z^j = \frac{1}{(2m - 1)!} \prod_{j=1}^{2m-2} (z + |\lambda_{m,j}|), \tag{6.6.6}$$

since $\lambda_{m,j} < 0$, $j = 1, \ldots, 2m - 2$ (see (6.4.16)). This result can be extended to any PF sequence as follows.

**Lemma 6.19.** *Let $\{a_j\}, j \in \mathbb{Z}_+$, be a sequence in $\ell^2$ satisfying $a_0 \neq 0$. Then $\{a_j\}$ is a PF sequence if and only if its symbol can be written as*

$$\sum_{j=0}^{\infty} a_j z^j = a_0 e^{\gamma z} \prod_{j=1}^{\infty} \frac{1 + \alpha_j z}{1 - \beta_j z}, \tag{6.6.7}$$

where $\gamma, \alpha_j, \beta_j \geq 0$ and

$$\sum_{j=1}^{\infty} (\alpha_j + \beta_j) < \infty.$$

**Example 6.20.** Consider the two-scale sequence $\{q_k\}$ of the $m^{\text{th}}$ order cardinal $B$-wavelet $\psi_m$ as given in (6.2.6). More generally, for each $r = 0, \ldots, m-1$, consider the sequence $\{q_{m,k}^{(r)}\}$ that governs its $r^{\text{th}}$ derivative $\psi_m^{(r)}$ as defined by (6.3.6). The symbol of $\{q_{m,k}^{(r)}\}$ has been shown in (6.3.29) to be

$$\sum_k q_{m,k}^{(r)} z^k = \sum_{k=0}^{3m-2+r} q_{m,k}^{(r)} z^k = \frac{2^{-m+r+1}}{(2m-1)!} E_{2m-1}(-z)(1-z)^{m+r},$$

and consequently, $q_{m,k}^{(r)}$, $k = 0, \ldots, 3m-2+r$, alternates in sign, and the symbol of $\{\tilde{q}_{m,k}^{(r)}\}$, $\tilde{q}_{m,k}^{(r)} := (-1)^k q_{m,k}^{(r)} = |q_{m,k}^{(r)}|$ becomes

$$\sum_k \tilde{q}_{m,k}^{(r)} z^k = \frac{2^{-m+r+1}}{(2m-1)!} E_{2m-1}(z)(1+z)^{m+r},$$

which is a polynomial with only negative zeros. Hence, by Lemma 6.19, $\{\tilde{q}_{m,k}^{(r)}\}$ is a PF sequence. ∎

Recall from Theorem 6.10 that $\{\tilde{q}_{m,k}^r\}$ can be computed by applying an LPTA. In general, we have the following result.

**Theorem 6.21.** *Let $\mathcal{P}$ be an LPTA as defined by (6.3.12) with $k_L(n)$, $k_R(n) \geq 0$ and $b_L(n)$, $b_R(n) > 0$, for all $n \in \mathbb{Z}_+$. Then each $\mathcal{P}(n)$ is a PF sequence.*

An important feature of PF sequences, and TP kernels in general, is their so-called "variation-diminishing" property. This is a "smoothing" effect in the sense that "oscillations", such as sign changes, of a sequence or function, gradually diminish when it is "filtered" by convolving with PF sequences or integrating with TP kernels.

**Definition 6.22.** *The number of strong (or actual) sign changes of a finite sequence **a** of real numbers, denoted by $S^-(\mathbf{a})$, is the count of sign changes of **a** after all the zero terms are deleted. The number of both strong and weak sign changes of this sequence **a**, denoted by $S^+(\mathbf{a})$, is the count of changes of signs when each of the (interior) zero terms of the sequence is considered as being positive or negative to give the maximum count. (Note: When a finite sequence is considered, the first and last terms of the sequence are supposed to be nonzero.) The number of strong sign changes of a continuous function $f$ with supp $f = [a, b]$ is defined to be*

$$S^-(f) = \sup\{S^-(\{f(x_0), \ldots, f(x_n)\}) \colon \ a < x_0 < \cdots < x_n < b, \quad n \in \mathbb{Z}_+\},$$
$$(6.6.8)$$

and the number of both strong and weak sign changes of the same $f$ is defined by

$$S^+(f) = \sup\{S^+(\{f(x_0), \ldots, f(x_n)\}): \ a < x_0 < \cdots < x_n < b, \quad n \in \mathbb{Z}_+\}. \tag{6.6.9}$$

**Remark.** In considering the count $S^+(f)$, we are only interested in those $f$ whose supports do not contain nontrivial intervals on which $f = 0$; that is, we assume $\overline{Z^c(f)} = \operatorname{supp} f$, where

$$\overline{Z^c(f)} = \text{closure of } \{x: \ f(x) \neq 0\}. \tag{6.6.10}$$

Otherwise, the count $S^+(f)$ should be considered on each component of supp $f$.

The following variation-diminishing property of cardinal $B$-splines can be established by applying the TP property.

**Theorem 6.23.** *Let $m \geq 2$ be any positive integer. Then*

$$S^-\left(\sum_{k=0}^n a_k N_m(\cdot - k)\right) \leq S^-(\{a_k\}), \quad a_0, a_n \neq 0. \tag{6.6.11}$$

As to the corresponding $B$-wavelets $\psi_m$, we will see that instead of possibly reducing the number of sign changes, a $B$-wavelet series always oscillates more often than its coefficient sequence. This property of "*complete oscillation*" of $\psi_m$, which is in contrast to the TP property of $N_m$, is what makes $\psi_m$ useful in applications to localization and measurement of irregularities such as singularities.

**Theorem 6.24.** *Let $m \geq 2$, and $d_0, d_n \neq 0$. Then*

$$S^-\left(\sum_{k=0}^n d_n \psi_m(\cdot - k)\right) \geq n + 3m - 2 \geq S^-(\{d_k\}) + 3m - 2. \tag{6.6.12}$$

When the zeros of a wavelet series are also counted as sign changes, one expects a larger lower bound. In this direction, we have the following result valid for linear $B$-wavelet series.

**Theorem 6.25.** *Let $d_0, d_n \neq 0$. Then*

$$S^+\left(\sum_{k=0}^n d_k \psi_2(\cdot - k)\right) \geq 2n + 4 - S^-(\{d_k\}). \tag{6.6.13}$$

Furthermore, if the support of this $B$-wavelet series is an interval and if $S^-(\{d_k\}) = 0$, then

$$S^+\left(\sum_{k=0}^n d_k \psi_2(\cdot - k)\right) = 2n + 4. \tag{6.6.14}$$

**Remark.** The result in (6.6.14) justifies the terminology of "complete oscillation".

**Proof of Theorem 6.24.** Recall from Theorem 6.3 in Section 6.1 that the (wavelet) space $W_0^m$ can be identified as the (multiresolution) subspace $V_1^{2m,0}$ of $V_1^{2m}$ via the $m^{\text{th}}$ order differential operator $D^m$.

However, instead of considering the relation $\psi_{I,m} = D^m \psi_{2m}$ there, let us introduce the $(2m)^{\text{th}}$ order cardinal spline

$$\psi_{2m}^*(x) = \frac{1}{2^{2m-1}} \sum_{j=0}^{2m-2} (-1)^j N_{2m}(j+1) N_{2m}(2x-j) \qquad (6.6.15)$$

that belongs to $V_1^{2m,0}$. It is easy to verify that

$$\{\Psi_{2m}^*(\cdot - k): \ k \in \mathbb{Z}\}$$

is a Riesz basis of $V_1^{2m,0}$. Furthermore, analogous to the relation $D^m \Psi_{2m} = \psi_{I,m}$, we also have

$$D\Psi_{2m}^* = \psi_m.$$

Indeed, applying (vii) in Theorem 4.3 $m$ times yields

$$(D^m \Psi_{2m}^*)(x) = \frac{1}{2^{m-1}} \sum_{j=0}^{2m-2} (-1)^j N_{2m}(j+1) N_{2m}^{(m)}(2x-j)$$

$$= \frac{1}{2^{m-1}} \sum_{j=0}^{2m-2} (-1)^j N_{2m}(j+1) \sum_{k=0}^{m} (-1)^k \binom{m}{k} N_m(2x-j-k)$$

$$= \sum_{\ell=0}^{3m-2} \left( \frac{(-1)^\ell}{2^{m-1}} \sum_{k=0}^{m} \binom{m}{k} N_{2m}(\ell-k+1) \right) N_m(2x-\ell)$$

$$= \sum_{\ell=0}^{3m-2} q_\ell N_m(2x-\ell) = \psi_m(x).$$

Hence, we may use $\Psi_{2m}^*$ in place of $\Psi_{2m}$ in Theorem 6.3. Now, consider the spline series

$$G(x) = \sum_{k=0}^{n} d_k \Psi_{2m}^*(x-k).$$

Since $d_0, d_n \neq 0$ and supp $\Psi_{2m}^* = [0, 2m-1]$, we see that

$$\text{supp } G = [0, n+2m-1]. \qquad (6.6.16)$$

Thus, it follows from the fact that

$$\begin{cases} G(k) = 0, & k \in \mathbb{Z}; \\ G^{(\ell)}(0) = G^{(\ell)}(n+2m-1) = 0, & \ell = 0, \ldots, 2m-2, \end{cases} \qquad (6.6.17)$$

that a family of points $\{x_j^{(\ell)}\}$, $j = 1, \ldots, n + 2m + \ell - 2$ and $\ell = 1, \ldots, m$, satisfying

$$
\begin{cases}
0 < x_1^{(1)} < 1 < x_2^{(1)} < 2 < \cdots < n + 2m - 2 < x_{n+2m-1}^{(1)} < n + 2m - 1 \\
0 < x_1^{(2)} < x_1^{(1)} < x_2^{(2)} < x_2^{(1)} < \cdots < x_{n+2m-1}^{(1)} < x_{n+2m}^{(2)} < n + 2m - 1 \\
\cdots\cdots\cdots\cdots\cdots\cdots\cdots\cdots\cdots\cdots\cdots\cdots\cdots\cdots\cdots\cdots\cdots\cdots\cdots \\
0 < x_1^{(m)} < x_1^{(m-1)} < x_2^{(m)} < x_2^{(m-1)} < \cdots < x_{n+3m-3}^{(m-1)} < x_{n+3m-2}^{(m)} < n + 2m - 1,
\end{cases}
$$

$$(6.6.18)$$

exists, such that $G^{(\ell)}$ has a (strong) sign change at each $x_j^{(\ell)}$, $j = 1, \ldots, n + 2m + \ell - 2$. In particular,

$$
G^{(m)}(x) = \sum_{k=0}^{n} d_k \psi_m(x - k)
$$

has a (strong) sign change at each of $x_1^{(m)}, \ldots, x_{n+3m-2}^{(m)}$. ∎

**Proof of Theorem 6.25.** To study the more delicate count $S^+$, we need a result from TP matrices, namely: if $A$ is any $p \times (n+1)$ TP matrix with $p > n$, then

$$
S^-(A\mathbf{v}) \leq \min(n, S^-(\mathbf{v})) = S^-(\mathbf{v}) \tag{6.6.19}
$$

for any $(n + 1)$-vector of real numbers. This fact is actually the key ingredient for establishing Theorem 6.23. In addition, if $\mathbf{v} = (v_0, \ldots, v_n)$ and $\tilde{\mathbf{v}} := (v_0, -v_1, v_2, \ldots, (-1)^n v_n)$, then it is easy to verify that

$$
S^+(\mathbf{v}) + S^-(\tilde{\mathbf{v}}) \geq n. \tag{6.6.20}
$$

To establish (6.6.13), we need the matrix

$$
A := \left[ (-1)^{j-2k+1} \psi_2 \left( \frac{j - 2k + 2}{2} \right) \right] \tag{6.6.21}
$$

$$
= \frac{1}{12}
\begin{bmatrix}
1 & & & & \\
6 & & & & \\
10 & 1 & & & \\
6 & 6 & & & \\
1 & 10 & & & \\
& 6 & \cdots & & \\
& 1 & & 1 & \\
& & & 6 & \\
& & & 10 & \\
& & & 6 & \\
& & & 1 &
\end{bmatrix}.
$$

It is also easy to verify this $(2n+5) \times (n+1)$ matrix $A$ is a TP matrix. Hence, by applying (6.6.19) and (6.6.20) to the vector $\mathbf{v} = (v_0, \ldots, v_{2n+4})$ with

$$v_j := \sum_{k=0}^{n} d_k \psi_2 \left( \frac{j-1}{2} - k \right),$$

we have

$$S^+(\mathbf{v}) \geq 2n + 4 - S^-(\tilde{\mathbf{v}}) \geq 2n + 4 - S^-(\{d_k\}).$$

This establishes (6.6.13).

Next, by using the two-scale relation of $\psi_2$ relative to $N_2$, the cardinal $B$-wavelet series $\Sigma d_k \psi_2(\cdot - k)$ can be written as a cardinal $B$-spline series in terms of $N_2(2 \cdot -k)$. Since the support of this series is an interval, we have, by applying yet another result on the TP of cardinal $B$-splines,

$$S^+ \left( \sum_{k=0}^{n} d_k \psi_2(\cdot - k) \right) \leq 2n + 4. \tag{6.6.22}$$

Hence, (6.6.14) follows from (6.6.13) and (6.6.22) when $S^-(\{d_k\}) = 0$.  ∎

In applications, it is quite essential to be precise about the oscillations of a $B$-wavelet series. For instance, the distribution of the zeros of a bandpass band-limited signal, which is governed by the "Nyquist rate", can be completely recovered from its isolated zeros (called "zero-crossings"), provided that certain conditions are satisfied. When a cardinal $B$-wavelet series is treated as a bandpass signal, however, it cannot be band-limited, being different from an entire function of exponential type. Nonetheless, when linear $B$-wavelets are used, we still have the following theorem. Analogous results for cardinal $B$-wavelets of higher orders are not available at this writing.

**Theorem 6.26.** *Let*

$$
\begin{cases}
f(x) = \displaystyle\sum_{j=0}^{n} d_j \psi_2(x - j); \\[2ex]
g(x) = \displaystyle\sum_{j=0}^{n} c_j \psi_2(x - j)
\end{cases}
$$

*be two linear spline-wavelet series with $\overline{Z^c}(f) = \overline{Z^c}(g) = [0, n+3]$ such that $S^-(\mathbf{c}) = 0$ and $g$ has only simple zeros. Then if $f$ and $g$ have the same zeros, $f$ must be a constant multiple of $g$.*

**Proof.** By assumption, we have $d_0 d_n \neq 0$ and $c_0 c_n \neq 0$. Choose $c$ such that $c_0 = cd_0$. Then

$$(g - cf)(x) = \sum_{j=1}^{n} (c_j - cd_j)\psi_2(x - j).$$

Suppose that $g - cf$ is not identically zero. Then we may assume, without loss of generality, that $\overline{Z^c}(g - cf) = [1, n+3]$. Since $(g - cf) \in V_1^2$, an application of the so-called Budan-Fourier Theorem affirms that the number of zeros $Z(g-cf)$, counting multiplicities, of $g-cf$ in the open interval $(1, n+3)$ does not exceed $2n + 2$, namely:

$$Z(g - cf) \leq 2n + 2. \tag{6.6.23}$$

On the other hand, since the restriction of $g$ to $(0,1]$ is a constant multiple of $\psi_2$, it has only a simple zero in $(0,1]$. Hence, it follows from Theorem 6.25 that

$$S^+(g)|_{(1,n+3)} = S^+(g)|_{(0,n+3)} - 1 \geq 2n + 4 - S^-(\mathbf{c}) - 1 = 2n + 3. \tag{6.6.24}$$

In addition, since $g$ has only simple zeros and $f$ has the same zeros as $g$, we have

$$Z(g)|_{(1,n+3)} \leq Z(g - cf). \tag{6.6.25}$$

Therefore, from (6.6.23) – (6.6.25), we obtain

$$2n + 3 \leq S^+(g)|_{(1,n+3)} \leq Z(g)|_{(1,n+3)} \leq Z(g - cf) \leq 2n + 2,$$

which is absurd. This completes the proof of the theorem.  ∎

# 7 Orthogonal Wavelets and Wavelet Packets

For obvious reasons, orthonormal (o.n.) bases are the most desirable bases of a Hilbert space. In particular, if an o.n. basis of $L^2(\mathbb{R})$ is generated by some $\mathcal{R}$-function $\psi$, then, being self-dual, $\psi$ is already a wavelet. Moreover, the two-scale sequence $\{q_n\}$ of $\psi$ relative to the scaling function $\phi$ is obtained from the two-scale sequence $\{p_n\}$ of $\phi$ simply by complex conjugation followed by alternation of signs and reverse in direction with a unit shift. For instance, as in (5.6.14), we may set

$$q_n = (-1)^n \bar{p}_{-n+1}.$$

In other words, essentially only one two-scale sequence governs both the multiresolution analysis (MRA) and its corresponding wavelet decomposition. What is more interesting is that by applying a mixture of $\{p_n\}$ and $\{q_n\}$ to produce two-scale relations with $\psi$, the wavelet spaces $W_n$ can be further decomposed orthogonally. The families of new orthogonal basis functions so produced are called "*wavelet packets*". This chapter is devoted to the analysis and construction of o.n. wavelets and their wavelet packets. In particular, the construction of compactly supported o.n. wavelets will be discussed.

## 7.1. Examples of orthogonal wavelets

A general framework of wavelets and their duals has been studied in Section 5.4, where the proposed strategy was to start with two admissible two-scale symbols $P = P_\phi$ and $G^* = G^*_{\tilde{\phi}}$, which are dual to each other. For the construction of o.n. wavelets, as we have seen in Section 5.6, $\phi$ is also necessarily self-dual, so that $G^* = P$ and the dual relation (5.4.7) becomes

$$|P(z)|^2 + |P(-z)|^2 = 1, \quad |z| = 1. \tag{7.1.1}$$

In addition, by Theorem 5.19, the general formulation of the two-scale symbol $Q$ for the o.n. wavelet $\psi$ relative to the scaling function $\phi$ is given by

$$c_0 z^{2n_0 - 1} \overline{P(-z)}, \quad |z| = 1, \tag{7.1.2}$$

where $c_0 = \pm 1$ and $n_0$ is an arbitrary integer. To be consistent with the choice in (5.6.13) and (5.6.14), let us set $c_0 = -1$ and $n_0 = 1$, so that

$$Q(z) = -z\overline{P(-z)}, \quad |z| = 1. \tag{7.1.3}$$

Hence, to construct an o.n. wavelet $\psi$, we must investigate the two-scale relations

$$\begin{cases} \hat{\phi}(\omega) = P(e^{-i\omega/2})\hat{\phi}\left(\dfrac{\omega}{2}\right); \\ \widehat{\psi}(\omega) = Q(e^{-i\omega/2})\hat{\phi}\left(\dfrac{\omega}{2}\right). \end{cases} \qquad (7.1.4)$$

Of course, if $\hat{\phi}$ or $P$ is known, then $\widehat{\psi}$, and hence $\psi$, is determined by applying the second relation in (7.1.4) using (7.1.3).

**Example 7.1.** Let $m$ be any positive integer and let $N_m$ denote the $m^{\text{th}}$ order cardinal $B$-spline. Then by Theorem 3.23 (see also (3.6.18)), the scaling function $\phi_m$ whose Fourier transform is given by

$$\hat{\phi}_m(\omega) = \frac{\widehat{N}_m(\omega)}{\left(\sum_k |\widehat{N}_m(\omega + 2\pi k)|^2\right)^{1/2}} = \frac{e^{-im\omega/2}\left(\frac{\sin\omega/2}{\omega/2}\right)^m}{(F_m(e^{-i\omega}))^{1/2}} \qquad (7.1.5)$$

is an o.n. scaling function in the sense that $\{\phi_m(\cdot - k)\colon k \in \mathbb{Z}\}$ is an o.n. basis of $V_0^m$. Here, $F_m$ is the generalized Euler-Frobenius Laurent polynomial relative to $N_m$ as defined in (6.1.13), and the structure of $F_m$ has been discussed in Section 6.4 (see (6.1.13) and Theorem 6.13). As a consequence, by applying (7.1.3)-(7.1.5), the Fourier transform $\widehat{\psi}$ of the o.n. wavelet $\psi$ that generates $\{W_j^m\}$ is seen to be

$$\widehat{\psi}(\omega) = -e^{-i\omega/2}\overline{P(e^{-i(\omega+2\pi)/2})}\hat{\phi}_m\left(\frac{\omega}{2}\right) \qquad (7.1.6)$$

$$= -e^{-i\omega/2}\overline{\hat{\phi}_m(\omega + 2\pi)}\hat{\phi}_m\left(\frac{\omega}{2}\right)\Big/\overline{\hat{\phi}_m\left(\frac{\omega}{2} + \pi\right)}$$

$$= -\left(\frac{4}{i\omega}\right)^m e^{-i\omega/2}\sin^{2m}\left(\frac{\omega}{4}\right)\left(\frac{F_m(-z)}{F_m(z^2)F_m(z)}\right)^{1/2},$$

where $z = e^{-i\omega/2}$. For $m = 1$, it is easy to see that the o.n. wavelet $\psi$ given by (7.1.66) is nothing but the Haar function $\psi_H$. ∎

Of course, if we already have an s.o. wavelet, then the orthonormalization procedure formulated in (3.6.18) readily yields an o.n. wavelet.

**Example 7.2.** Let $\psi$ be one of the s.o. cardinal spline-wavelets $\psi_{I,m}$, $\psi_m$, or $\widetilde{\psi}_m$, $m \geq 1$, introduced in the previous chapter. Then $\psi^\perp$, whose Fourier transform is given by (3.6.18), is an o.n. wavelet. For $\psi = \psi_{I,m}$, $\psi_m$, or $\widetilde{\psi}_m$, the results obtained in Chapter 6 can be easily applied to determine the Laurent series

$$\sum_k |\widehat{\psi}(\omega + 2\pi k)|^2.$$

(See (6.1.11)-(6.1.13), (6.2.5), and (6.2.9).)

In general, since we don't have any knowledge of the scaling function $\phi$ or its two-scale symbol $P = P_\phi$, we have to work very hard to come up with either one of them. Recall that our strategy is to construct $P$, and the next two sections will be devoted to this effort. To end this section, we give an example in which $\hat{\phi}$, instead of $P$, is constructed first.

**Example 7.3.** Let $0 < \varepsilon \leq \pi/3$, $0 < A < 1 < B < \infty$, and $N$ be an arbitrary positive integer. Select any $\hat{\eta} \in C^N(\mathbb{R})$ that satisfies the following conditions:

$$
\begin{cases}
\text{supp } \hat{\eta} = [-\pi - \varepsilon, \pi + \varepsilon]; \\
\hat{\eta}(\omega) = 1 \quad \text{for} \quad |\omega| \leq \pi - \varepsilon; \text{ and} \\
A \leq \sum_k |\hat{\eta}(\omega + 2\pi k)|^2 \leq B, \quad \omega \in \mathbb{R}.
\end{cases}
\tag{7.1.7}
$$

Then we can introduce a function $\phi$ whose Fourier transform is given by

$$
\hat{\phi}(\omega) := \frac{\hat{\eta}(\omega)}{\left( \sum_k |\hat{\eta}(\omega + 2\pi k)|^2 \right)^{1/2}}.
\tag{7.1.8}
$$

It is clear that $\hat{\phi}$ satisfies

$$
\text{supp } \hat{\phi} = [-\pi - \varepsilon, \pi + \varepsilon],
\tag{7.1.9}
$$

and that the function $P$ defined on the unit circle by

$$
P(e^{-i\omega}) := \sum_{k=-\infty}^{\infty} \hat{\phi}(2\omega + 4\pi k)
\tag{7.1.10}
$$

is in $C^N$. Hence, by integrating the Fourier series

$$
f(\omega) := P(e^{-i\omega}) = \frac{1}{2} \sum_{n=-\infty}^{\infty} p_n e^{-in\omega},
\tag{7.1.11}
$$

by parts $N$ times, we obtain

$$
p_n = \frac{1}{\pi} \int_0^{2\pi} f(\omega) e^{in\omega} d\omega = \left( \frac{i}{n} \right)^N \int_0^{2\pi} f^{(N)}(\omega) e^{in\omega} d\omega,
$$

so that

$$
p_n = O(|n|^{-N}), \quad |n| \to \infty.
\tag{7.1.12}
$$

In particular, we have $\{p_n\} \in \ell^2$. Next we show that

$$
\hat{\phi}(\omega) = P(e^{-i\omega/2}) \hat{\phi}\left( \frac{\omega}{2} \right), \quad \omega \in \mathbb{R},
\tag{7.1.13}
$$

by considering two separate cases.

(i) Suppose that $\omega \notin \operatorname{supp} \hat{\phi}$. Then we have

$$P(e^{-i\omega/2}) = \sum_{k \neq 0} \hat{\phi}(\omega + 4\pi k),$$

so that either $\hat{\phi}(\frac{\omega}{2}) = 0$, or else $P(e^{-i\omega/2}) = 0$. That is, both sides of (7.1.13) are zero in this case.

(ii) Suppose that $\omega \in \operatorname{supp} \hat{\phi}$. Then we can draw the following two conclusions. Firstly, since $0 < \varepsilon \leq \pi/3$, it follows that $\hat{\eta}(\frac{\omega}{2} + 2\pi k) = 0$ for all nonzero integers $k$, so that $\hat{\phi}(\frac{\omega}{2}) = \hat{\eta}(\frac{\omega}{2})/\hat{\eta}(\frac{\omega}{2}) = 1$. Secondly, we have $P(e^{-i\omega/2}) = \hat{\phi}(\omega)$. Hence, both sides of (7.1.13) are equal to $\hat{\phi}(\omega)$.

From the definition of $\hat{\phi}$ in (7.1.8), we already know that

$$\{\phi(\cdot - k)\colon k \in \mathbb{Z}\} \tag{7.1.14}$$

is an o.n. family (see Theorem 3.23), and that

$$\begin{cases} \hat{\phi}(0) = 1; \\ D^n \hat{\phi}(2\pi k) = 0, \quad 0 \neq k \in \mathbb{Z}, \quad n \in \mathbb{Z}_+. \end{cases} \tag{7.1.15}$$

In view of the Poisson Summation Formula (see (2.5.11)), this implies that (7.1.14) is also a partition of unity, namely:

$$\sum_{k=-\infty}^{\infty} \phi(x - k) = 1, \quad x \in \mathbb{R}. \tag{7.1.16}$$

Hence, to conclude that $\phi$ generates an MRA $\{V_n\}$ of $L^2(\mathbb{R})$, what remains to be verified is that for any $f \in L^2(\mathbb{R})$,

$$\|P_n(f) - f\|_2 \to 0, \quad n \to \infty, \tag{7.1.17}$$

where $P_n(f)$ denotes the $L^2(\mathbb{R})$ projection of $f$ onto $V_n$. For this purpose, we need to know the rate of decay of $\phi$. This is easy, for an $N$-fold integration by parts readily yields

$$\int_{-\infty}^{\infty} \hat{\phi}^{(N)}(\omega) e^{ix\omega} d\omega = (-ix)^N \int_{-\infty}^{\infty} \hat{\phi}(\omega) e^{ix\omega} d\omega = 2\pi (-ix)^N \phi(x)$$

which implies that

$$\phi(x) = O\left(\frac{1}{1 + |x|^N}\right). \tag{7.1.18}$$

Now, by using the kernel

$$K(x,y) = \sum_{k=-\infty}^{\infty} \phi(x-k)\overline{\phi(y-k)}, \qquad (7.1.19)$$

we can represent the projection $P_n(f)$ of $f$ as

$$(P_n f)(x) = 2^n \int_{-\infty}^{\infty} K(2^n x, 2^n y)f(y)dy.$$

Since (7.1.16) implies that

$$\int_{-\infty}^{\infty} K(x,y)dy = 1, \quad x \in \mathbb{R},$$

we may conclude that

$$\begin{aligned}
\|P_n f - f\|_2 &= 2^n \left\| \int_{-\infty}^{\infty} K(2^n \cdot, 2^n y)[f(y) - f(\cdot)]dy \right\|_2 \qquad (7.1.20) \\
&\leq \left\| \int_{-\infty}^{\infty} K(2^n \cdot, y)[f(2^{-n}y) - f(\cdot)]dy \right\|_2 \\
&\leq C \left\| \int_{-\infty}^{\infty} \frac{1}{1 + |2^n \cdot - y|^N} |f(2^{-n}y) - f(\cdot)|dy \right\|_2,
\end{aligned}$$

where the estimate

$$K(x,y) \leq \frac{C}{1 + |x-y|^N}, \qquad (7.1.21)$$

which follows from (7.1.18), has been used. By applying the generalized Minkowski inequality (which is the $L^2(\mathbb{R})$ analog of (2.4.3) and can be easily derived by using (2.1.1)), we obtain, from (7.1.20),

$$\|P_n f - f\|_2 \leq C \int_{-\infty}^{\infty} \frac{1}{1 + |y|^N} \|f(\cdot - 2^{-n}y) - f(\cdot)\|_2 dy. \qquad (7.1.22)$$

The rest of the proof is now standard, and is accomplished by breaking the integral into two parts. First, for each $\varepsilon > 0$, choose $M > 0$ so large that

$$\int_{|y| \geq M} \frac{1}{1 + |y|^N} dy < \varepsilon.$$

Then, since $f \in L^2(\mathbb{R})$, we have, for $|y| \leq M$,

$$\|f(\cdot - 2^{-n}y) - f(\cdot)\|_2 \to 0 \qquad (7.1.23)$$

uniformly, as $n \to \infty$. This establishes (7.1.17).  ∎

## 7.2. Identification of orthogonal two-scale symbols

The technical problem in the construction of scaling functions and wavelets is the identification of the admissible two-scale symbols $P = P_\phi$ (see Definition 5.4 and recall that our strategy in general is to start with a pair of dual two-scale symbols $P_\phi$ and $G_{\tilde\phi}^*$). Besides the cardinal spline-wavelets studied in Chapter 6 and the three examples of o.n. wavelets discussed in the previous section, there does not seem to be any procedure available other than constructing $\hat\phi$ as an infinite product

$$\hat\phi(\omega) = \prod_{k=1}^{\infty} P(e^{-i\omega/2^k}). \tag{7.2.1}$$

Although certain conditions imposed on the limit functions of such infinite products have been identified in Theorems 5.5 and 5.6, these conditions do not directly reflect on the two-scale symbols $P$ nor do they guarantee all the requirements for $\phi$ to generate an MRA. This section is concerned with a study of the Laurent series $P \in W$ that are two-scale symbols of some $\phi$. Since a complete characterization of $P$ does not seem to be feasible, we will be satisfied with a sufficient condition which can be easily applied. It is a continuation of our effort which began with Lemma 5.20. The distinction here is that special emphasis is placed on $\phi$ being an "*o.n. scaling function*", and by this we mean that $\phi$ not only generates an MRA of $L^2(\mathbb{R})$, but also satisfies the requirement

$$\langle \phi(\cdot - j), \phi(\cdot - k)\rangle = \delta_{j,k}, \quad j, k \in \mathbb{Z}. \tag{7.2.2}$$

Let us first recall from (5.1.12) and (7.1.1) the necessary conditions that a two-scale symbol $P$ of any o.n. scaling function $\phi$ must satisfy.

**Lemma 7.4.** *Let $P$ be a Laurent series in the Wiener Class $W$. If $P$ is a two-scale symbol of some scaling function $\phi$ which is o.n. in the sense of (7.2.2), then $P$ must satisfy*

$$P(1) = 1, \quad \text{and} \tag{7.2.3}$$

$$|P(z)|^2 + |P(-z)|^2 = 1, \quad |z| = 1. \tag{7.2.4}$$

Therefore, in view of (7.2.3), we are, as in (5.1.17), interested in those $P$ in $W$ that can be expressed as

$$\begin{cases} P(z) = \dfrac{1}{2}\sum_k p_k z^k = \left(\dfrac{1+z}{2}\right)^N S(z), \\[2mm] \text{where } N \text{ is some positive integer, and} \\[2mm] S \in W \text{ satisfies } S(1) = 1. \end{cases} \tag{7.2.5}$$

For any $S$ in (7.2.5), we write

$$\begin{cases} S(z) = \displaystyle\sum_k s_k z^k; \\[2mm] B := \displaystyle\max_{|z|=1} |S(z)|. \end{cases} \tag{7.2.6}$$

The objective of this section is to establish a useful condition on the factor $S$ which guarantees that $P$ is a two-scale symbol of some o.n. scaling function.

**Theorem 7.5.** *Let $P \in \mathcal{W}$ satisfy (7.2.4) and (7.2.5) for some $N \geq 1$, such that*

$$\sum_{k} |s_k| |k|^\varepsilon < \infty, \tag{7.2.7}$$

*for some $\varepsilon > 0$ and*

$$B < 2^{N-1}, \tag{7.2.8}$$

*where the notation in (7.2.6) is used. Then the infinite product*

$$g(\omega) := \prod_{k=1}^{\infty} P(e^{-i\omega/2^k}) \tag{7.2.9}$$

*converges to $g \in C(\mathbb{R}) \cap L^1(\mathbb{R}) \cap L^2(\mathbb{R})$ everywhere. Furthermore, the function $\phi \in L^2(\mathbb{R})$, with $\hat{\phi} = g$, is an o.n. scaling function that generates an MRA of $L^2(\mathbb{R})$.*

The proof of this theorem will depend on a sequence of lemmas.

**Lemma 7.6.** *Under the assumptions of (7.2.7) and (7.2.5) for some $N \geq 1$, the infinite product in (7.2.9) converges everywhere to a continuous function $g$.*

**Remark.** This lemma is different from Theorem 5.5 in that the hypothesis here is placed on $S$ instead of the finite products in (5.1.18).

**Proof.** From (7.2.7) and the relation (7.2.5), it is easy to see that $\{p_k\}$ also satisfies

$$\sum_{k} |p_k| |k|^\varepsilon < \infty; \tag{7.2.10}$$

and it follows from (7.2.10) that, for any $h > 0$,

$$|P(e^{-i(\omega+h)}) - P(e^{-i\omega})| \leq \frac{1}{2} \sum_{k} |p_k| |e^{-ikh} - 1|$$

$$\leq \frac{1}{2} \sum_{k} |p_k| \min(2, |k|h)$$

$$\leq |h|^\varepsilon \sum_{k} |p_k| |k|^\varepsilon.$$

(Here, without loss of generality, we consider $0 < \varepsilon \leq 1$.) Hence, as a function of $\omega$, $P(e^{-i\omega})$ is in the class $\operatorname{Lip} \varepsilon$ (see Definition 5.7). In addition, from the same derivation as that which led to (5.1.21), and the conclusion there (with the exception that $S$ is replaced by $P$ and $\alpha$ by $\varepsilon$), we may conclude that the infinite product in (7.2.9) converges. To prove that the limit function $g$ is continuous at each $\omega_0$, let us first assume that $g(\omega_0) \neq 0$. Under this assumption, we have

$$g(\omega) - g(\omega_0) = g(\omega_0)\{g(\omega)/g(\omega_0) - 1\}$$
$$= g(\omega_0)\left\{\exp\left[\sum_{k=1}^{\infty}\ln[1 - (1 - P(e^{-i\omega/2^k})/P(e^{-i\omega_0/2^k}))]\right] - 1\right\}$$
$$= g(\omega_0)\left\{\exp\left[O\left(\sum_{k=1}^{\infty}\frac{|\omega - \omega_0|^\varepsilon}{2^{k\varepsilon}}\right)\right] - 1\right\}$$
$$= o(1), \quad \omega \to \omega_0.$$

On the other hand, if $g(\omega_0) = 0$, then since $P(1) = 1$, there is a sufficiently large integer $k_0$, depending on $\omega_0$, such that the limit function

$$\tilde{g}(\omega) = \prod_{k=k_0}^{\infty} P(e^{-i\omega/2^k})$$

satisfies $\tilde{g}(\omega_0) \neq 0$. The derivation above already shows that $\tilde{g}$ is continuous at $\omega_0$, so that

$$g(\omega) = \tilde{g}(\omega_0) \prod_{k=1}^{k_0-1} P(e^{-i\omega/2^k})$$

is also continuous at $\omega_0$.  ∎

**Lemma 7.7.** *Under the assumptions of (7.2.8) and (7.2.5) for some $N \geq 1$, the limit function $g(\omega)$ in (7.2.9) satisfies*

$$|g(\omega)| \leq C\left(\frac{1}{1 + |\omega|}\right)^{1+\eta}, \quad \omega \in \mathbb{R}, \tag{7.2.11}$$

*for some $\eta > 0$.*

**Proof.** Let us return to the definition of $b_j$ in (5.1.18). In view of (7.2.8), we have

$$b_1 = \log_2 B_1 = \log_2 B < N - 1.$$

So, by choosing $n_0 = 1$ in Theorem 5.5, the conclusion in (5.1.20) of Theorem 5.5 yields (7.2.11), with

$$\eta := N - b_1 - 1. \quad \blacksquare$$

**Remark.** Let us pause for a moment to summarize what we have already proved and to give an outline of the rest of the proof of Theorem 7.5.

(i) As a consequence of Lemmas 7.6 and 7.7, we see that the infinite product in (7.2.9) converges everywhere to some function

$$g \in C(\mathbb{R}) \cap L^1(\mathbb{R}) \cap L^2(\mathbb{R}),$$

which will be called the infinite product itself. Hence, by Theorem 2.17, there is a unique $\phi \in L^2(\mathbb{R})$ whose Fourier transform $\hat{\phi}$ is this infinite product. Consequently, $\hat{\phi}$ satisfies:

$$\begin{cases} \hat{\phi}(\omega) = P(e^{-i\omega/2})\hat{\phi}\left(\dfrac{\omega}{2}\right), \text{ or} \\ \phi(x) = \sum_k p_k \phi(2x - k). \end{cases} \tag{7.2.12}$$

So, as usual, we may introduce

$$\phi_{j,k}(x) = 2^{j/2}\phi(2^j x - k), \quad j, k \in \mathbb{Z}, \tag{7.2.13}$$

and

$$V_j = \text{clos}_{L^2(\mathbb{R})}\langle \phi_{j,k} : k \in \mathbb{Z}\rangle. \tag{7.2.14}$$

By (7.2.12), it is clear that $\{V_j\}$ is a nested sequence of closed subspaces of $L^2(\mathbb{R})$; and to prove that it is an MRA of $L^2(\mathbb{R})$, we must show that the union of $V_j$ is dense in $L^2(\mathbb{R})$ and that $\phi$ generates a Riesz basis of $V_0$.

(ii) We shall first prove that $\phi$ generates an o.n. basis of $V_0$. To prove the density of the union of $V_j$, $j \in \mathbb{Z}$, we will be consistent with our approach discussed in Chapters 1 and 3, by constructing an o.n. wavelet basis of $L^2(\mathbb{R})$. This will imply not only

$$\text{clos}_{L^2(\mathbb{R})}\left(\bigcup_{j \in \mathbb{Z}} V_j\right) = L^2(\mathbb{R}), \tag{7.2.15}$$

but also

$$\bigcap_{j \in \mathbb{Z}} V_j = \{0\}. \tag{7.2.16}$$

(See Lemma 5.1.) From the study in Section 5.6, we know that a good candidate of an o.n. wavelet is

$$\psi(x) := \sum_{k \in \mathbb{Z}} q_k \phi(2x - k), \tag{7.2.17}$$

where

$$q_k := (-1)^k \bar{p}_{-k+1}. \tag{7.2.18}$$

Also, as usual, we set

$$\psi_{j,k}(x) = 2^{j/2}\psi(2^j x - k), \quad j, k, \in \mathbb{Z}. \tag{7.2.19}$$

Then the proof of Theorem 7.5 will be complete, provided we can show that the family

$$\{\psi_{j,k} : j, k \in \mathbb{Z}\} \tag{7.2.20}$$

is an o.n. basis of $L^2(\mathbb{R})$.

To prove that $\phi$ generates an o.n. basis (and hence, a Riesz basis) of $V_0$, we recall, from Theorem 3.23, that it is equivalent to showing that

$$\frac{1}{2\pi} \int_{-\infty}^{\infty} e^{ij\omega} |\hat{\phi}(\omega)|^2 d\omega = \delta_{j,0}, \quad j \in \mathbb{Z}. \tag{7.2.21}$$

**Lemma 7.8.** *Under the assumptions stated in Theorem 7.5, the infinite product $g = \hat{\phi}$ satisfies (7.2.21).*

**Proof.** The proof of this lemma is similar to the proof of one direction of Theorem 5.22. Indeed, by setting

$$g_n(\omega) := \left\{ \prod_{k=1}^{n} P(e^{-i\omega/2^k}) \right\} \chi_{[-2^n\pi, 2^n\pi]}(\omega), \tag{7.2.22}$$

we have

$$I_n := \frac{1}{2\pi} \int_{-\infty}^{\infty} e^{ij\omega} |g_n(\omega)|^2 d\omega = \frac{1}{2\pi} \int_{-2^n\pi}^{2^n\pi} \left( \prod_{k=1}^{n} |P(e^{-i\omega/2^k})|^2 \right) e^{ij\omega} d\omega,$$

which agrees with (5.4.22) with $G$ replaced by $\overline{P}$. Hence, from (5.4.25), we obtain

$$I_n = \frac{1}{2\pi} \int_{-\infty}^{\infty} e^{ij\omega} |g_n(\omega)|^2 d\omega = \delta_{j,0}, \quad j \in \mathbb{Z}; \tag{7.2.23}$$

and again by applying Lemma 5.20, we have

$$\frac{1}{2\pi} \int_{-\infty}^{\infty} e^{ij\omega} |\hat{\phi}(\omega)|^2 d\omega = \lim_{n \to \infty} I_n = \delta_{j,0},$$

for all $j \in \mathbb{Z}$. $\blacksquare$

In order to show that the family in (7.2.20) is an o.n. basis of $L^2(\mathbb{R})$, it is helpful to know the properties of the sequences $\{p_k\}$ and $\{q_k\}$, whose relation is given by (7.2.18).

**Lemma 7.9.** *Let $\{p_k\}$ and $\{q_k\}$ be defined by (7.2.5) and (7.2.18), where $P$ satisfies (7.2.3) and (7.2.4). Then these two sequences have the following properties:*

(i) $\sum_k p_k = 2$;

(ii) $\sum_k p_{k-2\ell} \overline{p}_{k-2m} = 2\delta_{\ell,m}$;

(iii) $\sum_k q_{k-2\ell} \overline{q}_{k-2m} = 2\delta_{\ell,m}$;

(iv) $\sum_k p_{k-2\ell} \overline{q}_{k-2m} = 0$; and

(v) $\sum_k \{p_{\ell-2k}\bar{p}_{m-2k} + q_{\ell-2k}\bar{q}_{m-2k}\} = 2\delta_{\ell,m}$,

for all $\ell, m \in \mathbb{Z}$.

**Proof.** Assertion (i) is equivalent to (7.2.3), assertion (ii) to (7.2.4), and in view of (7.2.18), (iii) is a trivial consequence of (ii). It is interesting to see that the "orthogonality" property (iv) can be easily derived by a simple change of indices. Indeed, we have

$$\sum_k p_{k-2\ell}\bar{q}_{k-2m} = \sum_k (-1)^k p_{k-2\ell}p_{-k+2m+1}$$

$$= \sum_j (-1)^{j+1}p_{-j+2m+1}p_{j-2\ell} = -\sum_k p_{k-2\ell}\bar{q}_{k-2m}.$$

Finally, to verify (v), we see that after applying (7.2.18) to change the $q$'s and $p$'s, the two sums in (v) simply form a partition of the summation in (ii), with one over the odd indices and the other over the even indices. ∎

As a consequence of the identity (v) in the lemma above, we have the following.

**Lemma 7.10.** *For all* $x \in \mathbb{R}$,

$$2\phi(2x - m) = \sum_k \{\bar{p}_{m-2k}\phi(x - k) + \bar{q}_{m-2k}\psi(x - k)\}, \quad m \in \mathbb{Z}. \tag{7.2.24}$$

**Proof.** Just apply (7.2.12), (7.2.17), and (v) in Lemma 7.9. ∎

The following decomposition formula will be useful.

**Lemma 7.11.** *For every* $f \in L^2(\mathbb{R})$,

$$\sum_k |\langle f, \phi_{j,k}\rangle|^2 = \sum_k \{|\langle f, \phi_{j-1,k}\rangle|^2 + |\langle f, \psi_{j-1,k}\rangle|^2\}, \quad j \in \mathbb{Z}. \tag{7.2.25}$$

**Proof.** Just apply Lemma 7.10 and (ii), (iii), and (iv) in Lemma 7.9. ∎

In addition to their nice "decomposition" properties as stated in Lemmas 7.10 and 7.11, the two families $\{\phi_{j,k}\}$ and $\{\psi_{j,k}\}$ also possess the following orthogonality properties as expected.

**Lemma 7.12.** *Under the assumptions stated in Theorem 7.5, the two families* $\{\phi_{j,k}\}$ *and* $\{\psi_{j,k}\}$ *satisfy the following:*

(i) $\langle \phi_{j,k}, \phi_{j,\ell}\rangle = \delta_{k,\ell}, \quad j, k, \ell \in \mathbb{Z};$

(ii) $\langle \phi_{j,k}, \psi_{j,\ell}\rangle = 0, \quad j, k, \ell \in \mathbb{Z};$ and

(iii) $\langle \psi_{j,k}, \psi_{\ell,m}\rangle = \delta_{j,\ell}\delta_{k,m}, \quad j, k, \ell, m \in \mathbb{Z}.$

**Proof.** Assertion (i) follows from Lemma 7.8 by applying Theorem 3.23 and a change of scale. To verify (ii), we simply apply (i), (7.2.12), (7.2.17), and Lemma 7.9, (iv). The same derivation, with the exception that (iii) instead of (iv) in Lemma 7.9, is applied, yields the assertion in (iii) for the situation $j = \ell$. For $j \neq \ell$, say $j > \ell$, we note from (7.2.17) that $\psi_{\ell,m} \in V_{\ell+1}$. Since $V_{\ell+1} \subset V_j$ and $\psi_{j,k}$ is orthogonal to $V_j$ by (ii), we see that $\psi_{j,k}$ is orthogonal to $\psi_{\ell,m}$. ∎

Hence, $\{\psi_{j,k}\}$ is an o.n. family in $L^2(\mathbb{R})$. As mentioned in an earlier remark, our approach to demonstrating the density of the union of $V_j$ in $L^2(\mathbb{R})$ is to verify that this family is an o.n. basis of $L^2(\mathbb{R})$. The standard procedure in harmonic analysis is to derive the "Parseval Identity" for $\{\psi_{j,k}\}$. This is what we will establish in the following lemma. Let us first remark that any o.n. family satisfies the "Bessel Inequality". The proof of this simple fact is identical to that of the Bessel Inequality for (trigonometric) Fourier series (see Theorem 2.18).

**Lemma 7.13.** *Under the assumptions stated in Theorem 7.5, the o.n. family* $\{\psi_{j,k}\}$, $j, k \in \mathbb{Z}$, *defined by (7.2.18) and (7.2.19), satisfies the following "Parseval Identity":*

$$\sum_{j,k\in\mathbb{Z}} |\langle f, \psi_{j,k}\rangle|^2 = \|f\|_2^2, \quad f \in L^2(\mathbb{R}). \tag{7.2.26}$$

**Proof.** Let $C_0^\infty$ denote the class of all functions which have compact supports and are infinitely differentiable. We will first establish (7.2.26) for all $f \in C_0^\infty$. As mentioned above, since $\{\phi_{j,k}: k \in \mathbb{Z}\}$ is an o.n. family for each $j$, we have the "Bessel Inequality":

$$\sum_{k\in\mathbb{Z}} |\langle f, \phi_{j,k}\rangle|^2 \leq \|f\|_2^2 < \infty, \quad j \in \mathbb{Z}.$$

Now, for any pair of positive integers $L$ and $M$, by summing both sides of (7.2.25) in Lemma 7.11 over $j = -L+1, \ldots, M$, and cancelling the common terms, we have

$$\sum_{j=-L}^{M-1} \sum_{k=-\infty}^{\infty} |\langle f, \psi_{j,k}\rangle|^2 = \sum_{k=-\infty}^{\infty} \{|\langle f, \phi_{M,k}\rangle|^2 - |\langle f, \phi_{-L,k}\rangle|^2\}. \tag{7.2.27}$$

Let us first take care of the second term on the right-hand side of (7.2.27). Since $f \in C_0^\infty$, there is some $K > 0$ such that supp $f \subset [-K, K]$. Now, by the Schwarz Inequality, we have

$$|\langle f, \phi_{-L,k}\rangle|^2 = \left| 2^{-L/2} \int_{-K}^{K} \phi(2^{-L}x - k)\overline{f(x)}\, dx \right|^2$$

$$\leq \left\{ 2^{-L} \int_{-K}^{K} |\phi(2^{-L}x - k)|^2 dx \right\} \|f\|_2^2$$

$$= \left\{ \int_{-K/2^L}^{K/2^L} |\phi(y - k)|^2 dy \right\} \|f\|_2^2,$$

so that

$$\sum_{k=-\infty}^{\infty} |\langle f, \phi_{-L,f}\rangle|^2 \leq \|f\|_2^2 \int_{B_L} |\phi(y)|^2 dy, \qquad (7.2.28)$$

for all sufficiently large $L$, where

$$B_L := \bigcup_{k\in\mathbb{Z}} [k - K/2^L, \quad k + K/2^L].$$

Since $\phi \in L^2(\mathbb{R})$ and the measure of $B_L \cap [-N, N]$ tends to zero as $L$ tends to infinity for any $N > 0$, the inequality in (7.2.28) yields

$$\lim_{L\to+\infty} \sum_{k=-\infty}^{\infty} |\langle f, \phi_{-L,k}\rangle|^2 = 0. \qquad (7.2.29)$$

(Incidentally, the result in (7.2.29) says that $\cap V_j = \{0\}$; see the remark made after the statement of Definition 5.2.) So, from (7.2.27), we have

$$\sum_{j=-\infty}^{M-1} \sum_{k=-\infty}^{\infty} |\langle f, \psi_{j,k}\rangle|^2 = \sum_{k=-\infty}^{\infty} |\langle f, \phi_{M,k}\rangle|^2, \quad f \in C_0^\infty. \qquad (7.2.30)$$

To study the quantity on the right-hand side of (7.2.30), we first observe, by applying the Parseval Identities, both for the Fourier transform and for the Fourier series, that

$$\sum_k |\langle f, \phi_{M,k}\rangle|^2$$

$$= \frac{1}{(2\pi)^2} \sum_k |\langle \hat{f}, \hat{\phi}_{M,k}\rangle|^2 \qquad (7.2.31)$$

$$= \frac{1}{(2\pi)^2} \sum_k 2^M \left| \int_0^{2\pi} \sum_{m=-\infty}^{\infty} \hat{f}(2^M(\omega + 2\pi m)) \overline{\hat{\phi}(\omega + 2\pi m)} e^{ik\omega} d\omega \right|^2$$

$$= \frac{2^M}{2\pi} \int_0^{2\pi} \left| \sum_{m=-\infty}^{\infty} \hat{f}(2^M(\omega + 2\pi m)) \overline{\hat{\phi}(\omega + 2\pi m)} \right|^2 d\omega$$

$$= \frac{2^M}{2\pi} \int_0^{2\pi} \left\{ \sum_{m,\ell} \hat{f}(2^M(\omega + 2\pi m)) \overline{\hat{f}(2^M(\omega + 2\pi\ell))} \right.$$

$$\left. \times \overline{\hat{\phi}(\omega + 2\pi m)} \, \hat{\phi}(\omega + 2\pi\ell) \right\} d\omega$$

$$= \frac{2^M}{2\pi} \int_{-\infty}^{\infty} \left\{ \sum_{n=-\infty}^{\infty} \hat{f}(2^M(\omega + 2\pi n)) \overline{\hat{f}(2^M\omega)} \, \overline{\hat{\phi}(\omega + 2\pi n)} \, \hat{\phi}(\omega) \right\} d\omega$$

$$= \sum_{\ell=-\infty}^{\infty} \frac{1}{2\pi} \int_{-\infty}^{\infty} \overline{\hat{f}(\omega)} \, \hat{f}(\omega + 2\pi\ell 2^M) \overline{\hat{\phi}(2^{-M}\omega + 2\pi\ell)} \, \hat{\phi}(2^{-M}\omega) d\omega$$

$$= \frac{1}{2\pi} \int_{-\infty}^{\infty} |\hat{\phi}(2^{-M}\omega)|^2 |\hat{f}(\omega)|^2 d\omega + R_M,$$

where $R_M$ is defined to be the sum over $\ell \neq 0$. Now, from (7.2.4) and the formulation of $\hat{\phi}$ as an infinite product of $P(e^{-i\omega/2^k})$, we have

$$|\hat{\phi}(\omega)| \leq 1, \quad \omega \in \mathbb{R}. \tag{7.2.32}$$

Hence, it follows that

$$|R_M| \leq \sum_{\ell \neq 0} \frac{1}{2\pi} \int_{-\infty}^{\infty} |\hat{f}(\omega)\hat{f}(\omega + 2\pi\ell 2^M)| d\omega. \tag{7.2.33}$$

Let us consider

$$F_M(\omega) := \sum_{\ell \neq 0} |\hat{f}(\omega + 2\pi\ell 2^M)|. \tag{7.2.34}$$

Since $f \in C_0^\infty$, it is clear that $\{F_M(\omega)\}$ is uniformly bounded on $\mathbb{R}$ and converges to $0$ uniformly on every compact set as $M$ tends to infinity. Hence, by observing that $\hat{f} \in L^1(\mathbb{R})$, it is easy to see that $R_M \to 0$, as $M \to +\infty$.

We now return to (7.2.31). First, recall from Lemmas 7.6 and 7.7 that $\hat{\phi} = g$ is continuous. In addition, by (7.2.3) and (7.2.9), we have $\hat{\phi}(0) = 1$. Hence, by using standard arguments, we may conclude that

$$\lim_{M \to +\infty} \frac{1}{2\pi} \int_{-\infty}^{\infty} |\hat{\phi}(2^{-M}\omega)|^2 |\hat{f}(\omega)|^2 d\omega = \frac{1}{2\pi} \|\hat{f}\|_2^2 = \|f\|_2^2. \tag{7.2.35}$$

Therefore, from the conclusions $R_M \to 0$, (7.2.35), (7.2.31), and (7.2.30), we have

$$\sum_{j,k \in \mathbb{Z}} |\langle f, \psi_{j,k} \rangle|^2 = \|f\|_2^2, \quad f \in C_0^\infty. \tag{7.2.36}$$

To extend (7.2.36) to all of $L^2(\mathbb{R})$, we use the fact that $C_0^\infty$ is dense in $L^2(\mathbb{R})$. So, for any $f \in L^2(\mathbb{R})$ and an arbitrary $\varepsilon > 0$, there is some $f_0 \in C_0^\infty$ such that $\|f - f_0\|_2 < \varepsilon$. Hence, by the Bessel Inequality,

$$\|\{\langle f - f_0, \psi_{j,k} \rangle\}\|_{\ell^2} \leq \|f - f_0\|_2 < \varepsilon,$$

whence

$$\left| \|\{\langle f, \psi_{j,k} \rangle\}\|_{\ell^2} - \|\{\langle f_0, \psi_{j,k} \rangle\}\|_{\ell^2} \right| \leq \|\{\langle f - f_0, \psi_{j,k} \rangle\}\|_{\ell^2} < \varepsilon. \tag{7.2.37}$$

Since we also have

$$\left| \|f\|_2 - \|f_0\|_2 \right| \leq \|f - f_0\|_2 < \varepsilon, \tag{7.2.38}$$

we infer (7.2.26) from (7.2.36)-(7.2.38). ∎

It is now easy to establish Theorem 7.5.

**Proof of Theorem 7.5.** Having established the Parseval Identity (7.2.26) in Lemma 7.13, we now use a standard argument to show that the o.n. family $\{\psi_{j,k}\}$ (established in Lemma 7.12, (iii)) is an o.n. basis of $L^2(\mathbb{R})$. Indeed, for any $f \in L^2(\mathbb{R})$ and any finite sequence $\{c_{j,k}\}$, we have

$$\|f - \Sigma c_{j,k}\psi_{j,k}\|_2^2 = \|f\|_2^2 - 2Re\,\Sigma \bar{c}_{j,k}\langle f, \psi_{j,k}\rangle + \Sigma |c_{j,k}|^2.$$

Hence, by choosing $c_{j,k} = \langle f, \psi_{j,k}\rangle$ for $|j| \leq N$ and $|k| \leq N$ and allowing $N$ to approach infinity, we have

$$f = \sum_{j,k} \langle f, \psi_{j,k}\rangle \psi_{j,k}, \quad f \in L^2(\mathbb{R}),$$

where the convergence is in $L^2(\mathbb{R})$. That is, we have proved that $\psi$ is an o.n. wavelet. Hence, by applying Lemma 5.1, we obtain (7.2.15), so that $\phi$ is a scaling function. ∎

## 7.3. Construction of compactly supported orthogonal wavelets

In view of the relationship between the two-scale sequences $\{p_k\}$ and $\{q_k\}$ as described by (7.2.18), in order to construct an o.n. wavelet $\psi$, it suffices to construct its corresponding o.n. scaling function $\phi$. The objective of this section is to describe a general procedure for constructing o.n. scaling functions and wavelets with compact supports. For simplicity, we only consider real-valued two-scale sequences $\{p_k\}$. According to Theorem 7.5 and (5.2.13), all we need is to identify those Laurent polynomials $S(z)$, corresponding to any given positive integer $N$, such that the conditions in (7.2.4)-(7.2.8) are satisfied.

More precisely, let $N$ be a positive integer and consider

$$P(z) = \left(\frac{1+z}{2}\right)^N S(z), \tag{7.3.1}$$

where $S(z)$ is a Laurent polynomial satisfying $S(1) = 1$. Since any finite Laurent series already satisfies (7.2.7), it suffices to identify those $S$ that satisfy (7.2.4) and

$$B := \max_{|z|=1} |S(z)| < 2^{N-1}. \tag{7.3.2}$$

To translate the condition (7.2.4), which is imposed on $P(z)$, into one that directly governs $S(z)$, we observe that since $S$ is a Laurent polynomial with real coefficients, $|S(e^{-i\omega})|^2$ is a cosine polynomial; and hence, we may write

$$|S(e^{-i\omega})|^2 = \widetilde{R}(\cos\omega), \tag{7.3.3}$$

where $\widetilde{R}$ is some (algebraic) polynomial with real coefficients. As in (5.6.37), the change of variables

$$\begin{cases} x = \dfrac{1 - \cos\omega}{2} = \sin^2\left(\dfrac{\omega}{2}\right); \\ R(x) = \widetilde{R}(\cos\omega) = \widetilde{R}(1 - 2x) \end{cases} \tag{7.3.4}$$

translates the condition (7.2.4) into

$$\left|\frac{1+e^{-i\omega}}{2}\right|^{2N}|S(e^{-i\omega})|^2 + \left|\frac{1-e^{-i\omega}}{2}\right|^{2N}|S(-e^{-i\omega})|^2 = 1,$$

or equivalently,

$$(1-x)^N R(x) + x^N R(1-x) = 1. \qquad (7.3.5)$$

According to (5.6.33) and (5.6.38), the general solution of (7.3.5) is given by

$$\begin{cases} R(x) = \displaystyle\sum_{k=0}^{N-1} \binom{N+k-1}{k} x^k + x^N T(x), \\ \text{where } T \text{ is a polynomial with } T(1-x) = -T(x). \end{cases} \qquad (7.3.6)$$

Let us summarize the above findings in the following, where

$$T_0(x) := T\left(\frac{1-2x}{2}\right).$$

**Lemma 7.14.** *Let $S$ be any Laurent polynomial that satisfies both (7.3.2) and*

$$|S(e^{-i\omega})|^2 = \sum_{k=0}^{N-1} \binom{N+k-1}{k} \left(\sin\frac{\omega}{2}\right)^{2k} + \left(\sin\frac{\omega}{2}\right)^{2N} T_0\left(\frac{\cos\omega}{2}\right), \quad (7.3.7)$$

*for some odd polynomial $T_0$. Then the Laurent polynomial*

$$\left(\frac{1+z}{2}\right)^N S(z) = \frac{1}{2}\sum_k p_k z^k, \qquad (7.3.8)$$

*where $S(1)$ is chosen to be 1, is the two-scale symbol of some compactly supported o.n. scaling function $\phi$ that generates an MRA of $L^2(\mathbb{R})$. Consequently,*

$$\psi(x) := \sum_k (-1)^k \bar{p}_{-k+1}\phi(2x-k) \qquad (7.3.9)$$

*is a compactly supported o.n. wavelet.*

**Remark.** If $S$ satisfies (7.3.7), then by setting $\omega = 0$, we have $|S(1)|^2 = 1$. So, the choice of $S$ to satisfy $S(1) = 1$ is easily achieved. The family of all odd polynomials $R_0(x)$, $|x| \leq \frac{1}{2}$, in (7.3.7) gives us some freedom in the construction of o.n. scaling functions. If choosing $R_0 = 0$ does not violate (7.3.2), then the corresponding scaling function has smallest support among those governed by (7.3.7). In the following, we will see that, indeed, the condition in (7.3.2) is satisfied by the choice $R_0 = 0$.

We need the following identity.

**Lemma 7.15.** *For all* $k, n \in \mathbb{Z}_+$,

$$\sum_{j=0}^{k} \binom{n+j}{j} = \binom{n+k+1}{k}. \tag{7.3.10}$$

**Proof.** This lemma is easily established by repeated applications of the identity

$$\binom{n+j+1}{j} = \binom{n+j}{j} + \binom{n+j}{j-1}.$$

Indeed, we have

$$
\begin{aligned}
\binom{n+k+1}{k} &= \binom{n+k}{k} + \binom{n+k}{k-1} \\
&= \binom{n+k}{k} + \left\{ \binom{n+k-1}{k-1} + \binom{n+k-1}{k-2} \right\} \\
&= \cdots = \sum_{j=0}^{k} \binom{n+j}{j}. \qquad \blacksquare
\end{aligned}
$$

The following result is a corollary of Lemma 7.14.

**Theorem 7.16.** *Let $N$ be any positive integer and $S(z)$ be any Laurent polynomial with real coefficients that satisfies*

$$|S(e^{-i\omega})|^2 = \sum_{j=0}^{N-1} \binom{N+j-1}{j} \left(\sin \frac{\omega}{2}\right)^{2j}, \tag{7.3.11}$$

*such that $S(1) = 1$. Then the Laurent polynomial in (7.3.8) is the two-scale symbol of a compactly supported o.n. scaling function $\phi$, and $\psi$, as defined by (7.3.9), is a compactly supported o.n. wavelet.*

**Proof.** By applying (7.3.10) with $n = k = N - 1$, we have

$$
\begin{aligned}
B^2 := \max_{|z|=1} |S(z)|^2 &\leq \sum_{j=0}^{N-1} \binom{N+j-1}{j} \\
&= \binom{2N-1}{N-1} = \frac{1}{2} \left\{ \binom{2N-1}{N-1} + \binom{2N-1}{N} \right\} \\
&< \frac{1}{2} \sum_{k=0}^{2N-1} \binom{2N-1}{k} = \frac{1}{2} \cdot 2^{2N-1} = 2^{2N-2},
\end{aligned}
$$

so that $B < 2^{N-1}$, and (7.3.2) is satisfied. $\qquad \blacksquare$

Thus, to construct compactly supported o.n. wavelets, the only technical problem is to solve (7.3.11) for $S(z)$. The following result, known as the Riesz Lemma, assures us that $S(z)$ always exists. We will give a "constructive" proof of this result in order to show how $S(z)$ can be derived.

**Theorem 7.17.** Let $a_0, \ldots, a_N \in \mathbb{R}$ with $a_N \neq 0$ such that

$$A(\omega) := \frac{a_0}{2} + \sum_{k=1}^{N} a_k \cos k\omega \geq 0, \quad \omega \in \mathbb{R}. \tag{7.3.12}$$

Then there exists a polynomial

$$B(z) = \sum_{k=0}^{N} b_k z^k \tag{7.3.13}$$

with real coefficients and exact degree $N$ that satisfies

$$|B(z)|^2 = A(\omega), \quad z = e^{-i\omega}. \tag{7.3.14}$$

**Proof.** Corresponding to the cosine polynomial $A(\omega)$, let us consider the algebraic polynomial

$$P_A(z) = \frac{1}{2} \sum_{k=-N}^{N} a_{|k|} z^{N+k}. \tag{7.3.15}$$

It is clear that $P_A$ satisfies

$$P_A(z) = z^N A(\omega), \quad z = e^{-i\omega}, \tag{7.3.16}$$

and

$$P_A(z) = z^{2N} P_A\left(\frac{1}{z}\right), \quad z \in \mathbb{C}. \tag{7.3.17}$$

Now, from the assumption that $a_N \neq 0$, we see that $P_A(0) \neq 0$, and by (7.3.17), it follows that all the zeros of $P_A$ occur in reciprocal pairs. In particular, any zero on the unit circle must have even multiplicity. Furthermore, since the coefficients of $P_A$ are real, all the complex zeros of $P_A$ also occur in conjugate pairs. That is, $P_A$ can be written as

$$P_A(z) = \frac{1}{2} a_N \left\{ \prod_{k=1}^{K} (z - r_k)(z - r_k^{-1}) \right\} \tag{7.3.18}$$

$$\times \left\{ \prod_{j=1}^{J} (z - z_j)(z - \bar{z}_j)(z - z_j^{-1})(z - \bar{z}_j^{-1}) \right\}$$

where $r_1, \ldots, r_K \in \mathbb{R} \backslash \{0\}$, $z_1, \ldots, z_J \in \mathbb{C} \backslash \mathbb{R}$, and $K + 2J = N$. Therefore, in view of the fact that

$$|(z - z_j)(z - \bar{z}_j^{-1})| = |z_j|^{-1} |z - z_j|^2, \quad z = e^{-i\omega}, \tag{7.3.19}$$

we have, by applying (7.3.12), (7.3.16), (7.3.18), and (7.3.19),

$$A(\omega) = |A(\omega)| = |P_A(z)|$$
$$= \frac{1}{2}|a_N| \prod_{k=1}^{K} |r_k^{-1}| \prod_{j=1}^{J} |z_j|^{-2}$$
$$\times \left| \prod_{k=1}^{K} (z - r_k) \prod_{j=1}^{J} (z - z_j)(z - \bar{z}_j) \right|^2 ,$$

where $z = e^{-i\omega}$. So, the polynomial

$$B(z) := \left( \frac{1}{2}|a_N| \prod_{k=1}^{K} |r_k^{-1}| \prod_{j=1}^{J} |z_j|^{-2} \right)^{\frac{1}{2}} \times \left\{ \prod_{k=1}^{K} (z - r_k) \prod_{j=1}^{J} (z - z_j)(z - \bar{z}_j) \right\}$$

(7.3.20)

has exact degree $K + 2J = N$ and satisfies (7.3.14). ∎

**Remark.** Observe that $B(z)$ is not unique, since we may select any zero from every reciprocal pair of zeros of $P_A(z)$ to formulate $B(z)$. In applying Theorems 7.15 or 7.16, it is necessary to normalize $B(z)$ so as to give $S(1) = 1$. In the following example, we will choose those zeros that do not lie in the open unit disk, and note that the normalization constant must be $-1$.

**Example 7.18.** Applying Theorems 7.16 and 7.17 with $N = 1$, we have

$$A(\omega) = |S(e^{-i\omega})|^2 = 1 + 2\sin^2\left(\frac{\omega}{2}\right) = \frac{4}{2} - \cos\omega,$$

so that the nonzero coefficients in (7.3.12) are $a_0 = 4$ and $a_1 = -1$. Now, by (7.3.15), we obtain

$$P_A(z) = \frac{1}{2}\{-1 + 4z - z^2\} = -\frac{1}{2}(z - (2 - \sqrt{3}))(z - (2 + \sqrt{3})).$$

If we choose the zero outside the unit circle, then we have, by applying (7.3.20),

$$S(z) = -B(z) = \frac{1}{\sqrt{2}} \frac{-1}{\sqrt{2 + \sqrt{3}}}((z - (2 + \sqrt{3})) \qquad (7.3.21)$$
$$= \frac{-1}{\sqrt{2}}\sqrt{2 - \sqrt{3}}\,(z - (2 + \sqrt{3}))$$
$$= -\frac{1}{2}\{(\sqrt{3} - 1)z - (\sqrt{3} + 1)\}.$$

Therefore, with the factor $S(z)$ in (7.3.21), the two-scale symbol is given by

$$P(z) = \left(\frac{1+z}{2}\right)^2 S(z) \tag{7.3.22}$$

$$= \frac{1}{8}(1+z)^2((1-\sqrt{3})z + (1+\sqrt{3}))$$

$$= \frac{1}{2}\left\{\frac{1+\sqrt{3}}{4} + \frac{3+\sqrt{3}}{4}z + \frac{3-\sqrt{3}}{4}z^2 + \frac{1-\sqrt{3}}{4}z^3\right\}.$$

The scaling function with the two-scale symbol given by (7.3.22) is called Daubechies' scaling function $\phi_3^D$ (see (5.2.6)).    ∎

To facilitate our discussion, let us introduce the following notation.

**Definition 7.19.** *For each integer $N \geq 2$, let $S_N(z)$ denote the solution $B(z)$, as given by (7.3.20), of the equation (7.3.11), formulated by selecting from each reciprocal pair of zeros of $P_A(z)$, the one that has larger magnitude, normalized such that $B(1) = 1$. Then the scaling function with two-scale symbol*

$$\left(\frac{1+z}{2}\right)^N S_N(z)$$

*will be denoted by $\phi_{N+1}^D$ and will be called the Daubechies scaling function of order $N + 1$.*

Next, we shall study the order of smoothness of the compactly supported o.n. Daubechies wavelets $\psi_m^D$ and their corresponding scaling functions $\phi_m^D$.

**Definition 7.20.** *Let $\gamma > 0$. A function $f \in L^2(\mathbb{R})$ is said to belong to the class $\widetilde{C}^\gamma$ if its Fourier transform $\hat{f}$ satisfies*

$$\int_{-\infty}^{\infty} |\hat{f}(\omega)|(1 + |\omega|)^\gamma d\omega < \infty. \tag{7.3.23}$$

In the following, we compare the class $\widetilde{C}^\gamma$ with the class $\text{Lip}^m \alpha$, where $m = [\gamma]$ is the largest integer not exceeding $\gamma$ and $\alpha = \gamma - m$ (see Definition 5.7). For convenience, we set

$$\text{Lip}^m 0 := C^m(\mathbb{R}). \tag{7.3.24}$$

**Lemma 7.21.** *For any $\gamma > 0$,*

$$\widetilde{C}^\gamma \subseteq \text{Lip}^m \alpha, \tag{7.3.25}$$

*with $m = [\gamma]$ and $\alpha = \gamma - m$.*

**Proof.** (i) Suppose $\gamma = 1$. Then for each $h > 0$, we have

$$\frac{f(x+h) - f(x)}{h} = \frac{1}{2\pi h} \int_{-\infty}^{\infty} \hat{f}(\omega)\{e^{i(x+h)\omega} - e^{ix\omega}\}d\omega. \tag{7.3.26}$$

Since

$$\left|\frac{e^{i(x+h)\omega} - e^{ix\omega}}{h}\right| \le |\omega|, \tag{7.3.27}$$

we can apply the Lebesgue Dominated Convergence Theorem to obtain

$$f'(x) = \frac{1}{2\pi}\int_{-\infty}^{\infty} (i\omega)\hat{f}(\omega)e^{ix\omega}\,d\omega, \quad f \in \tilde{C}.$$

Now, since $\omega\hat{f}(\omega) \in L^1(\mathbb{R})$, we have $f' \in C(\mathbb{R})$.

(ii) If $\gamma$ is a positive integer, then (7.3.25) can be established by induction.

(iii) Let $\alpha = \gamma - [\gamma] > 0$. Then for $f \in \tilde{C}^\gamma$ it follows from (ii) that $f \in C^m(\mathbb{R})$, where $m = [\gamma]$. Analogous to (7.3.26) and (7.3.27), we have

$$|f^{(m)}(x+h) - f^{(m)}(x)| \le \frac{1}{2\pi}\int_{-\infty}^{\infty} |\omega|^m|\hat{f}(\omega)|\min(|h\omega|, 2)d\omega$$

$$\le \frac{|h|^\alpha}{\pi}\int_{-\infty}^{\infty} |\omega|^\gamma|\hat{f}(\omega)|d\omega.$$

This implies that $f^{(m)} \in \text{Lip } \alpha$, or $f \in \text{Lip}^m\alpha$. ∎

We are now ready to establish the following.

**Theorem 7.22.** *There exists some positive number $\lambda$ such that $\phi_{m+1}^D$, $\psi_{m+1}^D \in \text{Lip}^{[\lambda m]}\alpha_m$, $\alpha_m := \lambda m - [\lambda m]$, for all integers $m \ge 2$.*

**Proof.** For $S = S_m$, let the quantity in (7.3.11) be denoted by $T_m$. Then we have, with $y = \sin^2\left(\frac{\omega}{4}\right)$,

$$B_2 := \max_{\omega \in \mathbb{R}} |S_m(e^{-i\omega})S_m(e^{-i\omega/2})|$$

$$= \max_{\omega \in \mathbb{R}} \left|T_m\left(\sin^2\frac{\omega}{2}\right) T_m\left(\sin^2\frac{\omega}{4}\right)\right|^{1/2}$$

$$= \max_{0 \le y \le 1} |T_m(4y(1-y))T_m(y)|^{1/2}.$$

Let us first observe that

$$\max_{0 \le y \le 1} T_m(y) < 2^{2(m-1)}$$

and that

$$T_m(y) \le \sum_{k=0}^{m-1} 2^{m+k-1}y^k \le 2^{m-1}m\max(1, (2y)^m).$$

Hence, for $0 \le y \le \frac{1}{2}$, we have

$$T_m(y)T_m(4y(1-y)) \le m2^{m-1}2^{2(m-1)} = m2^{3(m-1)}.$$

Also, for $y \geq \frac{1}{4}(2 + \sqrt{2})$ or $4y(1 - y) \leq \frac{1}{2}$, we have

$$T_m(y)T_m(4y(1 - y)) \leq 2^{2(m-1)}m2^{m-1} = m2^{3(m-1)}.$$

Finally, for $\frac{1}{2} \leq y \leq \frac{1}{4}(2 + \sqrt{2})$, we have

$$T_m(y)T_m(4y(1 - y)) \leq m^2 2^{4m-2} \left( \max_{0 \leq y \leq 1} [4y^2(1 - y)]^m \right) = m^2 2^{4m-2} \left( \frac{16}{27} \right)^m.$$

These estimates now yield

$$B_2 \leq m2^{2m-1} \left( \frac{16}{27} \right)^{m/2}.$$

Thus, it follows from (5.1.20) in Theorem 5.5, that

$$|\hat{\phi}_{m+1}(\omega)| = \left| \prod_{j=1}^{\infty} P_n(e^{-i\omega/2^j}) \right| \leq C(1 + |\omega|)^{[\ln m - m \ln \left( \frac{3\sqrt{3}}{4} \right)]/2 \ln 2}.$$

This exponent can be shown to be less than $-1$ for $m \geq 16$. For $m < 16$, one can estimate

$$B_1 := \max_{\omega \in \mathbb{R}} |S_m(e^{i\omega})| = \left[ \binom{2m - 1}{m} \right]^{1/2}$$

directly to get

$$|\hat{\phi}_{m+1}(\omega)| \leq C(1 + |\omega|)^{-1 - \eta m}$$

for some $\eta > 0$. An appeal to Lemma 7.21 now completes the proof of the theorem. ∎

**Remark.** Let $\alpha_m$ be the "largest" value for which $\phi_{m+1}^D \in \tilde{C}^{\alpha_m}$. Then we have

$$\lim_{m \to \infty} \frac{\alpha_m}{m} = 1 - \frac{1}{2} \log_2 3 \doteq 0.2075. \qquad (7.3.28)$$

The proof of this result is beyond the scope of this book.

## 7.4. Orthogonal wavelet packets

While the two-scale sequence $\{p_k\}$ of an o.n. scaling function $\phi$ contains all the information on $\phi$, the sequence $\{q_k\}$, defined by

$$q_k = (-1)^k \bar{p}_{-k+1},$$

completely characterizes its corresponding o.n. wavelet $\psi$. In what follows, let us use the notation:

$$\begin{cases} \mu_0(x) := \phi(x); \\ \mu_1(x) := \psi(x), \end{cases} \qquad (7.4.1)$$

and

$$\begin{cases} P_0(z) := P(z) = \dfrac{1}{2}\displaystyle\sum_k p_k z^k; \\[2mm] P_1(z) := Q(z) = \dfrac{1}{2}\displaystyle\sum_k q_k z^k = \dfrac{1}{2}\displaystyle\sum_k (-1)^k \bar{p}_{-k+1} z^k. \end{cases} \qquad (7.4.2)$$

Hence, the two-scale relations of the scaling function $\phi$ and its corresponding wavelet $\psi$ are given by

$$\begin{cases} \mu_0(x) = \displaystyle\sum_k p_k \mu_0(2x - k); \\[2mm] \mu_1(x) = \displaystyle\sum_k q_k \mu_0(2x - k), \end{cases} \qquad (7.4.3)$$

or equivalently,

$$\begin{cases} \hat{\mu}_0(\omega) = P_0(e^{-i\omega/2})\hat{\mu}_0\left(\dfrac{\omega}{2}\right); \\[2mm] \hat{\mu}_1(\omega) = P_1(e^{-i\omega/2})\hat{\mu}_0\left(\dfrac{\omega}{2}\right). \end{cases} \qquad (7.4.4)$$

This new notation is intended to facilitate the introduction of the following family of functions, called "wavelet packets". These functions give rise to o.n. bases which can be used to improve the performance of wavelets for time-frequency localization.

**Definition 7.23.** *The functions $\mu_n$, $n = 2\ell$ or $2\ell + 1$, $\ell = 0, 1, \ldots$, defined by*

$$\begin{cases} \mu_{2\ell}(x) = \displaystyle\sum_k p_k \mu_\ell(2x - k); \\[2mm] \mu_{2\ell+1}(x) = \displaystyle\sum_k q_k \mu_\ell(2x - k), \end{cases} \qquad (7.4.5)$$

*are called "wavelet packets" relative to the o.n. scaling function $\mu_0 = \phi$.*

Thus, the family $\{\mu_n\}$ is a generalization of the o.n. wavelet $\mu_1 = \psi$. To describe $\mu_n$, $n \in \mathbb{Z}_+$, via its Fourier transform, we need the dyadic expansion of $n \in \mathbb{Z}_+$, namely:

$$n = \sum_{j=1}^{\infty} \varepsilon_j 2^{j-1}, \quad \varepsilon_j \in \{0, 1\}. \qquad (7.4.6)$$

Observe that (7.4.6) is always a finite sum and that the expansion is unique. Indeed, if $2^{s_0-1} \le n < 2^{s_0}$, say, then we have $n = 2^{s_0-1} + n_1$ where $2^{s_1-1} \le n_1 < 2^{s_1}$ and $s_1 < s_0$; and iterating this procedure, we obtain $n = 2^{s_0-1} + 2^{s_1-1} + \cdots + 2^{s_k-1}$, where $1 \le s_k < \cdots < s_0$. That is, $\varepsilon_j = 1$ for $j = s_k, \ldots, s_0$, and $\varepsilon_j = 0$, otherwise.

**Theorem 7.24.** *Let $n$ be any nonnegative integer and let the dyadic expansion of $n$ be given by (7.4.6). Then the Fourier transform of the wavelet packet $\mu_n$ is given by*

$$\widehat{\mu}_n(\omega) = \prod_{k=1}^{\infty} P_{\varepsilon_k}(e^{-i\omega/2^k}), \quad \omega \in \mathbb{R}. \tag{7.4.7}$$

**Proof.** The Fourier transform equivalence of the two-scale relations (7.4.5) for the wavelet packets is given by

$$\begin{cases} \widehat{\mu}_{2\ell}(\omega) = P_0(e^{-i\omega/2})\widehat{\mu}_\ell\left(\dfrac{\omega}{2}\right); \\[2mm] \widehat{\mu}_{2\ell+1}(\omega) = P_1(e^{-i\omega/2})\widehat{\mu}_\ell\left(\dfrac{\omega}{2}\right). \end{cases} \tag{7.4.8}$$

In view of (7.4.4), we may proceed to prove (7.4.7) by induction on $n = 2\ell$ or $2\ell + 1$. Suppose that (7.4.7) holds for all $n$ with $0 \le n < 2^{s_0}$, and consider $2^{s_0} \le n < 2^{s_0+1}$. From the discussion presented above, we have

$$\varepsilon_j = \begin{cases} 1 & \text{for} \quad j = s_0 + 1 \\ 0 & \text{for} \quad j > s_0 + 1, \end{cases}$$

so that

$$n = \sum_{j=1}^{s_0+1} \varepsilon_j 2^{j-1}$$

and

$$\frac{n}{2} = \frac{\varepsilon_1}{2} + \sum_{j=1}^{s_0} \varepsilon_{j+1} 2^{j-1}.$$

As usual, let $[x]$ denote the largest integer not exceeding $x$, and observe that

$$n = 2\left[\frac{n}{2}\right] + \varepsilon_1. \tag{7.4.9}$$

Hence, from (7.4.8), we have

$$\widehat{\mu}_n(\omega) = P_{\varepsilon_1}(e^{-i\omega/2})\widehat{\mu}_{[n/2]}\left(\frac{\omega}{2}\right). \tag{7.4.10}$$

On the other hand, since

$$\left[\frac{n}{2}\right] = \sum_{j=1}^{s_0} \varepsilon_{j+1} 2^{j-1} < 2^{s_0},$$

it follows from the induction hypothesis that

$$\widehat{\mu}_{[n/2]}(\omega) = \prod_{j=1}^{\infty} P_{\varepsilon_{j+1}}(e^{-i\omega/2^j}). \tag{7.4.11}$$

Therefore, by combining (7.4.10) and (7.4.11), we obtain (7.4.7).  ∎

Next, we show that wavelet packets preserve the orthogonality property of the o.n. scaling function $\mu_0 = \phi$.

**Theorem 7.25.** Let $\phi$ be any o.n. scaling function and $\{\mu_n\}$ its corresponding family of wavelet packets. Then for each $n \in \mathbb{Z}$,

$$\langle \mu_n(\cdot - j), \mu_n(\cdot - k) \rangle = \delta_{j,k}, \quad j, k \in \mathbb{Z}. \tag{7.4.12}$$

**Proof.** Since $\mu_0 = \phi$ satisfies (7.4.12), we may proceed to prove (7.4.12) by induction. Suppose that (7.4.12) holds for all $n$ where $0 \le n < 2^{s_0}$, and consider $2^{s_0} \le n < 2^{s_0+1}$. Then as in the proof of Theorem 7.24, by applying (7.4.8) and (7.4.9), we have

$$
\begin{aligned}
&\langle \mu_n(\cdot - j), \mu_n(\cdot - k) \rangle \\
&= \frac{1}{2\pi} \int_{-\infty}^{\infty} |\widehat{\mu}_n(\omega)|^2 e^{i(k-j)\omega} d\omega \\
&= \frac{1}{2\pi} \int_{-\infty}^{\infty} |P_{\varepsilon_1}(e^{-i\omega/2})|^2 \left|\widehat{\mu}_{[n/2]}\left(\frac{\omega}{2}\right)\right|^2 e^{i(k-j)\omega} d\omega \\
&= \frac{1}{2\pi} \sum_{\ell=-\infty}^{\infty} \int_{4\pi\ell}^{4\pi(\ell+1)} |P_{\varepsilon_1}(e^{-i\omega/2})|^2 \left|\widehat{\mu}_{[n/2]}\left(\frac{\omega}{2}\right)\right|^2 e^{i(k-j)\omega} d\omega \\
&= \frac{1}{2\pi} \int_0^{4\pi} e^{i(k-j)\omega} |P_{\varepsilon_1}(e^{-i\omega/2})|^2 \sum_{\ell=-\infty}^{\infty} \left|\widehat{\mu}_{[n/2]}\left(\frac{\omega}{2} + 2\pi\ell\right)\right|^2 d\omega.
\end{aligned}
$$

Hence, by the induction hypothesis and Theorem 3.23, we obtain

$$
\begin{aligned}
&\langle \mu_n(\cdot - j), \mu_n(\cdot - k) \rangle \\
&= \frac{1}{2\pi} \int_0^{4\pi} e^{i(k-j)\omega} |P_{\varepsilon_1}(e^{-i\omega/2})|^2 d\omega \\
&= \frac{1}{2\pi} \int_0^{2\pi} e^{i(k-j)\omega} \left\{ |P_{\varepsilon_1}(e^{-i\omega/2})|^2 + |P_{\varepsilon_1}(-e^{-i\omega/2})|^2 \right\} d\omega \\
&= \frac{1}{2\pi} \int_0^{2\pi} e^{i(k-j)\omega} d\omega = \delta_{j,k},
\end{aligned}
$$

where one of the following two identities

$$
\begin{cases}
|P(z)|^2 + |P(-z)|^2 = 1; \\
|Q(z)|^2 + |Q(-z)|^2 = 1, \quad |z| = 1,
\end{cases} \tag{7.4.13}
$$

has been applied (see (7.2.4) and (7.1.3)). ∎

The orthogonality property between the o.n. scaling function $\mu_0 = \phi$ and its corresponding wavelet $\mu_1 = \psi$ also extends to wavelet packets, as asserted in the following theorem.

**Theorem 7.26.** *Let $\phi$ be any o.n. scaling function and $\{\mu_n\}$ its corresponding wavelet packets. Then*

$$\langle \mu_{2\ell}(\cdot - j), \mu_{2\ell+1}(\cdot - k) \rangle = 0, \quad j, k \in \mathbb{Z}, \quad \ell \in \mathbb{Z}_+. \tag{7.4.14}$$

**Proof.** Before we proceed, let us first record the identity

$$P_0(z)\overline{P_1(z)} + P_0(-z)\overline{P_1(-z)} = P(z)\overline{Q(z)} + P(-z)\overline{Q(-z)} = 0, \quad |z| = 1, \tag{7.4.15}$$

which is equivalent to the identity (iv) in Lemma 7.9. Hence, by applying (7.4.8), (7.4.15), and Theorems 7.25 and 3.23, we obtain, for all $j, k \in \mathbb{Z}$ and $\ell \in \mathbb{Z}_+$,

$$
\begin{aligned}
\langle \mu_{2\ell}(\cdot - j), & \mu_{2\ell+1}(\cdot - k) \rangle \\
&= \frac{1}{2\pi} \int_{-\infty}^{\infty} \widehat{\mu}_{2\ell}(\omega) \overline{\widehat{\mu}_{2\ell+1}(\omega)}\, e^{i(k-j)\omega}\, d\omega \\
&= \frac{1}{2\pi} \int_{-\infty}^{\infty} \left| \widehat{\mu}_\ell\left(\frac{\omega}{2}\right) \right|^2 P_0(z)\overline{P_1(z)}\, e^{i(k-j)\omega}\, d\omega \\
&= \frac{1}{2\pi} \int_0^{4\pi} \left\{ \sum_{m=-\infty}^{\infty} \left| \widehat{\mu}_\ell\left(\frac{\omega}{2} + 2\pi m\right) \right|^2 \right\} P_0(z)\overline{P_1(z)}\, e^{i(k-j)\omega}\, d\omega \\
&= \frac{1}{2\pi} \int_0^{4\pi} P_0(z)\overline{P_1(z)}\, e^{i(k-j)\omega}\, d\omega \\
&= \frac{1}{2\pi} \int_0^{2\pi} \left\{ P_0(z)\overline{P_1(z)} + P_0(-z)\overline{P_1(-z)} \right\} e^{i(k-j)\omega}\, d\omega \\
&= 0,
\end{aligned}
$$

where $z = e^{-i\omega/2}$. ∎

## 7.5. Orthogonal decomposition of wavelet series

Let $\{\mu_n\}$ be a family of wavelet packets corresponding to some o.n. scaling function $\mu_0 = \phi$. For each $n \in \mathbb{Z}_+$, consider the family of subspaces

$$U_j^n := \mathrm{clos}_{L^2(\mathbb{R})} \langle 2^{j/2}\mu_n(2^j \cdot - k) : k \in \mathbb{Z} \rangle, \quad j \in \mathbb{Z}, \quad n \in \mathbb{Z}_+, \tag{7.5.1}$$

generated by $\{\mu_n\}$. Recall that

$$\begin{cases} U_j^0 = V_j, & j \in \mathbb{Z}; \\ U_j^1 = W_j, & j \in \mathbb{Z}, \end{cases} \tag{7.5.2}$$

where $\{V_j\}$ is the MRA of $L^2(\mathbb{R})$ generated by $\mu_0 = \phi$ and $\{W_j\}$ is the sequence of orthogonal complementary (wavelet) subspaces generated by the wavelet $\mu_1 = \psi$. Then the orthogonal decomposition

$$V_{j+1} = V_j \oplus W_j, \quad j \in \mathbb{Z},$$

may be written as

$$U_{j+1}^0 = U_j^0 \oplus U_j^1, \quad j \in \mathbb{Z}. \tag{7.5.3}$$

In the following, we shall see that this orthogonal decomposition can be generalized from $n = 0$ to any $n \in \mathbb{Z}_+$.

**Theorem 7.27.** *Let $n$ be any nonnegative integer. Then*

$$U^n_{j+1} = U^{2n}_j \oplus U^{2n+1}_j, \quad j \in \mathbb{Z}. \tag{7.5.4}$$

**Proof.** From (7.4.5) in Definition 7.23, it is clear that $U^{2n}_j$ and $U^{2n+1}_j$ are subspaces of $U^n_{j+1}$. Furthermore, by Theorem 7.26, we see that these two subspaces are orthogonal to each other. So, it suffices to show that

$$\mu_n(2^{j+1}x - m) = \frac{1}{2} \sum_k \{\bar{p}_{m-2k}\mu_{2n}(2^j x - k) + \bar{q}_{m-2k}\mu_{2n+1}(2^j x - k)\} \tag{7.5.5}$$

holds for all $m \in \mathbb{Z}$.

To this end, we apply (7.4.5) to the right-hand side of (7.5.5) and simplify the expression by using the identity (v) in Lemma 7.9:

$$\frac{1}{2} \sum_k \{\bar{p}_{m-2k}\mu_{2n}(2^j x - k) + \bar{q}_{m-2k}\mu_{2n+1}(2^j x - k)\}$$

$$= \frac{1}{2} \sum_k \sum_\ell \{\bar{p}_{m-2k}p_\ell + \bar{q}_{m-2k}q_\ell\}\mu_n(2^{j+1}x - 2k - \ell)$$

$$= \frac{1}{2} \sum_k \sum_t \{\bar{p}_{m-2k}p_{t-2k} + \bar{q}_{m-2k}q_{t-2k}\}\mu_n(2^{j+1}x - t)$$

$$= \sum_t \left\{ \frac{1}{2} \sum_k [p_{t-2k}\bar{p}_{m-2k} + q_{t-2k}\bar{q}_{m-2k}] \right\} \mu_n(2^{j+1}x - t)$$

$$= \sum_t \delta_{t,m}\mu_n(2^{j+1}x - t) = \mu_n(2^{j+1}x - m).$$

This completes the proof of the theorem. ∎

The importance of an o.n. wavelet $\psi$ is that it generates an o.n. basis $\{\psi_{j,k}\}$, $j, k \in \mathbb{Z}$, of $L^2(\mathbb{R})$ in such a way that for each $j \in \mathbb{Z}$, the sub-family $\{\psi_{j,k}: k \in \mathbb{Z}\}$ is not only an o.n. basis of

$$W_j = \text{clos}_{L^2(\mathbb{R})} \langle \psi_{j,k}: k \in \mathbb{Z} \rangle,$$

but is also a time-window for extracting local information (on both magnitudes and locations) within the $j^{\text{th}}$ frequency band (or $j^{\text{th}}$ octave)

$$H_j := (2^{j+1}\Delta_{\hat{\psi}}, 2^{j+2}\Delta_{\hat{\psi}}], \tag{7.5.6}$$

where $\Delta_{\hat{\psi}}$ is the RMS bandwidth of the wavelet (see Sections 3.2 and 3.4, and particularly (3.4.1)–(3.4.5)). Observe that the width of the frequency band $H_j$ increases at higher frequency ranges. In the following, we will see that wavelet packets have the capability of partitioning the higher-frequency octaves to yield better frequency localization.

**Theorem 7.28.** For each $j = 1, 2, \ldots$,

$$
\begin{cases}
W_j = U_{j-1}^2 \oplus U_{j-1}^3; \\
W_j = U_{j-2}^4 \oplus U_{j-2}^5 \oplus U_{j-2}^6 \oplus U_{j-2}^7; \\
\quad \cdots \cdots \cdots \cdots \\
W_j = U_{j-k}^{2^k} \oplus U_{j-k}^{2^k+1} \oplus \cdots \oplus U_{j-k}^{2^{k+1}-1}; \\
\quad \cdots \cdots \cdots \cdots \\
W_j = U_0^{2^j} \oplus U_0^{2^j+1} \oplus \cdots \oplus U_0^{2^{j+1}-1}.
\end{cases}
\tag{7.5.7}
$$

Furthermore, for each $m = 0, \ldots, 2^k - 1$, $k = 1, \ldots, j$, and $j = 1, 2, \ldots$, the family

$$
\left\{ 2^{\frac{j-k}{2}} \mu_{2^k+m}(2^{j-k}x - \ell) \colon \ell \in \mathbb{Z} \right\}
\tag{7.5.8}
$$

is an orthonormal basis of $U_{j-k}^{2^k+m}$.

**Remark.** By using the $k^{\text{th}}$ orthogonal decomposition in (7.5.7), the $j^{\text{th}}$ frequency band $H_j$ is further partitioned into $2^k$ "*sub-bands*":

$$
H_j^{k,m}, \qquad m = 0, \ldots, 2^k - 1.
\tag{7.5.9}
$$

Of course, the o.n. basis in (7.5.8) of $U_{j-k}^{2^k+m}$ provides time-localization within the sub-band $H_j^{k,m}$, and the union of the family

$$
\{ H_j^{k,m} \colon m = 0, \ldots, 2^k - 1 \}
$$

is all of $H_j$.

**Proof.** The proof of (7.5.7) is simply a repeated application of (7.5.4) in Theorem 7.27, by setting $n = 1$ and recognizing that $U_j^1 = W_j$. That the family in (7.5.8) is an o.n. basis of $U_{j-k}^{2^k+m}$ is a consequence of (7.5.1) and (7.4.12) in Theorem 7.25. ∎

**Remark.** In view of the decomposition formula in (7.5.5), the decomposition of any wavelet series

$$
g_j(x) = \sum_n d_n^j \psi(2^j x - n) = \sum_n d_n^j \mu_1(2^j x - n)
\tag{7.5.10}
$$

into an orthogonal sum of wavelet packet components

$$
g_{j;k,m}(x) := \sum_n d_n^{j;k,m} \mu_{2^k+m}(2^{j-k}x - n), \quad m = 0, \ldots, 2^k - 1
\tag{7.5.11}
$$

(for any fixed value of $k$, $1 \leq k \leq j$) can be formulated as a *"tree"*, where each branch of the tree has two sub-branches. The same decomposition algorithm as given by (5.4.48) with

$$\begin{cases} a_n = \dfrac{1}{2}\bar{p}_n; \\ b_n = \dfrac{1}{2}\bar{q}_n = \dfrac{1}{2}(-1)^n p_{-n+1}, \end{cases} \tag{7.5.12}$$

can be used for decomposition at each branch. Of course, this tree decomposition algorithm should be designed to be adaptive; and in particular, if certain wavelet components are not as essential as others, then they should be made to have fewer branches by selecting smaller values of $k$. For reconstruction, the tree algorithm can be reversed by applying the reconstruction algorithm (5.4.49) with weight sequences $p_n$ and $q_n = (-1)^n \bar{p}_{-n+1}$.

For the "finest" orthogonal decomposition, each $k$ is chosen to be largest possible; that is, the last formula in (7.5.7) is used.

**Corollary 7.29.** *For each $j = 0, 1, 2, \ldots,$*

$$L^2(\mathbb{R}) = \bigoplus_{j \in \mathbb{Z}} W_j = \cdots \oplus W_{-1} \oplus W_0 \oplus U_0^2 \oplus U_0^3 \oplus \cdots . \tag{7.5.13}$$

Of course, the family

$$\{\psi_{j,k}, \ \mu_n(\cdot - k): \ j = \cdots, -1, 0; n = 2, 3, \ldots; \text{ and } k \in \mathbb{Z}\} \tag{7.5.14}$$

is an o.n. basis of $L^2(\mathbb{R})$.

# Notes

**Chapter 1.**

The standard reference for trigonometric series is Zygmund [9]. Other books on Fourier series that were helpful to us in the preparation of Chapters 1 and 2 are Bari [1], Helson [5], Katznelson [6], and Stein and Weiss [7].

The simplest orthogonal wavelet is the Haar function studied by A. Haar in [55]. The notion of integral wavelet transform (IWT), $(W_\psi f)(b, a)$, was first introduced by Grossmann and Morlet [54], although the techniques which are based on the use of translations and dilations can be traced back to Calderón [30] in the study of singular integral operators. The basic wavelet used to formulate the IWT is also called a "mother wavelet" in the wavelet literature. The formula for the recovery of any $f \in L^2(\mathbb{R})$ from its IWT $(W_\psi f)(b, a)$, $a, b \in \mathbb{R}$, can be found in [54].

The importance of the semi-discrete IWT, $(W_\psi f)(b, a)$, where $b \in \mathbb{R}$ and $a = 2^{-j}$, $j \in \mathbb{Z}$, in image compression was first pointed out by S. Mallat, and the notion of dyadic wavelets was also introduced by Mallat in his joint work with W.L. Hwang [60] and with S. Zhong [61, 62]. The stability condition on dyadic wavelets studied in Mallat and Hwang [60] can be viewed as a generalization of the Littlewood-Paley Identity. Inequalities of this type for frames and wavelets were studied in Daubechies [50] and Chui and Shi [37], and in particular, the identity was used by Chui and Shi [38] to characterize wavelets.

A general reference on signal processing is Oppenheim and Schafer [25], and a mathematical analysis of the subject is given in Chui and Chen [24]. Other references that deal with certain specialized but related topics in signal and image processing are Auslander, Kailath, and Mitter [22], and Rosenfeld [26].

That an $\mathcal{R}$-function is not necessarily an $\mathcal{R}$-wavelet (or wavelet) was already observed by Y. Meyer in the first volume in [21] and discussed by Daubechies in some detail in [50], where the results of Tchamitchian [68, 69] were used. The proof in this chapter was given by Chui and Shi in [37], following the ideas of Daubechies and Meyer.

The notion of multiresolution analysis (MRA) was first introduced by Meyer [63] and Mallat [58], and further developed by Mallat in [57, 59]. It was also Mallat [57-59] who constructed the wavelet decomposition and reconstruction algorithms using the MRA spaces. The presentation of these algorithms in this chapter follows [40, 43] in that the normalization constant $\sqrt{2}$ is absorbed by the basis functions to facilitate implementations. We remark that there is

245

some similarity between an MRA and the Laplacian pyramid algorithm due to Burt and Adelson [29].

The most comprehensive work on cardinal splines is [14] by Schoenberg, who was also responsible for the development of this subject. It was Meyer [63] and Lemarié [56] who noted that cardinal spline spaces give rise to MRA, although approximation theorists have also been considering subdivision schemes. (See Chui [11] and the references therein.) Corresponding to the $m^{\text{th}}$ order $B$-spline $N_m$, the $m^{\text{th}}$ order $B$-wavelet $\psi_m$ was introduced by Chui and Wang [43]. The dual $\tilde{\psi}_m$ of $\psi_m$ was also constructed in [43] in terms of the $m^{\text{th}}$ derivative of the shifted and scaled fundamental spline of order $2m$ introduced by Chui and Wang in [40].

A predictor-corrector algorithm can be devised to implement an IIR filter as an ARMA filter with poles lying both inside and outside the unit circle. Such algorithms can be optimized by incorporating them with noise processes (see Chui and Chen [23]). This procedure should apply to spline-wavelet decompositions without truncations.

**Chapter 2.**

There are many good references on Fourier transforms in the literature. The ones that were helpful to us in the preparation of Chapter 2 are Goldberg [4] and Titchmarsh [8].

Since functions in $L^p(\mathbb{R})$ or $L^p(0, 2\pi)$ are considered to be "equivalence classes" of functions, we are allowed to change the functions on sets of measure zero. In particular, in the statements of pointwise convergence, we always mean convergence to some representative of the equivalence class under consideration.

Again, the books [1, 5, 6, 7, 9] on harmonic analysis are good sources for further reading. This subject has a very rich history. We only include a very brief discussion on the development of pointwise convergence here. As early as 1876, du Bois-Reymond already showed the existence of a $2\pi$-periodic continuous function whose Fourier series diverges at some point. From this result, it is not difficult to prove the existence of a $2\pi$-periodic continuous function whose Fourier series is divergent on a dense subset of the real line $\mathbb{R}$. In 1923, Kolmogorov extended this observation and showed that the Fourier series of some $2\pi$-periodic continuous function may diverge almost everywhere. Three years later, he even extended this result and proved the existence of some $f \in L^1(0, 2\pi)$ with an a.e. divergent Fourier series. On the other hand, in 1966, Carleson proved that if $f \in L^2(0, 2\pi)$, then its Fourier series converges a.e. This very deep result was extended by Hunt in 1967 to every $f \in L^p(0, 2\pi)$, $p > 1$.

A somewhat extensive discussion of the Poisson Summation Formula is given in this chapter because several variations of this formula are used throughout this book.

**Chapter 3.**

The window Fourier transform was first introduced by Gabor [53] by using a Gaussian function as the window function. This is why it is also called the Gabor transform. In the engineering literature, this windowing process, which is not restricted to the use of a Gaussian function, is called a short-time Fourier transform (STFT) as well. See Daubechies [20, 50], Mallat [57], some of the chapters in the edited volumes [16, 17, 18, 19] and the lists of references therein. The reader is also referred to the same sources for material concerning the use of the standard deviation to define the radius, and hence width, of a window function as well as for discussions on the Uncertainty Principle. As mentioned earlier, when we refer to a function in $L^p(\mathbb{R})$, we actually mean a representative of an equivalence class of functions which are identical except on sets of measure zero. In particular, we always use a continuous representative as a window function whenever it is available. See also Champeney [2] for further discussions and references from the harmonic analysis point of view.

As mentioned in the notes for Chapter 1, the IWT was introduced by Grossmann and Morlet [54], where the admissibility condition (3.3.1) on the basic wavelets (also called the mother wavelets) was imposed. The semi-discrete version of the IWT has proved to be very useful in Mallat's work on image compression, first using zero-crossings and later using wavelet maxima (or local extrema) of the IWT on dyadic scale-levels. See the work of Mallat and Zhong [61, 62] and Mallat and Hwang [60]. In this regard, the stability condition (3.4.6) is crucial to the inversion formula (3.4.14) introduced in Mallat and Hwang [60]. The characterization of dyadic duals using the Littlewood-Paley identity is given in Chui and Shi [38].

The notion of frames was introduced by Duffin and Schaeffer [51] and studied in some detail by Daubechies [20, 50]. Example 3.18 is also attributed to Daubechies. The Interior Mapping Principle used in our discussion on the boundedness of the inverse map generated by the frame is standard in operator theory (see Vol. 1, page 57, of Dunford and Schwartz [3]). The stability property (3.5.18) of a frame was derived in Chui and Shi [37]. We also remark that Frazier and Jawerth [52] used dilation and translation in their work on $\varphi$-transforms.

Semi-orthogonal wavelets were introduced independently by Auscher [27] and Chui and Wang [40, 41, 43], while semi-orthogonal cardinal spline-wavelets were first constructed by Chui and Wang in [40, 43]. The orthonormalization procedure in (3.6.18) is due to Schweinler and Wigner [66] and is therefore called the Schweinler-Wigner o.n. procedure.

**Chapter 4.**

Spline analysis is an established subject. The reader is referred to de Boor [10], Nürnberger [13], Schoenberg [14], and Schumaker [15] for the univariate theory, and to Chui [11] for a multivariable investigation. A spline function with equally spaced (simple) knots is called a cardinal spline, and the reader is referred to Schoenberg [14, 65] for further study. In particular, the structure of Euler- Frobenius polynomials is documented in [14, 65].

The graphical display algorithm (Algorithm 4.7) introduced in Section 4.3 is an iteration of a standard subdivision scheme. It is presented here in order to motivate the introduction of the (wavelet) reconstruction algorithm in Chapter 5. The cardinal $B$-spline $B$-net algorithm (Algorithm 4.10) is the one-variable version of the box-spline graphical display algorithm developed in Chui and Lai [36].

The notion of quasi-interpolation is due to de Boor and Fix (see de Boor [10] and Schumaker [15]). The Neumann series approach was first introduced by Chui and Diamond in [35], where Theorem 4.13 was established. The characterization of quasi-interpolants presented here (see (4.5.35)-(4.5.36)) was given by Chui and Diamond [33]. The theory of interpolation by cardinal splines was developed by Schoenberg (see Schoenberg [14, 65]). The construction of local interpolation formulas presented in Section 4.6 by taking a "Boolean sum" of a completely local interpolation operator and a quasi-interpolation operator was introduced by Chui and Diamond in [34]. A tutorial study of this topic is given in Chui [32], and in more generality, in Chui [31].

## Chapter 5.

The notion of multiresolution analysis (MRA) was first introduced by Meyer [63] and Mallat [58], and further developed by Mallat in [57, 59]. Theorems 5.5 and 5.6 are due to Cohen [44]. The result on the support of a scaling function being given by the length of its two-scale sequence is due to Daubechies [49], and the scaling function defined in (5.2.6) was also given in [49]. The characterization of minimally supported scaling functions in terms of the non-existence of symmetric zeros was given by Chui and Wang [41], where the notion of generalized Euler-Frobenius (Laurent) polynomials was introduced and their properties, such as those in Theorem 5.10, were studied. The fact that the $m^{\text{th}}$ order B-spline is the only function that generates the MRA $\{V_j^m\}$ and has a finite two-scale sequence was also proved in [41]. The presentation of the direct-sum decomposition in Section 5.3 seems to be new, and the approach in Section 5.4 is a generalization of the work of Cohen, Daubechies, and Feauveau in [46]. In particular, the criterion (5.4.11) in Theorem 5.19 is new. The duality principle for spline-wavelets and, more generally, semi-orthogonal wavelets, was introduced in Chui and Wang [43] and [41], respectively, and the nonorthogonal (or biorthogonal) version is again found in Cohen, Daubechies, and Feauveau [46].

As mentioned earlier, wavelet decomposition and reconstruction algorithms were constructed in Mallat [58]. The formulation here follows [40, 43]. The importance of linear phase in filtering and the relation between linear phase and symmetry are well known in the engineering literature (see Oppenheim and Schafer [25]). Our discussion in Sections 5.5 and 5.6 is an extension of the results of Chui and Wang in [41].

The general solution of (5.6.39) of the dual relation (5.6.17) was given in Daubechies [49] for the orthonormal setting, and in Cohen, Daubechies, and Feauveau [46], in general. Compactly supported wavelets with compactly supported duals that meet the linear-phase requirement were also constructed by Cohen, Daubechies, and Feauveau in [46]. The filter bank approach was considered by Vetterli and Herley [70].

## Chapter 6.

The interpolatory spline-wavelets given in Theorem 6.1 are due to Chui and Wang [40], and the identification between the MRA subspaces $V_1^{2m,0}$ of $V_1^{2m}$ with the wavelet spaces $W_0^m$ was introduced in Chui and Wang [39]. The compactly supported cardinal spline-wavelets (or B-wavelets) given in (6.2.5), together with their duals in (6.2.10), were introduced by Chui and Wang in [43]. However, the presentation in Section 6.2 is quite different in that it makes use of the more general result, Theorem 5.19, from Chapter 5. The Pascal triangular algorithm (PTA) for computing B-wavelets and their derivatives was introduced in Chui and Wang [42].

The presentation of properties of the Euler-Frobenius polynomials follows Schoenberg [65], while the material on error analysis in spline-wavelet decomposition is taken from Chui and Wang [42]. The most comprehensive reference on total positivity is Karlin [12], and additional information on Pólya frequency (PF) sequences can be found in Schoenberg [14]. Theorem 6.21 on certain linear PTA's producing PF sequences was proved in Chui and Wang [39], where the notion of complete oscillation was first introduced and its relation to zero-crossings explored.

## Chapter 7.

The first nontrivial wavelet was constructed by Strömberg [67], using spline functions. The Meyer wavelets [64], as studied in Example 7.3, are o.n. wavelets with compactly supported Fourier transforms. The o.n. spline wavelets given in Example 7.1 are usually called the Battle-Lemarié wavelets, since they were constructed independently by Battle [28] and Lemarié [56] using different methods. However, none of these o.n. wavelets has compact support. Based on the structure of MRA, Daubechies [49] was the first to construct compactly supported o.n. wavelets. Hence, her construction was based on the identification of o.n. scaling functions as discussed in Section 7.2, although the presentation there is somewhat different from that in [49]. The construction of the Daubechies wavelets also depends on the Riesz Lemma as stated in Theorem 7.17. Theorem 7.22 was also proved by Daubechies [49], while the result stated in (7.3.28) was established by Cohen and Daubechies in [45].

Wavelet packets, also called "wave packets" by Coifman and Meyer, were introduced in Coifman, Meyer, Quake, and Wickerhauser [47]. See also Coifman, Meyer, and Wickerhauser [48] for further information.

# References

**Books**

*Fourier Analysis and Operator Theory*

1. Bari, N. K., *A Treatise on Trigonometric Series*, Macmillan, 1964.
2. Champeney, D. C., *A Handbook of Fourier Theorems*, Cambridge University Press, Cambridge, 1987.
3. Dunford, N. and J. Schwartz, *Linear Operators*, Interscience, New York, 1958.
4. Goldberg, R. R., *Fourier Transforms*, Cambridge University Press, 1965.
5. Helson, H., *Harmonic Analysis*, The Wadsworth & Brooks/Cole Mathematics Series, Addison-Wesley, 1983, reprinted by Wadsworth, 1991.
6. Katznelson, Y., *An Introduction to Harmonic Analysis*, John Wiley and Sons, 1968, reprinted by Dover, New York, 1976.
7. Stein, E. M. and G. Weiss, *Introduction to Fourier Analysis on Euclidean Spaces*, Princeton University Press, 1971.
8. Titchmarsh, E. C., *Introduction to the Theory of Fourier Integrals* (Second Edition), Oxford University Press, 1948.
9. Zygmund, A., *Trigonometric Series* (Second Edition), in two volumes, Cambridge University Press, 1959.

*Spline Analysis*

10. de Boor, C., *A Practical Guide to Splines*, Applied Mathematical Sciences, Vol. 27, Springer-Verlag, 1978.
11. Chui, C. K., *Multivariate Splines*, CBMS-NSF Series in Applied Math. #54, SIAM Publ., Philadelphia, 1988.
12. Karlin, S., *Total Positivity*, Stanford University Press, Stanford, CA, 1968.
13. Nürnberger, G., *Approximation by Spline Functions*, Springer-Verlag, New York, 1989.
14. Schoenberg, I. J., *Cardinal Spline Interpolation*, CBMS-NSF Series in Applied Math. #12, SIAM Publ., Philadelphia, 1973.
15. Schumaker, L. L., *Spline Functions: Basic Theory*, Wiley-Interscience, New York, 1981.

*Wavelet Theory and Applications*

16. Beylkin, G., R. Coifman, I. Daubechies, S. Mallat, Y. Meyer, L. Raphael, and B. Ruskai, (eds.), *Wavelets and Their Applications*, Jones and Bartlett, Cambridge, MA, 1992.
17. Chui, C. K., (ed.), *Approximation Theory and Functional Analysis*, Academic Press, Boston, 1991.
18. Chui, C. K., (ed.), *Wavelets: A Tutorial in Theory and Applications*, Academic Press, Boston, 1992.
19. Combes, J. M., A. Grossmann, and P. Tchamitchian, (eds.), *Wavelets: Time-Frequency Methods and Phase Space*, Springer-Verlag, New York, 1989; Second Edition, 1991.
20. Daubechies, I., *Wavelets*, CBMS-NSF Series in Appl. Math., SIAM Publ., Philadelphia, 1992.
21. Meyer, Y., *Ondelettes et Opérateurs*, Vol. I and Vol. II, Hermann, Paris, 1990. Also, Meyer, Y. and R. R. Coifman, Vol. III, 1991.

*Signal and Image Processing*

22. Auslander, L., T. Kailath, and S. Mitter, (eds.) *Signal Processing I: Signal Processing Theory*, The IMA Volumes in Mathematics and Its Applications #22, Springer-Verlag, New York, 1990.
23. Chui, C. K. and G. Chen, *Kalman Filtering with Real-Time Applications*, Springer Series in Information Sciences #17, Springer-Verlag, New York, 1987; Second Edition 1991.
24. Chui, C. K. and G. Chen, *Signal Processing and Systems Theory - Selected Topics*, Springer Series in Information Sciences #26, Springer-Verlag, New York, 1992.
25. Oppenheim, A. V. and R. W. Schafer, *Discrete-Time Signal Processing*, Prentice Hall Signal Proc. Series, Prentice Hall, Englewood Cliffs, New Jersey, 1989.
26. Rosenfeld, A., (ed.), *Multiresolution Image Processing and Analysis*, Springer Series in Information Sciences #12, Springer-Verlag, New York, 1984.

**Papers**

27. Auscher, P., Ondettes fractales et applications, Thèse de Doctorat, University Paris-Dauphine, 1989.
28. Battle, G., A block spin construction of ondelettes, Part I: Lemarié functions, *Comm. Math. Phys.* **110** (1987), 601–615.
29. Burt, P. J. and E. H. Adelson, The Laplacian pyramid as a compact image code, *IEEE Trans. Comm.* **31** (1983), 482–540.
30. Calderón, A. P., Intermediate spaces and interpolation, the complex method, *Studia Math.* **24** (1964), 113–190.
31. Chui, C. K., Construction and applications of interpolation formulas, in *Multivariate Approximation and Interpolation*, W. Haussmann and K. Jetter (eds.), ISNM Series Math., Birkhaüser Verlag, Basel, 1990, 11–23.

32. Chui, C. K., Vertex splines and their applications to interpolation of discrete data, in *Computation of Curves and Surfaces*, W. Dahmen, M. Gasca, and C. A Micchelli (eds.), Kluwer Academic Publishers, 1990, 137–181.

33. Chui, C. K. and H. Diamond, A characterization of multivariate quasi-interpolation formulas and applications, *Numer. Math.* **57** (1990), 105–121.

34. Chui, C. K. and H. Diamond, A general framework for local interpolation, *Numer. Math.* **58** (1991), 569–581.

35. Chui, C. K. and H. Diamond, A natural formulation of quasi-interpolation by multivariate splines, *Proc. Amer. Math. Soc.* **99** (1987), 643–646.

36. Chui, C. K. and M. J. Lai, Computation of box splines and *B*-splines on triangulations of nonuniform rectangular partitions, *Approx. Th. and Its Appl.* **3** (1987), 37–62.

37. Chui, C. K. and X. L. Shi, Inequalities of Littlewood-Paley type for frames and wavelets, CAT Report #249, Texas A&M University, 1991.

38. Chui, C. K. and X. L. Shi, On a Littlewood-Paley identity and characterization of wavelets, CAT Report #250, Texas A&M University, 1991.

39. Chui, C. K. and J. Z. Wang, An analysis of cardinal spline-wavelets, *J. Approx. Theory*, to appear.

40. Chui, C. K. and J. Z. Wang, A cardinal spline approach to wavelets, *Proc. Amer. Math. Soc.* **113** (1991), 785–793.

41. Chui, C. K. and J. Z. Wang, A general framework of compactly supported splines and wavelets, *J. Approx. Theory*, to appear.

42. Chui, C. K. and J. Z. Wang, Computational and algorithmic aspects of cardinal spline-wavelets, CAT Report #235, Texas A&M University, 1990.

43. Chui, C. K. and J. Z. Wang, On compactly supported spline wavelets and a duality principle, *Trans. Amer. Math. Soc.*, 1991, to appear.

44. Cohen, A., Ondelettes, Analyses multirésolutions et traitement numerique du signal, Doctoral Thesis, Univ. Paris-Dauphine, 1990.

45. Cohen, A. and I. Daubechies, Nonseparable bidimensional wavelet bases, AT&T Bell Laboratories, 1991, preprint.

46. Cohen, A., I. Daubechies, and J. C. Feauveau, Bi-orthogonal bases of compactly supported wavelets, *Comm. Pure and Appl. Math.*, 1991, to appear.

47. Coifman, R., Y. Meyer, S. Quake, and M. V. Wickerhauser, Signal processing and compression with wave packets, in *Proceedings of the Conference on Wavelets*, Marseilles, Spring 1989.

48. Coifman, R. , Y. Meyer, and M. V. Wickerhauser, Wavelet analysis and signal processing, 1991, preprint.

49. Daubechies, I., Orthonormal bases of compactly supported wavelets, *Comm. Pure and Appl. Math.* **41** (1988), 909–996.

50. Daubechies, I., The wavelet transform, time-frequency localization and signal analysis, *IEEE Trans. Inform. Theory* **36** (1990), 961–1005.

51. Duffin, R. J. and A. C. Schaeffer, A class of nonharmonic Fourier series, *Trans. Amer. Math. Soc.* **72** (1952), 341–366.
52. Frazier, M. and B. Jawerth, Decomposition of Besov spaces, *Indiana University Math. J.* **34** (1985), 777-799.
53. Gabor, D., Theory of communication, *J. IEE (London)* **93** (1946), 429–457.
54. Grossmann, A. and J. Morlet, Decomposition of Hardy functions into square integrable wavelets of constant shape, *SIAM J. Math. Anal.* **15** (1984), 723–736.
55. Haar, A., Zur Theorie der orthogonalen Funktionensysteme, *Math. Ann.* **69** (1910), 331–371.
56. Lemarié, P. G., Ondelettes à localisation exponentielles, *J. Math. Pure et Appl.* **67** (1988), 227–236.
57. Mallat, S., A theory of multiresolution signal decomposition: the wavelet representation, *IEEE Trans. Pattern Anal. Machine Intell.* **11** (1989), 674–693.
58. Mallat, S., Multiresolution representation and wavelets, Ph.D. Thesis, University of Pennsylvania, Philadelphia, PA, 1988.
59. Mallat, S., Multiresolution approximations and wavelet orthonormal bases of $L^2(\mathbf{R})$, *Trans. Amer. Math. Soc.* **315** (1989), 69–87.
60. Mallat, S. and W. L. Hwang, Singularity detection and processing with wavelets, 1991, preprint.
61. Mallat, S. and S. Zhong, Reconstruction of functions from the wavelet transform local maxima, 1990, preprint.
62. Mallat, S. and S. Zhong, Wavelet transform maxima and multiscale edges, in *Wavelets and Their Applications*, G. Beylkin, R. Coifman, I. Daubechies, S. Mallat, Y. Meyer, L. Raphael, and B. Ruskai (eds.), Jones and Bartlett, Cambridge, MA, 1991.
63. Meyer, Y., Ondelettes et fonctions splines, *Séminaire EDP*, École Polytechnique, Paris, December 1986.
64. Meyer, Y., Principe d'incertitude, bases Hilbertiennes et algèbres d'opérateurs, *Séminaire Bourbaki* **662** (1985-1986).
65. Schoenberg, I. J., Contributions to the problem of approximation of equidistant data by analytic functions, *Quart. Appl. Math.* **4** (1946), 45–99, 112–141.
66. Schweinler, H. C. and E. P. Wigner, Orthogonalization methods, *J. Math. Phys.* **11** (1970), 1693–1694.
67. Strömberg, J. O., A modified Franklin system and higher order spline systems on $\mathbf{R}^n$ as unconditional bases for Hardy spaces, in *Proc. Conf. in Honor of Antoni Zygmund*, Vol. II, W. Beckner, A. P. Calderón, R. Fefferman, and P. W. Jones (eds.), Wadsworth, NY, 1981, 475–493.
68. Tchamitchian, Ph., Biorthogonalité et théorie des opérateurs, *Rev. Math. Iberoamericana*, to appear.

69. Tchamitchian, Ph., Calcul symbolique sur les opérateurs de Caldéron-Zygmund et bases inconditionnelles de $L^2(\mathbb{R}^n)$, *C. R. Acad. Sc. Paris* **303** (1986), 215–218.
70. Vetterli, M. and C. Herley, Wavelets and filter banks: theory and design, *IEEE ASSP*, 1992, to appear.

Although wavelet analysis is a relatively new subject, there is already a vast literature devoted to both its theory and applications as well as several related areas. The list of references given here only reflects what is covered in this book and discussed in the Notes. A much more extensive bibliography is included in the second volume, *Wavelets–A Tutorial in Theory and Applications*, of this book series.

# Subject Index

**A**

Admissible
two-scale symbol 124, 125, 130
data 101
Algebra 89, 140, 141
Analog signal 6, 7, 23, 37, 49, 50,
53, 55, 57, 61, 159
Antisymmetry 15, 18, 21, 22, 162,
168, 185
Approximate identity 28
Associative 28
Autocorrelation function 32, 47,
48, 134
Autoregressive-moving average
(ARMA) 21, 246
sequences 21

**B**

$B$-net 95-98, 100, 194, 248
$B$-net algorithm 98, 194, 248
$B$-spline 17, 18, 54, 56, 59, 81, 85,
87, 90, 92, 95, 98, 100, 101,
110, 111, 117, 122,123, 129,
130, 140, 160, 165, 166, 168,
176-180, 183-185, 188, 192,
194, 200, 201, 202, 204, 207,
209, 213, 216, 246, 248
$B$-wavelet 184, 185, 188, 193, 194,
199, 204, 207, 210, 211, 213,
246
Backward differences 83
Banach space 24, 36
Banach-Steinhaus Theorem 72, 73
Band-limited signal 214
Bandwidth 62
Basic function 2, 3
Basic wavelet 5, 7-11, 59, 60, 62,
64-66, 81, 120, 156, 157, 185,
186, 245
Battle-Lemarié wavelets 249
Bernstein polynomial 95-97

Bernstein polynomial operator 95,
96
Bessel Inequality 39, 40, 42, 70,
226, 228
Bi-orthogonality 16
Binary dilation 4, 5
Blending operation 115
Blurred 19
Boolean sum 248
Bounded linear
functional 107
local spline operator 103
operator 33, 69, 105
Bounded variation 44, 46, 48

**C**

Cardinal $B$-spline 17, 56, 85, 98,
100, 101, 110, 123, 129, 130,
140, 160, 165, 168,177, 178,
180, 183, 185, 192, 194, 202,
204, 209, 248
Cauchy sequence 34, 40
center 7, 8, 11, 50, 54, 55, 60-63,
65, 86
frequency 60, 62
Circle group 23
Commutative 28, 79
Compactly supported 15, 18, 21,
46, 48, 54, 56, 84, 120, 132,
140, 165, 168-172, 176, 182,
183, 187, 188, 199, 215, 229-
231, 234, 248, 249
Complete oscillation 207, 208, 210,
211
Compressed data 18
Concave upward 95
Constant-Q 62
Control net 95
Convolution 17, 23, 27, 28, 30, 38,
40, 47, 56, 102, 104, 107, 140,
184

identity 28, 30

## D

Daubechies scaling functions 129, 234

Daubechies wavelets 234

Decomposition algorithm 199, 201

Decomposition relation 19, 142, 145, 156-158, 199

Delta function 27, 30

Differential operator 83, 182, 211

Digital signal 23, 37

Direct sum 15, 16, 19, 20, 119, 141, 144

Direct-sum decomposition 120, 142, 145, 248

Dirichlet kernel 38, 40

Dirichlet-Jordan Test 44

Discrete Fourier transform 37-40, 42, 137, 160

Distortion 18, 21, 160

Dominated Convergence Theorem 25, 33, 127, 235

Downsampling 20, 159

Dual 9-14, 16, 23, 67, 68, 70, 72, 74, 75, 77-80, 119, 120, 122, 146, 148, 149, 151, 152, 154, 155, 168-172, 174, 176, 182, 184, 185, 215, 220, 246, 248

Dual scaling function 171

Duality principle 149, 156, 248

Dyadic
  dual 67, 68, 80
  expansion 237, 238
  translation 4

Dyadic wavelet 11, 65-68, 73, 74

## E

Error analysis 195, 199, 202, 249

Euler-Frobenius polynomial 48, 111, 112, 134, 171, 182, 183,

195, 199, 200, 202, 203, 208

## F

Fatou's lemma 33

Fejér kernel 40

Filtered 209

Finite
  energy 6, 37, 50, 159
  two-scale relations 128, 134

Finite impulse response (FIR) 21

Fourier analysis 1, 6, 23, 28

Fourier
  coefficients 2, 5, 37-40, 42, 43, 47
  series 38-40, 45, 48, 217, 226
  transform 5-7, 11, 23, 25-28, 30, 32-40, 42, 44, 47, 49-52, 54-56, 59, 67, 68, 73, 75-77, 91, 111, 112, 122, 126, 130, 135, 137, 143, 144, 160, 161, 216, 217, 223, 227, 234, 237, 238, 247

Frames 11, 49, 68, 81, 155, 245, 247

Frequency band 7, 8, 60, 62, 65, 120, 160, 241, 242

Frequency window 8, 55, 62

Frequency-domain 6

Frequency-localization 8

Fubini Theorem 28, 31

Fundamental
  cardinal spline function 110, 112, 178
  spline 110, 246

## G

Gabor transform 49-55, 59, 247

Gabor window 53

Gaussian function 26, 27, 49, 50, 54, 56-58, 63, 77, 185, 186, 247

Generalized Euler-Frobenius Laurent polynomial 134, 169,

180, 183, 200, 216

Generalized Euler-Frobenius polynomial 171

Generalized linear phase 160-165, 168, 170-172, 174

Generalized Minkowski Inequality 37, 41, 219

Graphical display 81, 90, 92-95, 128, 133, 187, 248

**H**

Haar
  function 5, 54, 70, 170, 171, 216, 245
  wavelet 165, 177

Heaviside unit step 26

Height of the time-frequency window 53

Hertz 57

Hilbert space 24, 37, 83, 119, 215

**I**

Image compression 245, 247

Impulse response 21

Impulse train 194

Infinite impulse response (IIR) 21, 246

Inner product 2, 4, 6, 24, 34, 36, 37, 79

Integral dilation 2

Integral shifts 3

Integral wavelet transform (IWT) 5-14, 49, 51, 59, 60, 65, 74, 79, 156, 157, 185, 245

Interpolatory graphical display algorithm 90, 92, 93, 95, 133, 187

Inverse formula 9

Inverse Fourier transforms 23, 49

Isometric 3, 42

**K**

Knot sequence 82, 84, 89, 92, 100, 114, 159, 181, 192

**L**

$\ell^2$-linear combination 3

$L^2(\mathbf{R})$-closure 84, 85

Laplacian pyramid algorithm 246

Laurent polynomial 48, 102, 111, 134, 169, 170, 172-174, 180, 183, 200, 206, 216, 229, 230, 231

Laurent series 124, 140, 146-148, 183, 201, 202, 216, 220, 229

Lebesgue integration theory 25, 32

Lebesgue theory 1, 23, 34

Leibniz Rule 87

Limit in the mean of order two 34

Linear functional 30, 107

Linear Pascal triangular algorithm (LPTA) 189, 209

Linear phase 8, 61, 160-165, 168, 170-172, 174, 248

Linear-phase filtering 22, 120, 159, 168

Littlewood-Paley Identity 245, 247

**M**

Matrix operator 71

Measurable function 46

Meyer wavelets 249

Measure zero 24, 246, 247

Minkowski Inequality 24, 37, 41, 219

Modified Euler-Frobenius polynomial 203

Modulus of continuity 41, 43, 124, 125

Mother wavelet 245

Moving average 20, 21, 93, 100, 159, 194, 199

Multiresolution analysis (MRA) 16-18, 119-121, 128, 134, 140, 145, 147, 151, 159, 169, 177, 215, 245, 248, 249

**N**

Nested sequence 17, 18, 84, 85, 90, 119-121, 177, 223
Neumann series 102, 248
Norm 3, 4, 6, 24, 33
Nyquist rate 213

**O**

Octave 60, 62, 120, 160, 241
On-line 81, 105, 110
One-sided degree 134
Open Mapping Theorem 72, 73
Orthogonal
    sum 15, 120, 242
    wavelet 4, 5, 7, 13-15, 74, 75, 165, 236, 245
Orthonormal (o.n.) 16
    basis 4, 120, 242
    family 75, 127, 165, 166, 168
Orthonormalization procedure 78, 216, 247
Oscillation 189, 207, 208, 210, 211, 249

**P**

Parseval Identity 2, 6, 8, 32, 34, 36, 40, 42, 43, 47, 52, 54, 55, 58, 59, 61-63, 66, 76, 77, 205, 226, 229
Partial sum 38-40
Pascal triangular algorithm (PTA) 188, 189, 249
Phase-shift 8, 61, 65
Piecewise continuous 1, 23, 82
Poisson Summation Formula 23, 45, 101, 111, 122, 135, 195, 218, 246

Polynomial spline 81
Polynomial-reproduction 115
Pólya frequency (PF) sequences 208, 209
Predictor-corrector algorithm 246

**Q**

Quantization 21
Quasi-interpolation 105-110, 113-116, 177, 248
    operator 105-108, 113-116, 248

**R**

Real-time 18, 49, 81, 92, 105, 110, 114, 128, 133, 177
Reciprocal polynomial 134
Reconstruction
    algorithm 21, 91, 128, 146, 156, 158, 160, 200, 243, 248
    formula 10, 11, 70
Recover 26
Recursive scheme 130, 131, 188, 191, 192
Reflection 35, 47, 65, 100
Riemann-Lebesgue Lemma 25, 28
Riesz basis 12-15, 17, 71, 73, 74, 77-79, 85, 90, 93, 121-123, 127, 135, 142, 144-146, 181, 182, 184, 202, 211, 223, 224
    bound 89
Riesz Lemma 231, 249
Riesz-Fischer Theorem 39
Root mean square (RMS) bandwidth 55, 241
Root mean square (RMS) duration 50, 55

**S**

Sample space 156, 159, 177
Sampling rate 68-71, 73, 74, 80
Scaling function 17, 19, 21, 91, 119, 121, 122, 127-130, 135, 136,

140, 147, 159, 162, 163, 164,
165, 168-171, 177, 182, 187,
193, 215-217, 220, 221, 229-
231, 234, 236-240, 248
Scaling process 106
Schwarz Inequality 24, 32, 54, 57,
226
Self-dual 14, 75, 78, 215
Semi-orthogonal (s.o.) 15, 16, 18,
75, 77-79, 169-171, 177, 182,
183, 216
Semi-orthogonal wavelet 15, 75
Shape 95
Short-time Fourier transform
(STFT) 49, 56, 59, 60, 247
Single-frequency 4
Sinusoidal wave 3
Skew-symmetric 161, 164, 165,
168, 170, 171, 176
Small wave 7, 10
Sobolev space 105
Spectral information 6, 7, 23, 49,
50, 55, 120
Spectrum 6, 8, 49, 50, 53, 57
Square-integrable 1-3, 32
Square-summable sequences 12
Stability condition 11, 66, 67, 69,
71, 245, 247
Sub-band 242
Superposition 2
Support 18, 48, 49, 54, 56, 84, 86,
110, 119, 132, 134, 136, 137,
139, 140, 162, 165, 169, 170,
183, 184, 210, 213, 230, 248,
249
Symbol 4, 21, 37, 48, 76, 102-104,
111, 122, 124, 125, 127, 129-
131, 134-137, 139-141, 146,
147, 160, 162, 163, 165, 169,
170, 172-174, 176, 179, 180,

182, 183-185, 189, 193, 194,
200, 209, 210, 215, 217, 220,
230, 231, 234
Symmetric 14, 15, 18, 21, 22, 81,
86, 94, 102, 111, 120, 135-
137, 139, 140, 161, 162, 164,
165, 168, 170, 171, 176, 183,
185, 197, 206, 248
Symmetric zero 135

**T**
Thresholding 160
Time window 7
Time-domain 6, 23, 49, 57
Time-frequency analysis 1, 6, 8-10,
23, 49, 62, 74, 81, 146
Time-frequency localization 10, 22,
54, 56, 186, 237
Time-localization 7, 49, 50, 59, 242
Time-scale coordinate 14
Toeplitz matrix 208
Total positivity (TP) 81, 207, 208,
209, 249
   kernel 208, 209
   matrix 208, 213
Tree decomposition algorithm 243
Two-scale relation 90-92, 122, 125,
128, 129, 132-136, 140, 165,
170, 171, 179, 187, 213
   sequence 119, 122, 129, 130, 162-
165, 170, 171, 179, 182, 209,
215, 236, 248
   symbol 122, 124, 125, 127, 129-
131, 135-137, 139-141, 146,
147, 163, 165, 169, 170, 172,
173, 179, 180, 182-184, 193,
215, 217, 220, 230, 231, 234

**U**
Uncertainty Principle 49, 50, 54,
56, 57, 81, 247

Unconditional basis 48, 71

**V**

Variation-diminishing 209, 210

**W**

Wave packets 249
Wavelet
    analysis 1, 6, 23
    coefficient 5
    decomposition 19, 20, 91, 156,
        157, 177, 199, 215, 245, 248,
        249
    maxima 247
    series 4-6, 13, 14, 49, 74, 79, 120,
        155, 160, 194, 207, 210, 213,
        240, 242
Wavelet packets 215, 236-240, 242,
        249
Weierstraß Theorem 40
Weight 21, 29, 33, 94, 100, 199,
        200, 243
Width 7, 8, 50, 51, 53, 55, 56, 60,
        62, 241, 247
Width of the time-frequency win-
        dow 53, 56
Wiener Class 140, 142, 220
Window
    function 7-9, 11, 49-56, 59, 61,
        247
    inverse Fourier transform 52

**Z**

Zero-crossings 207, 213
Zoom-in and zoom-out 7, 9

# Appendix

In order to be able to implement the wavelet decomposition and reconstruction algorithms as described by (5.4.48) and (5.4.49), we need the weight sequences $\{a_k\}, \{b_k\}, \{p_k\},$ and $\{q_k\}$. Besides, the reconstruction (or two-scale) sequences $\{p_k\}$ and $\{q_k\}$ can be used to graph the scaling function $\phi$ and wavelet $\psi$ (see (5.2.11), (5.2.14)–(5.2.17), and (5.3.4)). In what follows, we consider the cardinal B-splines $\phi = N_m$ and B-wavelets $\psi = \psi_m$ of order $m$. Recall that their two-scale sequences

$$\begin{cases} p_k = p_{m,k} \; ; \\ q_k = q_{m,k} \; , \end{cases}$$

as given in (6.3.3)–(6.3.4), are particularly simple, and their decomposition sequences

$$\begin{cases} a_k = a_{m,k} \; ; \\ b_k = b_{m,k} \end{cases}$$

can be computed by using (6.5.1)–(6.5.2). Since these sequences are symmetric, it is sufficient to compute half of the values. More precisely, we have

$$\begin{cases} a_{m,k} = a_{m,m-k} \; ; \\ b_{m,k} = b_{m,3m-2-k} \end{cases} \qquad (A.1)$$

and

$$\begin{cases} p_{m,k} = p_{m,m-k} \; ; \\ q_{m,k} = q_{m,3m-2-k} \; , \end{cases} \qquad (A.2)$$

for all $k \in \mathbb{Z}$. In practice, particularly in cardinal spline interpolation, linear and cubic splines are used most frequently. We therefore give the values of $a_k = a_{m,k}$, $b_k = b_{m,k}$, and $p_k = p_{m,k}$, $q_k = q_{m,k}$ for $m = 2$ and 4 in Tables A.1 and A.2. The reader is reminded to refer to (A.1) and (A.2) for the other "half" of the sequences.

|   | $m = 2$ | | $m = 4$ | |
|---|---------|---|---------|---|
| $k$ | $a_k$ | $b_{k+1}$ | $a_{k+1}$ | $b_{k+4}$ |
| 1 | 0.683012701892 | 0.866025403784 | 0.893162856314 | -1.475394519892 |
| 2 | 0.316987298108 | -0.316987298108 | 0.400680825467 | 0.468422596633 |
| 3 | -0.116025403784 | -0.232050807569 | -0.282211870811 | 0.742097698477 |
| 4 | -0.084936490539 | 0.084936490539 | -0.232924626134 | -0.345770890775 |
| 5 | 0.031088913246 | 0.062177826491 | 0.129083571218 | -0.389745580800 |
| 6 | 0.022758664048 | -0.022758664047 | 0.126457446356 | 0.196794277304 |
| 7 | -0.008330249198 | -0.016660498395 | -0.066420837387 | 0.207690838380 |
| 8 | -0.006098165652 | 0.006098165652 | -0.067903608499 | -0.106775803373 |
| 9 | 0.002232083545 | 0.004464167091 | 0.035226101674 | -0.111058440711 |
| 10 | 0.001633998562 | -0.001633998561 | 0.036373586989 | 0.057330952254 |
| 11 | -0.000598084983 | -0.001196169967 | -0.018815686621 | 0.059433388390 |
| 12 | -0.000437828595 | 0.000437828595 | -0.019473269356 | -0.030709700871 |
| 13 | 0.000160256388 | 0.000320512777 | 0.010066747520 | -0.031811811318 |
| 14 | 0.000117315818 | -0.000117315818 | 0.010424052187 | 0.016440944687 |
| 15 | -0.000042940569 | -0.000085881139 | -0.005387929819 | 0.017028029466 |
| 16 | -0.000031434679 | 0.000031434678 | -0.005579839208 | -0.008800839839 |
| 17 | 0.000011505891 | 0.000023011782 | 0.002883979478 | -0.009114745138 |
| 18 | 0.000008422897 | -0.000008422897 | 0.002986784625 | 0.004710957034 |
| 19 | -0.000003082990 | -0.000006165980 | -0.001543728719 | 0.004878941541 |
| 20 | -0.000002256905 | 0.0000022569054 | -0.001598768083 | -0.002521687975 |
| 21 | 0.000000826079 | 0.0000016521587 | 0.000826326663 | -0.002611601542 |

**Table A.1.** Spline-wavelet decomposition sequences.

|   | $m = 2$ | | $m = 4$ | |
|---|---------|---|---------|---|
| $k$ | $p_k$ | $q_{k+1}$ | $p_{k+1}$ | $q_{k+4}$ |
| 1 | $\frac{2}{2}$ | $\frac{20}{4!}$ | $\frac{6}{8}$ | $-\frac{24264}{8!}$ |
| 2 | $\frac{1}{2}$ | $-\frac{12}{4!}$ | $\frac{4}{8}$ | $\frac{18482}{8!}$ |
| 3 | | $\frac{2}{4!}$ | $\frac{1}{8}$ | $-\frac{7904}{8!}$ |
| 4 | | | | $\frac{1677}{8!}$ |
| 5 | | | | $-\frac{124}{8!}$ |
| 6 | | | | $\frac{1}{8!}$ |

**Table A.2.** Reconstruction (or two-scale) sequences.

# WAVELET ANALYSIS AND ITS APPLICATIONS

CHARLES K. CHUI, SERIES EDITOR

1. Charles K. Chui, *An Introduction to Wavelets*
2. Charles K. Chui, ed., *Wavelets: A Tutorial in Theory and Applications*